I0105314

WARLIKE AND PEACEFUL SOCIETIES

Warlike and Peaceful Societies

The Interaction of Genes and Culture

Agner Fog

OpenBook Publishers

https://www.openbookpublishers.com

© 2017 Agner Fog

This work is licensed under a Creative Commons Attribution 4.0 International license (CC BY 4.0). This license allows you to share, copy, distribute and transmit the work; to adapt the work and to make commercial use of the work providing attribution is made to the authors (but not in any way that suggests that they endorse you or your use of the work). Attribution should include the following information:

Agner Fog, *Warlike and Peaceful Societies: The Interaction of Genes and Culture*. Cambridge, UK: Open Book Publishers, 2017, http://dx.doi.org/10.11647/OBP.0128

In order to access detailed and updated information on the license, please visit https://www.openbookpublishers.com/product/657#copyright

Further details about CC BY licenses are available at http://creativecommons.org/licenses/by/4.0/

All external links were active at the time of publication unless otherwise stated and have been archived via the Internet Archive Wayback Machine at https://archive.org/web

Digital material and resources associated with this volume are available at https://www.openbookpublishers.com/product/657#resources

Every effort has been made to identify and contact copyright holders and any omission or error will be corrected if notification is made to the publisher.

ISBN Paperback: 978-1-78374-403-9
ISBN Hardback: 978-1-78374-404-6
ISBN Digital (PDF): 978-1-78374-405-3
ISBN Digital ebook (epub): 978-1-78374-406-0
ISBN Digital ebook (mobi): 978-1-78374-407-7
DOI: 10.11647/OBP.0128

Cover image: War memorial, Whitehall, London. Photo by Aimee Rivers (2011). Flickr, https://www.flickr.com/photos/sermoa/5361099123, CC BY-SA 2.0
Cover design: Anna Gatti

All paper used by Open Book Publishers is SFI (Sustainable Forestry Initiative), PEFC (Programme for the Endorsement of Forest Certification Schemes) and Forest Stewardship Council(r)(FSC(r) certified.

Printed in the United Kingdom, United States, and Australia
by Lightning Source for Open Book Publishers (Cambridge, UK)

Contents

1. Introduction

All through recorded history we have seen extreme differences between different human societies. Some societies in some periods have been warlike and cruel beyond comprehension, while other societies in other times and places have been remarkably peaceful and tolerant.

Societies that are warlike, hierarchical, and intolerant with strict discipline are called *regal*. Societies that are peaceful, egalitarian, and tolerant are called *kungic*. Many societies are something in between these two extremes. The present book offers a new, groundbreaking theory that explains this extreme variability in social organization and culture based on evolutionary theory.

It has often been discussed whether such dramatic differences in human behavior are due to genetic differences or cultural norms. One aspect that is often missing in the genes-versus-culture debate is that genes may code for flexibility. Our genes enable us to behave differently under different conditions. This is called *phenotypic plasticity*.[1]

The theory presented here demonstrates that humans have a plasticity that enables us to adapt to different conditions of war or peace. Warlike or regal behavior has been adaptive under conditions of collective danger that were sometimes present in our evolutionary past, while conditions of collective security that were present at other times and places made peaceful or kungic behavior optimal from an evolutionary point of view. In other words, the potentials for both warlike and peaceful behavior are present in our genetic makeup.

---

1   Bateson and Gluckman (2011, p. 31)

© 2017 Agner Fog, CC BY 4.0   https://doi.org/10.11647/OBP.0128.01

Nobody is born a devil or a saint. Depending on our living conditions, we may become authoritarian and belligerent or peace-loving and tolerant. The theory presented here explains a likely evolutionary mechanism behind this flexible psychology and analyzes the conditions that make us either strident or docile.

This theory, which will be called *regality theory*, can answer many burning questions about both individual and collective behavior: Why have so many tyrants fought cruel and unnecessary wars? Why do many people support their tyrants? Why do some people hate foreigners while other people readily embrace them? Why have people used their apparently peaceful religion to justify some of the worst atrocities in history? And why have other people dedicated their lives to the most unselfish charitable causes based on the very same religions? Why do some militants commit acts of terrorism against innocent people? And why can a few acts of terrorism that cause a limited amount of harm lead to dramatic changes in the political climate, while other events that cause much more harm have no noticeable political effect?

The remarkable differences between warlike and peaceful societies are reflected in many characteristics of culture, including aspects that have no obvious relationship with war and peace, such as art preferences and sexual morals. This book explores such side effects as well and presents statistical evidence in support of the theory.

1.1. A different kind of social science

'Scientific genius is extinct', wrote Dean Simonton in *Nature* a few years ago. In his view, the only kind of scientific progress we see today is marginal improvements within old paradigms that have already been thoroughly explored. Revolutionary new ideas either do not occur or fail to be acknowledged.[2] Scientists who are trained in one particular paradigm are unlikely to understand and accept a new, radically different paradigm.[3] While everybody hails interdisciplinary research, the reality today is that many scientists guard their own scientific

---

2   Simonton (2013)
3   Kuhn (1962, chapter 12)

territory. Scientists today have little freedom to choose their own subjects of research. The highly competitive funding system is more likely to support old research areas than radically new ones because it is controlled by established scientists through the peer review system.[4]

Most scientists start by specializing in one particular scientific paradigm and then search for problems that this paradigm can be applied to. The present book reflects the opposite approach. It starts with a problem and then searches for paradigms that can contribute to solving the problem. This includes paradigms from the natural sciences, such as evolutionary biology and ecology, as well as from the social sciences, such as anthropology, history, political science, economics, and cultural studies.

Unfortunately, there is much animosity and little mutual understanding between the natural and the social sciences. Many regard it as impossible to establish something similar to the laws of the natural sciences for social phenomena.[5] Evolutionary theories of human behavior are rejected by many sociologists on those grounds,[6] and some particularly fashionable branches of social studies are aversive to any search for causal regularities in social and cultural systems.[7] This is not a good starting point for bridging the gap between the social and the natural sciences. Fortunately, authors in other branches of the social sciences have strongly defended the study of social phenomena based on solid scientific principles.[8] We have to rely in particular on those social science traditions that explicitly search for regularities; for example, comparative historical analysis[9] and social systems theory.[10]

Too many studies of social phenomena have focused on an isolated phenomenon, using a single theoretical framework that allows only a single type of explanation. Such studies cannot account for the rich complexity of human culture and social developments. We need a

---

4   Becher and Trowler (2001), Lucas (2006), van Arensbergen, van der Weijden and van den Besselaar (2014)

5   Hayek (1967)

6   Horowitz, Yaworsky and Kickham (2014), O'Malley (2007)

7   Beed and Beed (2000)

8   Kincaid (1996, chapter 3)

9   Mahoney and Thelen (2015)

10   Richardson, G. (1991)

social science that combines the insights of many different scientific disciplines to better understand the interactions between individual and collective action, between planned and unintended developments, between human action and structural causes, between endogenous and exogenous factors, and so on. The present book strives towards this goal of multicausal explanations.

The large number of different scientific disciplines involved in this book makes it impossible to go into deep details for each discipline. For example, the text does not go into details with historical examples, but instead discusses the specific aspects of historical events that are relevant to the theoretical discussion. Readers who want to go deeper into a particular subject are referred to the literature references (chapter 10).

In sciences like physics and mathematics, a theory is called 'beautiful' if it can solve a broad range of problems using one simple formula and if it can be applied to problems other than the one that prompted the development of the theory. Regality theory is a beautiful theory in this sense. What started as an attempt to explain morals by cultural selection ends up as an evolutionary psychology theory of collective action that can explain a broad range of phenomena: individual characteristics such as authoritarianism, xenophobia, or tolerance; social phenomena such as political hierarchy, bellicosity, discipline, or egalitarianism; and even cultural phenomena such as religiosity, music genres, and architectural style. Regality theory is not a 'grand theory', though. It can contribute to the explanation of many interesting phenomena, but it needs to be combined with other theories in order to fully explain these phenomena.

Some branches of social studies readily mix science and ideology. That is a dangerous course. Regality theory is useful for explaining many different political phenomena, and the theory may be useful for guiding political decisions, but this must be a one-way interaction. The present book is based on the principle that science may influence politics but politics should not influence science. The fundamental science should be immune to political and ideological influences even if the research should reveal politically inconvenient truths.

1.2. Overview of the book

This book relies on many different scientific disciplines from both the natural sciences and the social sciences. Many concepts are briefly explained because one cannot expect the reader to be competent in such a broad range of different disciplines, but you should still be prepared to look up unfamiliar words and concepts. The good news is that it is not necessary to read and understand all the chapters in order to get a basic understanding of the theory.

Chapter 2.1 gives a short introduction to regality theory. It is necessary to read this chapter first in order to understand the rest of the book. You may read the remaining chapters in any way you like. You may focus on the chapters that are most relevant to your field of interest and skip other chapters or read them later. There are cross-references throughout the book where one chapter relates to another.

Chapter 2.2 explains the evolutionary mechanism that regality theory is based on. This is the theoretical justification for the theory. Chapters 2.3, 2.4, and 2.5 contain further discussion of evolutionary aspects of the theory. Chapter 2.6 relates the theory to common cultural phenomena.

Chapter 3 and its subsections discuss how regality theory can benefit from contributions from other scientific disciplines. Regality theory cannot stand alone. We are dealing with social, cultural, and psychological phenomena that are influenced by a complex interplay of many different causes and mechanisms. Such a complex system cannot be described adequately by a single theory. We need to look into such diverse disciplines as ecology, demography, anthropology, history, political science, economics, social psychology, cultural studies, media studies, and many others in order to get a full understanding of the complex social phenomena of warlike and peaceful behavior. The subsections of chapter 3 discuss relevant findings from a number of disciplines that can be combined with regality theory to provide a more complete understanding of the social, cultural, and psychological phenomena we want to study.

Many different academic traditions have made observations about different kinds of societies and cultures or different kinds of personalities and psychological reactions that have important similarities with regality theory. However, they have done so without the same degree of theoretical understanding of fundamental causes and mechanisms. Chapter 3 also discusses how such findings from other areas of study can be integrated with regality theory.

Chapter 4 looks at different theories about the causes of war and peace as well as the dynamic processes behind different kinds of violent conflicts.

Chapter 5 and 6 analyze various aspects of contemporary cultures. Chapter 5 looks at a number of economic factors that produce regal cultures, perhaps unintentionally, through fear and collective danger. The commercial mass media profit from fear. Economic instability causes insecurity and conflicts. Changing economic conditions have changed the patterns of war and violent conflicts so that proxy war, insurgency, and terrorism have mainly replaced conventional interstate war.

Chapter 6 looks at cases where fear and conflict are used intentionally as strategic weapons by powerful nations as well as by smaller insurgent groups.

The psychological plasticity that regality theory describes has its origins in a distant evolutionary past, and we cannot be certain that it is still adaptive (in the evolutionary sense) in a modern setting. We may gain more insight by looking at non-modern cultures that are more similar to our evolutionary past. Chapter 7 describes a number of ancient cultures, ranging from the most peaceful to the most warlike, and living under very different ecological environments. In connection with each culture is a discussion of how it relates to the predictions of regality theory.

While examples are useful for illustrating a theory, we cannot rule out the possibility that the agreement between a theory and a few examples is just a coincidence. Chapter 8 contains a number of statistical tests to distinguish between random coincidence and significant correlations. Various methods are used for testing the predictions of regality theory at both the individual level and the level of whole societies in both modern and ancient non-industrial societies.

Chapter 9 concludes the book with a discussion and summary of the findings and possible applications of regality theory.

2. The Theory of Regal and Kungic Cultures

2.1. In a nutshell: 'regal' and 'kungic' explained

It is easy to observe that some cultures are warlike and totalitarian while other cultures are peaceful and tolerant.[1] It is more difficult to explain why. Regality theory seeks to explain such cultural differences as adaptations to the different levels of danger and conflict that societies are exposed to.[2] Nobody is born belligerent or peaceful, according to this theory. Instead, humans have evolved a psychological plasticity that shapes our personalities to fit the environments we live in. This psychological mechanism makes people prefer a strong leader and strict discipline in the event of war or other collective danger.

The mechanism explained here is an interplay between genes and culture. The genes code for a flexibility that allows the psychological sentiments of each person to respond to the level of war and the need for collective action. The zeitgeist and culture adapts to these sentiments in such a way that the society becomes well prepared to meet any external threats.

Fighting in war is hard and dangerous, and it would be more attractive for the individual not to fight and to let others do the fighting. This is the well-known collective action problem. Regality theory proposes

---

1   Russell (1972)
2   Fog (1999, p. 91)

© 2017 Agner Fog, CC BY 4.0   https://doi.org/10.11647/OBP.0128.02

that the collective action problem can be overcome by installing a strong leader who can reward brave warriors and punish defectors. If the leader has enough support, then he[3] can coerce everybody to fight and let everybody benefit from the collective fighting. If no leader has enough support, then nobody will fight and everybody will suffer from the resulting collective defeat. But in neither situation will an individual have to fight alone and let others free ride on the benefits he makes for his group. Everybody will benefit from having a strong leader in the case of war, and therefore everybody should desire a strong leader when facing collective danger.

However, a strong leader is a disadvantage in the absence of war, because a tyrannical leader can exploit his followers and suppress their freedom. Therefore, it is advantageous to have a psychological plasticity that makes us prefer a strong leader in the event of war, but not in the event of peace. Regality theory proposes that such a plasticity has evolved by natural selection. People will prefer a strong leader and strict discipline when the probability of war or other collective danger is perceived to be high, while people will prefer an egalitarian society with more lax discipline when there is no collective danger.

If the majority of the members of a tribe or other group desire a strong leader and strict discipline, then, surely, this will be what they get. They will develop a hierarchical political structure and a very punitive system of discipline. It has been observed that this affects not only the political structure but many other aspects of the culture as well. People will develop a strong feeling of tribal or national identity, and their world view will be more polarized between friends and enemies. Tolerance of strangers and deviants will go down. Religion will be used as a means to keep people in line. And, perhaps most surprisingly, it has been observed that styles of art and music will gradually change so as to achieve psychological congruence with the sociopolitical structure and the world view.

Such a culture is called *regal*. We will use the word 'regal' to denote the psychological preferences of the individuals as well as the political structure and the culture and artifacts that are characteristic of a society with frequent wars, threats of war, or other collective dangers

---

3   Chapter 2.5 explains why most war leaders in history were men.

that require collective action. The opposite of regal is *kungic*. A kungic culture is peaceful, egalitarian, and tolerant. The characteristics of regal and kungic cultures will be explained in more detail in chapter 2.6.

The word 'regal' comes from the Latin *regalis*, which means 'royal'. The word 'kungic' is coined after the !Kung bushmen, who have the most kungic culture found in the present study. War and other collective dangers (perceived or real) that push a society towards a more regal structure are called *regalizing factors*, and this process is called *regalization*.

2.2. Evolutionary basis for regality theory

It has often been observed that people prefer a strong leader and a strong social group in times of crisis,[4] and a number of scientists have independently suggested that this may be an adaptive response to the need for collective action.[5] However, so far there has been little discussion of why this would be adaptive.

The new theory proposed here relies on a psychological mechanism that makes people prefer a strong leader in times of intergroup conflict but not in times of peace and safety. Such a mechanism could be adaptive because it reduces or eliminates the free rider problem in collective fighting.

Several mechanisms have been proposed to explain the phenomenon called parochial altruism—the fact that people are willing to fight for their social group despite the fitness costs.[6] The most important evolutionary explanations that have previously been proposed include kin selection,[7] group selection,[8] reciprocal selection,[9] altruistic punishment,[10] prestige,[11] sexual selection (women are attracted to brave warriors),[12] the opportunity of successful warriors to profit from the

---

4   Hastings and Shaffer (2008), Jugert and Duckitt (2009), Ladd (2007)
5   Fog (1997), Navarrete, Kurzban, Fessler and Kirkpatrick (2004), van Vugt (2006), Hastings and Shaffer (2008), Kessler and Cohrs (2008), Glowacki and von Rueden (2015)
6   Bowles and Gintis (2011), Nowak (2006)
7   Thayer (2004)
8   Crofoot and Wrangham (2010), Lehmann and Feldman (2008), Thayer (2004)
9   Tooby and Cosmides (1988, 2010)
10   Boyd, Gintis, Bowles and Richerson (2003)
11   Glowacki and Wrangham (2013)
12   Van der Dennen (1995), Wrangham (1999), Glowacki and Wrangham (2013)

spoils and to mate with captured women from the losing group,[13] and cultural group selection.[14]

It is a common characteristic of these proposed mechanisms that the effects are relatively weak, and perhaps too weak to compensate for the extremely high fitness costs of fighting.[15] The fitness gain in the form of increased mating opportunities does not necessarily go to the people that have run the highest risks; and the mechanism of punishing defectors involves the additional collective action problem of who should bear the costs of being the punisher.[16]

The alternative explanation proposed here is a mechanism that may have been important in the evolution of collective fighting in prehistory. In the event of war, or imminent war, the members of a social group will show a psychological preference for having a strong leader and a social system with strict discipline. If enough members of the group express these preferences, then the group will soon develop a hierarchical political structure with a strong and powerful leader who can command group members to fight, devise a strategy, reward brave warriors, and punish defectors.

There is an important difference between being willing to fight for one's social group and being willing to support a strong leader. The altruistic individual who volunteers to fight for his group will run a high personal risk, while all the non-fighting members of his group will benefit from his bravery. As the cost to the individual warrior is likely higher than his share of the group-level gain, this behavior will not be promoted by simple natural selection. But the strategy of supporting a strong leader is different. If only a few members of the group desire a strong leader, then there will be no strong leader and no collective fighting. If enough group members support a strong leader, then this leader will be able to dominate everybody, including the minority that do not support him, and command them to fight. Thus, it is possible for the group to suppress the fitness advantage of free riding by installing a strong leader.[17] The individual who shows the preference for a strong

---

13   Van der Dennen (1995), Chagnon (1990), Choi (2007), Glowacki and Wrangham (2015)
14   Henrich (2004)
15   Bradley (1999)
16   Glowacki and von Rueden (2015), Fowler (2005)
17   Glowacki and von Rueden (2015), Hooper, Kaplan and Boone (2010)

leader will have to carry the costs of fighting, but he will also enjoy the benefits of everybody else fighting. Either everybody fights or nobody fights—there is no place for free riders. The group-level benefit of everybody fighting in a coordinated way could very well be sufficiently high to outweigh the individual fitness costs of fighting, even when the benefit is divided between all the group members.

For promoting a complex task such as fighting, where extraordinary above-average skills are particularly valuable, we can expect that a system including both reward and punishment will be more efficient than a system based on punishment alone. A system based on punishment only would make warriors deliver the minimum performance necessary to avoid punishment; and defectors might even avoid punishment if they could convincingly fake illness. A punishment system could possibly evolve by other mechanisms if the costs of punishing are sufficiently low.[18] However, we would expect rewards to be considerably more costly to deliver than punishments and require a higher payback to evolve.

We can imagine a Stone Age scenario like this: a tribal people experiences frequent conflicts with a neighbor tribe. This makes the people prefer a strong leader. Such a leader emerges, and his people trust and support him. He will lead the battles, devise strategies, and appoint people to various tasks. He may deliver rewards and punishments himself, or he may delegate this task to persons of intermediate rank. Rewards are particularly important for making it attractive for warriors to fight to the best of their abilities. Brave warriors may be rewarded with better food, weapons, protection, and other resources and— perhaps most importantly—with prestige.[19] A high prestige gives the brave warrior access to an attractive wife and perhaps multiple wives. This translates directly to biological fitness. Cowards who do not fight wholeheartedly will get a bad reputation and low prestige. This will give them a disadvantage in social exchanges and a disadvantage in the search for a mate. Such a system gives the best fighters the highest rewards and compensates for the risks of injury or death. The chances of winning a war against a neighbor tribe are increased as a result. The whole group is likely to support the leader, because everybody benefits

18 Fowler (2005)
19 Glowacki and Wrangham (2013), von Rueden, Gurven and Kaplan (2010)

from the increased chances of winning wars and there is no way of achieving the same result without a strong leader.

There is a trade-off between the benefit of being part of a strong and powerful political organization with a strong leader and the cost of repression within this organization.[20] This balance is likely to be tipped in a peaceful environment where the need for collective protection is low. The individual would have no reason to submit to a strong leader in this case. On the contrary, the individual would most likely see his own fitness reduced by a despotic male leader who could take advantage of everybody else and even monopolize a large number of women.[21] Therefore, the optimal strategy for the individual must be to have a flexible psychology, showing a preference for strong leadership and strict discipline when intergroup conflicts are frequent or expected, and a preference for an egalitarian social structure when intergroup conflicts are perceived to be unlikely.[22] The group-level effect of this psychological flexibility is that the higher the level of intergroup conflict, the more the group will invest in a strong organization that strengthens its ability to organize collective fighting.

This is the basic hypothesis of regality theory. A high level of intergroup conflict or perceived collective danger will activate a psychological desire for a strong leader in the group members, and the group will develop a hierarchical structure as a result. The opposite situation is a group living under safe and peaceful conditions where there is no neighbor group to fight with. People in this situation will not accept a strong leader who limits their freedom. A leader who is too strict will lose the support of his people and will not be able to stay in power. The group will develop an egalitarian structure as a result.

To recapitulate, a regal group is a group that has developed strong organization, discipline, and fighting spirit as a response to conflict or danger. A kungic group is a group that has adjusted to a peaceful and safe environment. The words 'regal' and 'kungic' are also used for the individual psychological preferences that lead to strong or weak group organization, respectively.

---

20 Summers (2005)
21 Betzig (2008)
22 Gavrilets and Fortunato (2014)

The regal and kungic forms of social organization can be considered the extremes of a continuous scale, where most societies are placed somewhere near the middle of this scale. We can call this the regality scale or the regal-kungic scale. The regality of a culture is determined by the frequency and severity of intergroup conflicts and other dangers that require collective action.

It might be problematic to assign numerical values to the regal-kungic scale when dealing with very different cultures under different historical and environmental conditions. In many cases, it is more useful to use it as a relative scale. For example, we may prefer to say that culture A is more regal than culture B instead of saying that culture A is regal or culture B is kungic. Likewise, it can be useful to follow a particular culture over time and see if it is getting more regal or more kungic.

2.3. An evolutionarily stable strategy

Some scientists have proposed that altruistic punishment may promote cooperation in human societies. One or more altruists in the group will bear the costs of punishing defectors. A recently published model of evolutionary game theory indicates that conformity, cooperation, and altruistic punishment in a social group are likely to be stronger when the group is under threat than when it is not.[23]

Other models in evolutionary game theory show that cooperative punishment is more stable than punishment administered by voluntary individuals, and it has been suggested that a punishing institution (policeman) might be evolutionarily stable in genetic or cultural evolution.[24] Regality theory proposes that this policeman can be replaced by a leader (who may appoint a policeman). The leader is rewarded with the fitness advantage of being a leader, and he can punish anybody who does not support him. This overcomes the collective action problem in the theory of altruistic punishment. Regality theory also allows the leader to administer rewards, which would be hard to explain by other theories because of the high costs of rewarding.

---

23   Roos, Gelfand, Nau and Lun (2015)
24   Sigmund, de Silva, Traulsen and Hauert (2010), Jaffe and Zaballa (2010)

We will now discuss whether a strong leader is necessary to make group members fight for their group in times of war. We will first consider the hypothetical situation where there is no strong leader but where group members are willing to fight for their group because they have a genetic predisposition to do so. Members who fight for their group are called altruists, while members who do not fight are called egoists. A group mostly of altruists is likely to win over a group mostly of egoists. This is called *group selection*. However, there is also selection within the group. The altruists have a considerable risk of dying in battle, while the egoists survive. Therefore, there will be more and more egoists for each generation. A group that contains only altruists and no egoists can thrive and grow, but it is vulnerable to invasion by egoists. In other words, group selection is not effective if there is more than a negligible rate of migration into the group.[25] We know from history and anthropology that conquered groups are rarely completely massacred. Some members of a losing group, especially women and children, are likely to survive and join the winning group. If the losing group in our hypothetical scenario contains egoists, then some of these egoists will survive and enter the winning group and eventually outcompete the altruists.

We will now consider a second scenario where there is a strong leader supported by the majority of group members. This is the scenario that regality theory is based on. The leader can reward brave warriors and punish defectors who do not fight for their group. We will assume that these rewards and punishments are strong enough to compensate for the fitness costs of fighting. For example, we can assume that the bravest warriors get the most attractive wives and therefore have many children, while the cowards get less attractive wives and therefore have fewer children. Or perhaps the bravest warriors get multiple wives while the cowards get none. Most warriors will fight to the best of their abilities in order to get the most rewards.

A group with a strong leader who can organize this kind of reward and punishment is likely to win battles against less organized groups that have no strong leader. The successful group will win more territory, which benefits all members of the group. Therefore, it will be attractive for all members of the group to support the leader.

---

25   West, el Mouden and Gardner (2011)

Our first scenario (group selection) was not stable because it was vulnerable to invasion by egoists. We will now discuss whether the second scenario (regality) is vulnerable to invasion by individuals who do not support the strong leader. If the majority of group members support the leader, then the minority that do not support the leader will be forced to fight anyway, and they will also be punished for not supporting the leader. History is full of examples of tyrannical leaders who punish anybody who does not support them. Therefore, there is no fitness advantage to not supporting the leader. On the contrary, there is likely severe persecution. If the non-supporters form a majority strong enough to overthrow the strong leader and put a weak leader in his place, then the whole group will be weakened and be less likely to win wars. This benefits neither the supporters nor the non-supporters. The conclusion must be that a group with a strong leader in wartime is not vulnerable to invasion by non-supporters. Therefore, regality is an evolutionarily stable strategy in a conflict-prone environment.

The preference for a strong leader is not activated in a permanently peaceful and safe environment, according to regality theory. The genotype that is not activated in phenotype is not subject to selection but only to random genetic drift under these conditions.

The prediction from regality theory is that people will show regal psychological reactions in the face of any collective danger that affects them directly and that requires collective efforts to overcome. Due to the weakness of group selection, we will expect the regal reaction of people to be much weaker in the case of dangers that affect only unrelated group members.

2.4. The behavior of the leader

So far we have discussed which strategy is most fit for ordinary group members. Now we will look at the role of the leader and discuss how we can expect the leader to act from a selfish fitness-maximizing point of view. It is no surprise that people are willing to be leaders. There is a large fitness advantage to being a leader or having a high position in the hierarchy of a successful group.[26] A powerful leader of a hierarchical

---

26   Summers (2005), Betzig (2008)

organization is typically able to take advantage of everybody else to benefit himself and his family.[27] Many of the most powerful leaders in history have assembled enormous wealth and large numbers of wives or concubines. Of course, everything must have been on a smaller scale in prehistory, but even among chimpanzees and other social animals there is a large advantage to being the alpha male.[28]

There are also costs to being a leader. The leader may have to take a frontline role in battles, and there is a real risk of being killed by an enemy group, by a rival for the leadership position, or by rebels who think that the leader is too despotic. A leader can be expected to make higher sacrifices or take higher risks in intergroup conflicts than low-ranking members because he has more at stake.[29] Nevertheless, we can assume that the fitness benefits of being a leader are much higher than the costs.

Based on this, we can expect the fitness-maximizing strategies of leaders to be very different from the strategies of followers. A typical survival strategy for a low-ranking individual could involve being an agreeable person, making friendships and alliances, and helping friends in need in the hope that they will later return the favor.[30] In contrast, we can expect the optimal strategy for a leader to involve doing everything to consolidate and increase his power, to weaken rivals for the position, to amass resources and wealth for himself and his family, and to have as many wives and concubines as he can get away with. The only thing that limits his despotism is the risk of losing the support of his followers. A person of intermediate rank or a person with chances of becoming a leader will be likely to use any strategy that can enhance his rank.

Psychological research confirms that people of high rank behave differently from people of low rank. Wealthy and high-ranking people of both sexes behave more egoistically and are more likely to cheat or behave unethically than other people.[31] They tend to feel entitled to their position.[32] They tend to take side with other high-ranking people

---

27   Anderson and Willer (2014), Padilla, Hogan and Kaiser (2007)
28   Boesch, Kohou, Néné and Vigilant (2006)
29   Gavrilets and Fortunato (2014), Johnson (2015)
30   Kiyonari, Tanida and Yamagishi (2000)
31   Piff et al. (2012), Trautmann, van de Kuilen and Zeckhauser (2013), Bendahan, Zehnder, Pralong and Antonakis (2014), Gino and Pierce (2009)
32   Piff (2014)

in conflicts.[33] They have a higher tendency to sexual infidelity,[34] they are more self-sufficient,[35] and they have less empathy for other people.[36] The reduced empathy is not entirely bad, however. It enables the leader to make more rational decisions that give higher weight to collective interests than to the interests of single individuals.[37]

A leader can exploit his followers to enrich himself, the more so the more power he has. The power of the leader is weakened if the followers can easily leave the group and join another group with a more agreeable leader. This explains why the most despotic leaders in history have appeared in large agrarian societies from which it was difficult for peasants to escape.[38]

The ability of a leader to exploit his followers is higher the more regal his group is. We can therefore expect leaders to try to increase the regality of their group by exaggerating dangers to the group and by fighting unnecessary wars.[39] Statistical studies of wars through history show such a strong connection between empires and war that we may assume that emperors need to fight wars to maintain their empires.[40] Powerful leaders may even fabricate enemies or fight fictitious dangers to maintain and consolidate their power. For example, the Inquisition tried to uphold the threatened monopoly of power of the Catholic Church in the Renaissance through the persecution of heretics and witches.[41] More examples of such fabricated dangers are discussed in chapter 6.

2.5. Why are most warriors and chiefs men?

The reader may have noticed that I am referring to warriors and leaders as 'he'. There is a reason for this. Throughout history, most warriors have been men and most leaders of warring societies have been men. Obviously, culture and tradition plays a role here, but there is more to

---

33 Lammers and Yang (2012)
34 Lammers et al. (2011)
35 Vohs, Mead and Goode (Goode 2006)
36 Kraus et al. (2012), Haslam and Loughnan (2014)
37 Côté, Piff and Willer (2013)
38 Price and van Vugt (2014)
39 Price and van Vugt (2015)
40 Eckhardt (1992, p. 184)
41 Ben-Yehuda (1980)

it than that. A growing amount of research indicates that the traditional division of labor between the sexes has biological roots. Many of the differences between men and women are connected with the Darwinian pursuit of reproductive success. The reproductive success of a man is limited mainly by his access to mating with women. There is practically no limit to how many children a man can sire if he can get enough women to cooperate. The situation for women is very different. The number of children that a woman can give birth to is limited mainly by her physiology and energy uptake, while the number of sexual partners plays only a minor role. Therefore, the reproductive strategies of men and women are very different and this leads to many conflicts of interest.[42]

Through most of our evolutionary history, our ancestors lived as hunters and gatherers, where more men than women hunted big animals, and where more women than men gathered fruits and roots and hunted small animals. Investigations of hunter-gatherer societies have found that the hunting of big game is not the most efficient way of getting food. There may be several reasons why men hunt, but showing off appears to be among the most important ones. This can be explained by the so-called *costly signaling theory*. A successful hunt proves that a man is strong and smart and therefore an attractive mating partner. Successful hunters have higher prestige and status than other men, and this translates into reproductive success.[43] Anthropologists have found that good hunters had higher prestige and more children than other men in all of the societies investigated.[44] A similar strategy for women would probably not increase their reproductive success.

There are similar reasons why men go to war. In the Stone Age, fighting and hunting were related activities that required some of the same skills and tools.[45] Brave warriors have high prestige, and there is reason to believe that this gives them a reproductive advantage.[46] The opportunity for capturing women from an enemy group further contributes to the fitness of warriors.[47] On the other hand, there are

42   Geary (2010), Chapman (2015)
43   Von Rueden and Jaeggi (2016)
44   Bird, R. (1999), Smith, E. (2004)
45   LeBlanc (2014)
46   Lehmann and Feldman (2008)
47   Van der Dennen (1995, p. 328)

examples where any advantage to the individual warrior is outweighed by the increased mortality.[48] If some male warriors die, then there will be an excess of women in the group, and this will lead to an increased reproductive fitness of the surviving men if polygamy is allowed. The total number of children produced by the group will be almost the same if a few men are lost. In other words, there is a fitness loss to the unlucky warriors who die but a fitness gain to the survivors, so the average fitness of the men in the group is almost unaffected by the deaths of a few warriors. If all men in a group fight and the deaths are randomly and unpredictably distributed among them, then natural selection can still favor fighting, because the risk of dying is offset by the chance of getting an extra wife if you survive. This would certainly not be the case if women were warriors. A woman dying is a lost opportunity for reproduction for the whole group, and the fitness of the surviving women would not be increased much by polyandry.[49]

Men are physically stronger than women on average, and the differences between the sexes are particularly marked in skills that are relevant to hunting and fighting, such as throwing distance. It has been suggested that such differences are the result of evolutionary forces that have favored these skills in men more than in women.[50]

It was argued above that social rewards are necessary for making people fight for their group. Men are more sensitive than women to social rewards because they have more potential for gaining fitness.[51] We can therefore assume that it has been easier to persuade men than women to fight throughout our evolutionary history. Psychological experiments confirm that men are more willing than women to make sacrifices for their group in situations of intergroup conflict, and this confirms the so-called *male warrior hypothesis*.[52] Of course, there are also practical reasons behind the tradition that war is the domain of men rather than women. Women in hunter-gatherer societies often breastfeed their babies for several years, and it would be unwise to carry an unweaned baby to the battlefield.

---

48 Beckerman et al. (2009)
49 Van der Dennen (1995, p. 325)
50 Geary (2010, p. 290)
51 Geary (2010)
52 Van Vugt, de Cremer and Janssen (2007), McDonald, Navarrete and van Vugt (2012)

In wartime, it seems logical to choose an experienced warrior as leader, and this would normally mean a man. But even in peaceful societies we can observe that most leaders are men. There is a large fitness advantage to having high status, and this advantage is higher for men than for women because the reproductive success of men is more variable.[53] We can therefore assume that men are willing to work hard and make large sacrifices in order to increase their social status, and more so than women. This is confirmed by anthropological evidence. Almost all known societies have more male than female leaders.[54]

The advantage of being a leader is higher in regal than in kungic societies. We can therefore predict that regal cultures will be more male-dominated than kungic cultures. Psychological studies have found that people prefer a masculine leader in times of intergroup conflict, while they prefer a feminine leader in situations of within-group competition. This confirms the traditional roles of men as war leaders and women as peace brokers.[55]

Cultural theorists have often mentioned examples of cultures with unusual sex roles to prove their theory that sex roles and male dominance are culturally determined. Sociologist Steven Goldberg investigated these examples by studying the original ethnographic sources, and he found that in all of these cases there are more men than women in influential positions. This supports the theory that men are willing to sacrifice more to increase their status than women are.[56] However, it would be foolish to deny the huge cultural differences in the level of male dominance. Proponents of cultural explanations have emphasized cultural differences, while proponents of biological explanations have ignored them. Here, regality theory may actually contribute to resolving this long-standing disagreement. Regal societies are generally more male-dominated than kungic societies, as the examples in chapter 7 show. We can therefore confirm that there are some cultures with high male domination and other cultures with more equality between the sexes, and that these cultural differences can be explained to a large

---

53   Von Rueden, Gurven and Kaplan (2010), von Rueden and Jaeggi (2016)
54   Goldberg (1993)
55   Van Vugt and Spisak (2008), Spisak and Dekker (2012)
56   Goldberg (1993, chapter 2)

extent by differences in the level of war or collective danger, according to regality theory.

While women are rarely engaged in direct combat, they may contribute to warfare by other means. The outcome of a war is important for the entire group, women as well as men. We can therefore expect women as well as men to support a strong leader when this is necessary for success in war. Regality theory applies to women and men alike, and we can expect everybody to desire a strong leader in times of war or collective danger.

2.6. Cultural effects of regal and kungic tendencies

Studies have revealed many interesting differences between warlike and peaceful cultures. Some of these differences are obvious, others are quite surprising. Here we will discuss some of the cultural tendencies that are characteristic of regal and kungic cultures, respectively, according to regality theory.[57]

When most or all members of a society desire a strong leader, it is hardly surprising that they tend to build a hierarchical political system with a powerful leader at the top. We can also predict that they will develop a strict system of discipline and punishment. These developments may be due to psychological preferences, cultural selection, or rational decision making, and most likely a combination of all three.

Military success requires a strong morale and group spirit or fighting spirit. Regal societies tend to develop a strong feeling of group identity and a world view of friends versus enemies, while some of the most kungic cultures do not even have a name for their own social group. Likewise, regal societies tend to be quite xenophobic and intolerant of all kinds of deviants, while kungic groups are very tolerant.

The ideology, philosophy, and religion of regal societies are typically used as tools for strengthening the morale and group spirit. For example, the ideology of a regal society may state that individuals exist for the benefit of the society, while kungic societies tend to have the opposite ideology, namely that the society exists for the benefit of the individuals.

---

57 Fog (1999, p. 101)

Political and religious leaders in regal societies often support each other to strengthen their power if they are not in fact the same person. Emperors often claim to have divine status. Religion is often used as a means to discipline people, for example by threatening supernatural punishment or promising rewards after death.[58] The religious world of supernatural beings typically reflects or emphasizes important aspects of the social structure of the mundane world.[59]

Regal societies typically also have strict discipline in the area of sexuality. Strict sexual morals may force young people to marry early and have many children since alternative (non-reproductive) outlets for their sexual drive are prohibited.[60] The strict sexual morals do not, however, prevent high-ranking men from having multiple wives or concubines. Kungic societies typically have more permissive sexual morals and lower birth rates.

A possible consequence of the regal ideology that individuals exist for the benefit of the society is that the rate of suicide is low. People do not have the right to take their own lives. This, of course, does not preclude suicide for culturally prescribed reasons, such as shame or self-sacrifice in battle. We can expect kungic societies to have a higher rate of the kind of suicide that Émile Durkheim has called 'anomic suicide'.[61]

Interestingly, the differences between regal and kungic cultures in social structure and worldview are also reflected in art, fiction, music, architecture, and other forms of art. People tend to prefer psychological congruence between the different aspects of their culture, and this also applies to artistic taste. Various forms of art are efficient means for communicating ideological values and cultural unity.[62] Musical style, in particular, has been observed to correlate with social structure, lifestyle, personality, and political preferences.[63] It cannot be ruled out, though, that some of the observed correlations are due to cultural diffusion.[64]

---

58 Watts et al. (2015), McNamara, Norenzayan and Henrich (2014)
59 Moor, Ultee and Need (2007)
60 Garcia and Kruger (2010), van Ussel (1970)
61 Durkheim (1897)
62 Sütterlin (1998)
63 Lomax (1968), Delsing, ter Bogt, Engels and Meeus (2008), North and Hargreaves (2007), Zweigenhaft (2008)
64 Erickson (1976)

Regal cultures tend to produce pictorial art that is highly perfectionist and embellished with endless repetition of meticulous ornamentation and thus reflects the glory of gods or kings. Regal fiction often glorifies gods or kings with clear distinctions between good and evil, friends and enemies. The architecture of regal societies is often particularly conspicuous: large and pompous palaces and religious buildings with luxurious ornamentation and oversized gates that make visitors feel humble (see figure 1). Regal music is also highly embellished and sometimes pompous.

Figure 1. Example of regal architecture. Cologne Cathedral (Kölner Dom). Built 1248–1880. Photo by Tobi 87, 2009.[65]

---

65   CC BY 3.0, https://commons.wikimedia.org/wiki/File:K%C3%B6lner_Dom.jpg

The architecture, art, and music of kungic cultures is less rule-bound and more individualistic, with appreciation of fantasy and innovativeness and a broad range of themes (see figure 2).

Figure 2. Example of kungic architecture. Residential buildings, Bispebjerg Bakke, Copenhagen. Built 2004–2007. Photo by Agner Fog, 2017.

There is a systematic asymmetry in the human cultural heritage. Regal societies are sometimes quite intolerant of art that is not congruent with their culture, and they may even destroy art from previous more kungic periods. For example, the government in Nazi Germany systematically destroyed what they called 'degenerate art'[66] and the Taliban in Afghanistan destroyed the great Buddha statues in Bamiyan.[67] Kungic cultures (including our own modern culture) on the other hand, are very tolerant and even admiring of foreign art and often go to great lengths to preserve the magnificent art and architecture of previous more regal times.

It appears that there was a similar difference between regal and kungic art in prehistoric times. Kungic cultures have produced smaller artifacts of perishable materials, while regal cultures have typically produced large and impressive artifacts of durable materials and perhaps destroyed any remaining artifacts of previous kungic times. This effect most likely causes a systematic sampling bias in the archaeological record.[68]

---

66 Goggin (1991)
67 Francioni and Lenzerini (2006)
68 Fog (2006)

The cultural characteristics that are typical for regal and kungic societies are listed in table 1. It has been observed that societies can reshape these characteristics, not only as a response to changing threats of war but also as a response to other dangers that threaten the social group as a whole, such as economic crisis, famine, natural disasters,[69] and even imaginary dangers such as witches and devils.[70] It is therefore possible that the observed psychological response is a general mechanism of adaptation to the level of danger that threatens the social group as a whole, or perhaps even to any problem that requires collective effort to solve.[71] The effects of dangers to the individual may be different, as discussed in chapter 3.3.

Regal societies	Kungic societies
A hierarchical political system with a strong leader	A flat and egalitarian political system
Strong feelings of national or tribal identity	High individualism
Strict discipline and punishment of deviants	Lax discipline and high tolerance of deviants
Xenophobia	Tolerance of foreigners
The world is seen as full of dangers and enemies	The world is seen as peaceful and safe with little or no distinction between us and them
Belief that individuals exist for the benefit of society	Belief that society exists for the benefit of individuals
Strict religion	Religion has little or no disciplining power
Strict sexual morals	High sexual freedom
High birth rate	Low birth rate
Low parental investment, i.e. short childhood and low education	Long childhood and education
Low suicide rate (except for culturally prescribed reasons)	High rate of anomic suicide
Art and music is perfectionist, highly embellished, and follows specific schemes	Art and music express individual fantasy with appreciation of individuality and innovativeness

Table 1. Regal and kungic cultural indicators.

---

69   Ember and Ember (1992), Kirch (1984)
70   Ben-Yehuda (1990, p. 123)
71   Fog (1999)

3. Contributions from Other Theories

3.1. Influence of the environment: Contributions from ecological theory

In ecology and niche theory, the *competitive exclusion principle* says that complete competitors cannot coexist indefinitely.[1] While this principle has mostly been applied to the areas of ecology and economics, other aspects of niche theory have been successfully applied to eco-cultural specialization.[2] We cannot expect two social groups in close proximity to live in peace if they are adapted to the same environment and depend on the same resources. The two competing groups may merge, separate, differentiate, or fight. But they may not coexist indefinitely unless something prevents them from fighting, such as geographic barriers, technical difficulties, or third party intervention.

The competitive exclusion principle applies to humans as well as to animals with territorial groups. It has been observed that chimpanzees often attack and kill members of neighbor groups and gradually steal their territory.[3] The intensity of intergroup conflict among chimpanzees increases with the population density and the number of males.[4] The

---

1   Hardin (1960)
2   Banks et al. (2006), Maffi (2005)
3   Boesch (2010)
4   Wilson et al. (2014)

© 2017 Agner Fog, CC BY 4.0 https://doi.org/10.11647/OBP.0128.03

closely related species, the bonobo, is much more peaceful, although it is very similar to the chimpanzee in other respects. The reason why the chimpanzee is violent while the bonobo is peaceful may be that there is a patchy distribution of food north of the Congo River, where the chimpanzees live, but a more scattered distribution south of this river, where the bonobos live. A concentration of food or other resources in small patches leads to *contest competition*, where the strongest individuals get the most. A more scattered distribution of food leads to *scramble competition*, where the individual that finds a piece of food first will get it. There is no reason to fight over access to food in an environment of scramble competition.[5]

Chimpanzees and bonobos are our closest relatives among the animals. Humans are equally closely related to both species because the evolutionary split between chimpanzees and bonobos occurred later than the split between humans and the great apes. This has led many scientists to speculate whether human nature is violent like the chimpanzee or peaceful like the bonobo.[6]

Anthropologists have turned to ethnographic and archaeological evidence in order to find out whether warfare was common among early humans, but the evidence is elusive. The few hunter-gatherer groups that have survived long enough to be studied by anthropologists live in marginal areas where the population density is too low for large-scale warfare and where there are no defendable resources to fight over.[7] The victims of warfare among prehistoric nomadic hunters and gatherers were unlikely to be buried, so archaeological traces of injured skeletons may be hard to find.[8]

Some scientists have claimed that humans are peaceful by nature and that the limited evidence of prehistoric violence can be explained as small-scale feuds and raids.[9] Others claim that lethal intergroup violence has been common throughout the evolutionary history of humans.[10]

---

5   White (2013)
6   Wrangham (2012)
7   Gat (2006, p. 14)
8   LeBlanc (2014), Lahr et al. (2016)
9   Fry (2009), Ellingson (2001)
10   Keeley (1996), Guilaine and Zammit (2008), Allen and Jones (2014)

Figure 3. Prehistoric cave painting showing warfare. Bhimbetka, India.
Photo by Nikhil2789, 2008.[11]

An increasing amount of evidence indicates that there might have been violent intergroup fighting throughout human prehistory.[12] A study of nomadic Australian aborigines has found that the level of violence was independent of the population density,[13] while studies in many other areas have found that the level of violence depends on ecological and environmental factors such as the concentration and defendability of resources.[14] Mass killings took place mainly in sedentary cultures and most markedly in connection with agriculture or otherwise defendable resources.[15] Recent archaeological findings show evidence of mass killing among hunter-gatherers near a fertile lakeshore in the late Pleistocene or early Holocene.[16]

---

11 CC BY-SA 3.0, https://commons.wikimedia.org/wiki/File:Cave_Painting-_War_Bhi mbetika_caves.jpg
12 Lahr (2016), Keeley (1996), Guilaine and Zammit (2008), Allen and Jones (2014), Gat (2015)
13 Gat (2015)
14 Boone (1992), Thorpe (2003)
15 Boone (1992), Martin and Frayer (1997, p. 334)
16 Lahr (2016)

Nomadic hunter-gatherers would flee more often than fight, and thus rarely die in a battlefield but more likely die from malnutrition and diseases after fleeing to an inferior territory. Systematic studies of nomadic hunter-gatherer groups show that intergroup violence is common. A review of published examples of peaceful hunter-gatherer groups finds that most of these cases can be explained by their isolation, pacification, or being surrounded by cultures with a different ecology. Typically, the peaceful hunter-gatherer groups were surrounded by agricultural societies whom they would never attack.[17] This confirms the competitive exclusion principle: Hunter–gatherer groups surrounded by other hunter-gatherers will fight, at least occasionally, while hunter-gatherer groups living in their own niche surrounded by groups with a different means of subsistence can exist peacefully. While the peaceful hunter-gatherer societies had little or no intergroup violence, they had plenty of interpersonal violence.[18] The distinction between group-internal and external violence is important here.

We can now answer the long-debated question of whether humans are violent or peaceful by nature: it depends on the environment. A sedentary culture concentrated around defendable resources invites conflict, while a nomadic lifestyle in an environment of sparse resources leads to scramble competition rather than fighting. Regality theory posits that humans have a flexible psychology that allows fast adaption to a peaceful or warlike environment and culture.

It may be possible to roughly predict the degree of intergroup conflict for a particular culture if we study the ecology, mode of subsistence, available technology, and geography. We will expect conflicts to be unlikely for a social group that has adapted to its own specialized niche, but likely for a group that depends on the same niche as a nearby neighbor group. Conflicts can be impeded if traveling is difficult because of geographic barriers or if it is technically difficult to collect and transport sufficient food and water for supporting a troop of warriors.

If food is sparse, and consequently the population density is low, then it will be difficult to assemble a sufficiently large group of warriors to attack an enemy, the warriors will have a long way to travel, and it

---

17   Wrangham and Glowacki (2012)
18   Kelly (2000)

will be difficult to supply enough food for them. Some people find it counterintuitive that low food supply should lead to peace. However, we have to distinguish between a low but stable food supply and a fluctuating food supply. If the food supply is permanently low but stable, then the population density will necessarily be low. Imagine a landscape where food is sparsely distributed and people live in small villages or camps far from each other. How would it be possible to assemble enough warriors from allied neighbor groups to attack an enemy, travel together to the distant enemy territory, and provide and transport enough food and other necessities for the traveling troops? The logistic problems simply make large-scale war impossible in the absence of technological means for food preservation and transport. In any case, there would be little reason for warfare among nomadic peoples in sparsely populated areas, because they would have few possessions worth plundering and territories would be too large to defend.[19]

However, if food is plentiful or concentrated in rich and fertile patches, then the population density will soon become high, geographical distances between enemy groups are likely to be shorter, and it will be easier to organize larger political groups. If, furthermore, the food supply is fluctuating and unpredictable, then there will be occasional periods of famine where contest competition prevails and people fight over the insufficient supply of food. Anthropologists and archaeologists have found evidence of higher levels of conflict connected with settlements in fertile areas such as river valleys. Along with the higher levels of conflict came also alliance formation, peacemaking efforts, and exchange of prestige goods.[20]

In conclusion, we predict that the level of intergroup conflict will be low in areas where food is sparse or where mountains, dense vegetation, aridity, or other environmental factors make traveling difficult. On the other hand, we can expect frequent wars where food production is efficient and concentrated in defendable patches, and where there are efficient means of traveling and food preservation. It has been observed that efficient food production and food storage is connected

---

19 Roscoe (2014), Fry and Söderberg (2013)
20 Dye (2013)

with conformity,[21] which we may interpret as a sign of regality. The predictability of the food supply is also important. Unpredictable famine and natural disasters are factors likely to cause war.[22]

The anthropologist Kirk Endicott has suggested that it might be possible to predict whether a population is violent or peaceful based on the environment. As an example, he describes the Moriori of the Chatham Islands near New Zealand, who changed from violent to peaceful after living a few hundred years in isolation.[23] We will test the feasibility of this kind of prediction in chapters 7 and 8.

3.2. Nature or nurture: Evolution of sociality

Collaboration between biologists, archaeologists, and anthropologists has led to new insights into human social behavior. The modern human species has evolved a large brain in parallel with a more complex social organization. The brain size of our ancestors has tripled over a period of two million years. It is generally believed that the increased brain capacity has been necessary for dealing with more complex social structures, for advanced language, and for developing culture.[24]

Culture is information that is transmitted from person to person through teaching, observation, and imitation. Culture can evolve just like genes evolve, but cultural evolution is much more efficient than genetic evolution because it can involve goal-directed innovation and intelligent problem-solving. New inventions can also be transmitted from any person to any other person, unlike genes, which are transmitted only from parent to child.[25] While many animals are able to learn from conspecifics, the human *capacity for culture* is far more complex than that of any animal. The huge advantage that cultural evolution gives has only been possible through the evolution of a large and efficient brain.[26]

Neuroscientists have found that specific regions of the human brain are involved in various aspects of social behavior and cognition, such as empathy, cooperation, identification with an arbitrary group,

---

21 Berry (1967)
22 Ember and Ember (1992)
23 Endicott (2013)
24 Dunbar, Gamble and Gowlett (2010)
25 Richerson and Boyd (2008, p. 60), Fog (1999, p. 52)
26 Richerson and Boyd (2008, p. 7)

distinction between in-group and out-group, in-group favoritism, group competition, and recognition of social status.[27] A hormone and neurotransmitter called oxytocin, which is involved in many aspects of social behavior in animals and humans, has been found to increase in-group favoritism in humans.[28] These findings support the theory that social organization, hierarchy, and intergroup conflict are important factors that have influenced human evolution and shaped the functioning of our brains. They also show that culture plays an important role, because it is demonstrated that humans are able to identify easily with an arbitrary group of people of mixed races.[29]

Several studies have confirmed that test persons show more cooperation and in-group favoritism when given oxytocin.[30] Interestingly, the effects of oxytocin are different for men and women. The hormone increases the perception of competition in men and the perception of kinship in women. This finding throws new light on the male warrior hypothesis, according to which men are more adapted for fighting and competition than women are.[31]

However, oxytocin is not unambiguously connected with regality, because it does not increase punitiveness.[32] It would be naive to think that we can find a single biological signal for regality. What the new findings of neuroscience tell us, however, is that behaviors that are relevant for regality theory, such as group identification, ethnocentrism, and intergroup conflict, can be influenced by biological signals without the persons being conscious of any such influence. We can therefore reject the theory that violent conflict is caused solely by culture and rational decision-making.[33] While biological processes have a strong influence on social behavior, the opposite is also true. There is plenty of evidence that social processes and cultural differences can influence the human brain and hormonal processes.[34] The role of culture in the shaping of regal psychological reactions is further discussed in chapter 4.1.

---

27   Cikara and van Bavel (2014)
28   De Dreu et al. (2011)
29   Cikara and van Bavel (2014)
30   De Dreu (2012), Sheng et al. (2013), Stallen et al. (2012)
31   Fischer-Shofty, Levkovitz and Shamay-Tsoory (2013)
32   Krueger et al. (2013)
33   Durrant (2011)
34   Kim and Sasaki (2014)

3.3. Fertility: Contributions from life history theory

The regal versus kungic culture dimension has an interesting connection with the *r* versus *K* strategy dimension in evolutionary ecology, which applies to both animals and humans. An *r-strategy* means that individuals start early to reproduce, have many children, and care little for each child. A *K-strategy* means a high age at first reproduction, few children, and a high investment in the care and upbringing of each child. The *r* versus *K* strategy parameter (also called 'fast' versus 'slow' strategy) is a simplification of a more complex set of parameters in biological life history theory, but this simplification will be sufficient for the current purpose.[35]

Humans have a typical *K*-strategy compared with most animals.[36] This strategy is not completely fixed, however. Recent research has shown that there is some room for individual differences and adjustment to the environment. Several studies have found that humans choose a more *r*-like strategy when they live in an environment where the mortality and morbidity of adults is high. A more *K*-like strategy is chosen where the mortality is low, where resources are predictable and defendable, and where the population density is near the carrying capacity of the environment.[37] Economic factors and education also influence the strategy.[38]

While the *r* versus *K* life history theory sees reproductive strategy from the point of view of the individual, the regal versus kungic culture theory is more concerned with a social-level perspective. The optimal strategy from the perspective of the social group in times of war is to produce many children and to raise them as quickly as possible to become fierce warriors. In times of peace, the optimal strategy from the group's perspective is to produce few children in order to avoid overexploitation of the environment and ecological collapse. Group selection theory has not provided a satisfactory explanation of why

---

35 Stearns (1992, p. 207)
36 Mueller (1997)
37 Belsky, Schlomer and Ellis (2012), Ellis, Figueredo, Brumbach and Schlomer (2009), Griskevicius, Delton, Robertson and Tybur (2011), Low (1990), McAllister, Pepper, Virgo and Coall (2016)
38 Shenk (2009)

reproduction is limited, but life history theory seems to provide at least part of the explanation.

In times of war, mortality is high and individuals will choose an *r*-like strategy. In times of peace and stability, we can expect the population density to match the carrying capacity of the environment, and we can expect to see a *K*-like strategy. The interesting observation is that there is a fairly good agreement between the interests of the group according to regality theory and the interests of the individual according to the *r*/*K* life history theory. There is some degree of synergy between the two mechanisms and we will expect a positive correlation between the regal/kungic culture dimension and the *r*/*K* life history strategy dimension. The names regal and kungic were in fact chosen to reflect the resemblance with *r*/*K* life history theory, although the analogy should not be taken too far.

There is one important difference between the predictions of the two theories. Regality theory predicts that fertility will go up as a response to collective danger that requires collective action, while the *r*/*K* life history theory predicts that fertility will go up as a response to any danger, including dangers that affect only the individual.

We will return to the connection between regality theory and *r*/*K* theory in chapter 4.1.

3.4. Contributions from political demography

Political developments, including war and peace, depend on demographic factors. Many momentous historical events are related to the so-called *demographic transition*, which is illustrated in figure 4. Throughout most of human history, the birth rate and death rate have been almost equally high, so that the total population was constant or growing only slowly. At a certain time in history, the death rate began to decline due to improvements in sanitation, hygiene, medicine, nutrition, and living conditions. The higher life expectancy gave parents confidence that they did not need so many children, and after several decades the birth rate began to decrease as well. In Europe, the birth rate began to fall rapidly in the 1960s. In the 1990s, the birth rate had fallen to the same level as the death rate, and population growth stagnated in Europe.[39] This pattern of demographic transition has since been repeated in other parts of the world, including Asia and the Americas, and we are now beginning to see a demographic transition in Africa as well. We can see in figure 5 that the population size has reached a peak in Europe and that the growth rate is tapering off in Asia, America, and Oceania. The growth rate has hardly begun to decrease in Africa.[40]

The fact that the death rate begins to fall first while the birth rate decreases only several decades later has the consequence that we see a very rapid population growth in the intermediate period, where the death rate is low and the birth rate is still high.[41]

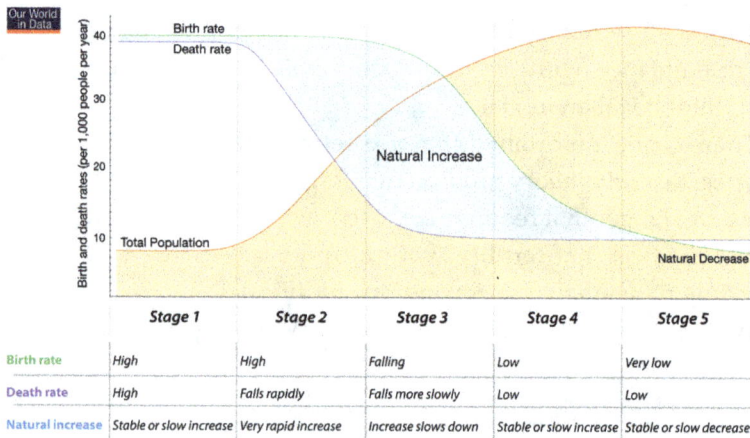

	Stage 1	Stage 2	Stage 3	Stage 4	Stage 5
Birth rate	High	High	Falling	Low	Very low
Death rate	High	Falls rapidly	Falls more slowly	Low	Low
Natural increase	Stable or slow increase	Very rapid increase	Increase slows down	Stable or slow increase	Stable or slow decrease

Figure 4. Demographic transition. By Max Roser, 2016.[42]

The demographic transition can be interpreted as a shift in reproductive strategy from *r*-strategy to *K*-strategy, relatively speaking, as explained in chapter 3.3. It is necessary to have many children as long as the death rate is high. As the death rate falls and improvements in living conditions make it possible to feed more people, the population grows until the new carrying capacity has been reached. Rapid population growth does not

39   Goldstone (2012)
40   Green (2012)
41   Goldstone (2012)
42   CC BY-SA 4.0, https://commons.wikimedia.org/wiki/File:Demographic-Transition OWID.png

necessarily lead to violent conflicts,[43] but the new situation with a larger population and growing urbanization makes life more competitive. The optimal strategy for parents in this new crowded environment is to have fewer children and to invest more in the education of these few children—in other words, a more *K*-like strategy. When these children grow up, they have to spend their most fertile years competing for social positions rather than raising large families.[44]

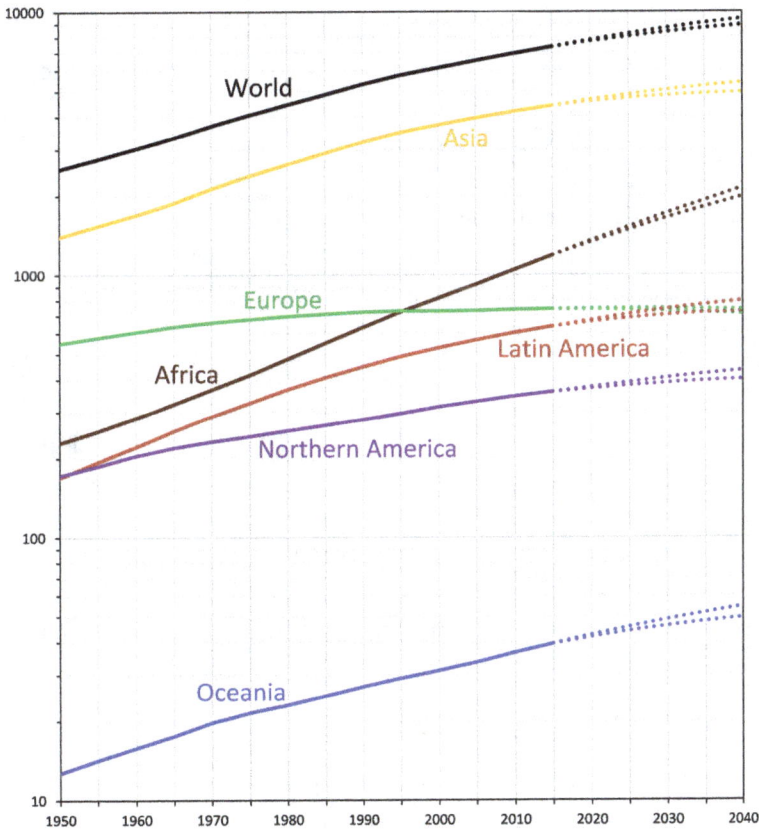

Figure 5. World population by year. Millions, logarithmic scale.
By Agner Fog, 2017. Data from UN.[45]

43   Urdal (2005)
44   Heinsohn (2003, p. 46)
45   United Nations Data Retrieval System. 'World Population Prospects: The 2015 Revision', http://data.un.org/

A period of rapid population growth leaves a cohort of people that is much more numerous than the parent generation. The bulge in the population pyramid that is seen when a large cohort reaches maturity is called a *youth bulge*. The young people in a youth bulge are likely to have problems finding a suitable social position. If tradition dictates that the family inheritance goes to the oldest son, then the subsequent sons are likely to feel superfluous and they will have to fight hard to get a position that matches their expectations. This causes political instability. Disinherited, jobless, unmarried young men are often willing to take high risks in their attempts to find their place in life because they have nothing to lose.[46] Such youth bulges of frustrated young men have fueled many violent conflicts in history. In fact, violent conflicts can hardly be fought without a surplus of venturesome young men.[47]

The European population could not have colonized half the world if they had not had a surplus of young men. In fact, most of the major wars, conquests, and revolutions in European history happened after periods of rapid population growth.[48]

When a large surplus of young people combined with economic stagnation leads to underemployment and low wages, the likely reaction is social discontent and cycles of rebellion and repression.[49] This can destabilize any political system. The frustrated young people will always be able to find a political or religious ideology that can justify their need to fight against a social system that has no place for them.[50] This can lead to revolution and rebellion against a dictatorship, but it can just as well lead to rebellion against a democratic system. Statistical studies show that civil conflicts are much more likely to break out, and democracies to become unstable, in times with large youth cohorts.[51]

The sharply declining fertility rate in the last stage of a demographic transition provides a window of opportunity for economic development called a *demographic dividend*. Savings increase as the number of people

---

46   Wilson and Daly (1985)
47   Goldstone (2012), Heinsohn (2003)
48   Heinsohn (2003, p. 72)
49   Urdal (2006), Cincotta and Doces (2012)
50   Heinsohn (2003, p. 40)
51   Cincotta and Leahy (2007), Cincotta and Doces (2012), Weber (2013)

of working age increases and the number of dependents decreases. This is likely to lead not only to economic prosperity but also to political stability and democracy.[52]

We can conclude that a youth bulge can drive political changes in any direction, depending on the conditions. Youth bulges have been connected with ethnic conflicts[53] and even large territorial conflicts such as the two world wars,[54] but also with rebellions that have their origin in social and economic factors, such as the French Revolution, the Iranian Revolution, and the Arab Spring.[55] It appears that a surplus of fearless young people is necessary for the regal process of building an empire as well as for the kungic process of breaking it down.

3.5. World view and personality: Authoritarianism theory

The psychological characteristics that we call regal have a striking similarity with the phenomenon that social psychologists call *authoritarianism,* and in fact many of the findings of the current study could possibly be explained with the theory of authoritarianism.

Authoritarianism was originally regarded as a mainly fixed part of a person's personality that is formed in childhood.[56] Later studies contradict this and find that authoritarianism is influenced by threats, danger, and uncertainty.[57] Interestingly, it is found that threats to a group that the person belongs to have stronger effect than threats to the person as an individual, and that this increases group cohesion, exactly as predicted by regality theory.[58] However, the evidence is somewhat mixed on the question of individual versus collective threat. Some studies show no authoritarian effect on the test persons unless

---

52 Lechler (2015)
53 Esty et al. (1995), Goldstone et al. (2000)
54 Hvistendahl (2011)
55 LaGraffe (2012), Yair and Miodownik (2016), Apt (2014, chapter 3)
56 Adorno, Frenkel-Brunswik, Levinson and Sanford (1950)
57 Van Hiel, Pandelaere, and Duriez (2004)
58 Jugert and Duckitt (2009), Feldman and Stenner (1997), Duckitt and Fisher (2003), Hogg (1992, p. 35)

they are personally affected by the (collective) threat.[59] There are also disagreements among theorists about the role of predispositions and the direction of causality.[60]

Given the importance of group threat and collective action, it is peculiar that most studies regard authoritarianism as an individual phenomenon and only a few studies treat it as a group phenomenon.[61] There are many different theories about how authoritarianism is generated. Some of these theories are in accordance with the predictions of regality theory, others are not.

A newer version of authoritarianism theory distinguishes between two measures called *right-wing authoritarianism* (RWA) and *social dominance orientation* (SDO), each of which is connected with certain characteristic world views.[62] RWA is linked with the view of the world as a dangerous place, and SDO is linked with the view of the world as a competitive jungle where might is right. RWA leads to social conformity and SDO leads to tough-mindedness, according to this theory, and both lead to negative attitudes towards out-groups.[63]

RWA and SDO are often regarded as two different forms of authoritarianism, where RWA represents *authoritarian submission*, while SDO represents *authoritarian dominance*.[64] Both lead to discrimination against minorities but for different reasons. People with high RWA discriminate because of fear, while people with high SDO discriminate because this gives them a higher status than those they discriminate against.[65]

The RWA theory is particularly relevant for regality theory because the view of the world as a dangerous place in RWA theory can be regarded as very similar to perceived collective danger in regality theory. People with high RWA make society more hierarchical, while people with high SDO use this hierarchy to their personal advantage.

---

59   Asbrock and Fritsche (2013)
60   Cohrs (2013)
61   Duckitt (1989), Stellmacher (2005)
62   Duckitt (2001), Perry (2013)
63   Duckitt (2001), Sibley, Wilson and Duckitt (2007), Sibley and Duckitt (2008), McFarland (2010)
64   Van Hiel, Pandelaere, and Duriez (2004), Altemeyer (1998)
65   Altemeyer (1998)

Research in these theories has led to results that are in good agreement with regality theory because they link social conformity, punitiveness, and xenophobia to collective danger. Evidently, authoritarianism theory and regality theory are two different paradigms looking at the same phenomenon. However, many of the predictions of regality theory that are tested in the present study would have been difficult to make from authoritarianism theory because the latter has more focus on individual psychology than on social and cultural structures.

The area of authoritarianism research is rich in experimental studies.[66] Many of these studies have found that authoritarian responses can be elicited by quite small stimuli, such as making known threats more salient or even by asking the test persons to imagine a particular threatening scenario. The experimental methods that have been developed in the area of authoritarianism research might be useful for testing the effects of different kinds of threats in connection with regality theory.

My main criticism of authoritarianism theory relates to its poor theoretical foundation. Authoritarianism theory has its roots in psychoanalytic theories, which are not falsifiable,[67] though modern versions such as RWA and SDO theories have little or no connection to psychoanalysis. The theoretical model behind traditional authoritarianism theory has limited empirical support, and the theory has often been criticized for political bias.[68] The term 'right-wing authoritarianism' is ill chosen because of its inherent political bias, and also because the concept of right-wing ideology makes sense only in a certain cultural context. In fact, some studies have found the same kind of authoritarianism among communists, who by definition must be called left wing.[69] One researcher even made the contradictory finding that some persons who scored high on the right-wing authoritarianism scale also scored high on a 'left-wing authoritarianism' scale.[70]

---

66   Duckitt (2013)
67   Popper (1963, p. 37)
68   Stellmacher and Petzel (2005), Sibley and Duckitt (2008), Eckhardt (1991), Ray, J. (1989)
69   McFarland, Ageyev and Abalakina (1993)
70   Altemeyer (1998)

While many authoritarianism theorists regard certain political ideologies as undesired psychological aberrations, regality theory sees the same ideologies as adaptive responses to perceived collective danger (or, at least, would-be adaptive in the environment of evolutionary adaptation). The evolutionary theory is less likely to lead to biased and ethnocentric thinking, and we should bear this in mind in our choice of terminology. While authoritarianism theory certainly needs revision and a change of terminology, it builds on a research tradition that has produced many interesting experimental results that may be valuable when reinterpreted in the light of regality theory.[71]

3.6. Contributions from other social psychological theories

There are several other theories that resemble authoritarianism theory but avoid some of the problems mentioned above.

The influential *realistic group conflict theory* finds that hostility between groups is likely to occur when their goals conflict or when they are competing for the same resources.[72]

Integrated threat theory finds that several different kinds of threat can lead to hostility towards outgroups. Realistic group conflict is one of them, symbolic threat is another. Symbolic threat occurs, for example, when immigrants with different values, norms, and beliefs are perceived to threaten one's own culture. Some cases of prejudice are caused mostly by realistic conflicts, while other cases are caused mostly by symbolic conflicts. Other kinds of threats include negative stereotyping and intergroup anxiety, which may be regarded as independent factors in some models.[73]

The theory of *group-based control restoration* finds that people tend to show in-group favoritism in times of social or personal crisis. The proposed mechanism requires that the in-group is relevant to the problem at hand and that the person feels a lack of personal control.[74]

---

71   Duckitt (2015)
72   Jackson (1993)
73   Riek, Mania and Gaertner (2006), González et al. (2008)
74   Fritsche, Jonas and Kessler (2011)

This is in accordance with regality theory in predicting that people want to strengthen their group when facing threats that they cannot handle alone. An interesting question is whether the discrimination against out-groups is directed specifically against a certain out-group that is blamed for the threat, or whether there is a more unspecific discrimination against any group of deviants. In fact, there is evidence for both possibilities.[75] For example, the fear of Islamic terrorists has led to widespread hostility against Arabs and Muslims in general, but also to more punitive attitudes towards perpetrators of totally unrelated crimes such as car theft and rape, as well as to more authoritarian parenting.[76] In the light of regality theory, we may see this as an indication that the regal psychological response is only partially goal-directed towards a specific danger. To some degree, the regal response seems to be an unspecific reaction of strengthening the in-group against any type of danger. A clear example of an unspecific response is homophobia. Many studies have found a strong correlation between authoritarian attitudes and hostility towards homosexuals, even though homosexuals in general pose no threat to the authoritarian person.[77]

People who are uncertain about their social identity are likely to look for a prototypical leader in order to define their identity, according to the *social identity theory of leadership*.[78] People with a high degree of self-uncertainty are more likely to join radical extremist groups, according to *uncertainty–identity theory*. The essence of this theory is that uncertainty about oneself and one's identity motivates people to join groups with high entitativity: clear boundaries, internal homogeneity, and possibly a hierarchical structure and strong leadership.[79] Most of the literature is unclear about what kinds of uncertainties have this effect. All the examples of uncertainty mentioned in the uncertainty–identity theory literature are threatening uncertainties, such as economic problems, ethnic conflicts, and natural disasters.[80] A recent study focusing mainly

---

75   Greenaway, Louis, Hornsey and Jones (2014)
76   Fritsche, Jonas and Kessler (2011)
77   Basow and Johnson (2000)
78   Hogg, van Knippenberg, and Rast (2012)
79   Hogg (2014)
80   Hogg, Meehan, and Farquharson (2010), Hogg and Adelman (2013), Rast, Hogg, and Giessner (2013), Schoel et al. (2011)

on economic problems finds that both individual and collective economic problems have this effect, but the measure of individual uncertainty is very coarse.[81]

This takes us back to the already well-known connection between threat and authoritarianism. Uncertainty and threat are known to be connected with political conservatism, according to the so-called *uncertainty–threat model*.[82] It appears that uncertainty and threat are better at explaining conservative and authoritarian extremist groups than anti-authoritarian extremist groups. Additional factors that have been suggested to explain extremism in minority groups include perceived injustice, group threat, perceived illegitimacy of authorities, and a feeling of being disconnected from the majority of society.[83]

Terror management theory is based on the assumption that people become anxious when thinking about their own future death. Anything that reminds people of their own mortality will provoke a psychological defense that emphasizes their social group and a world view which gives meaning to life and death.[84] Critics of this theory have found that the same psychological reaction is seen when threats unrelated to mortality are made salient. This supports an evolutionary explanation related to group defense (in accordance with regality theory) rather than the terror management theory.[85] Another study finds that terrorism salience, but not mortality salience, invokes a group defense response. This finding supports *system justification theory*, but not terror management theory.[86] A meta-analytical review of the evidence finds that mortality salience has the same effect as other threats to the world view, but the effect of mortality salience is delayed whereas the effect of other threats is immediate.[87]

A recent theory about the connection between conflict and political attitudes is the *stress-based model of political extremism*. This model is based on studies of protracted conflicts such as the Israeli–Palestinian conflict and the conflict in Northern Ireland. The studies find that

---

81   Kakkar and Sivanathan (2017)
82   Jost et al. (2007)
83   Doosje, Loseman and Bos (2013)
84   Greenberg and Arndt (2011)
85   Navarrete, Kurzban, Fessler and Kirkpatrick (2004)
86   Ullrich and Cohrs (2007)
87   Martens, Burke, Schimel and Faucher (2011)

a population exposed to political violence will develop symptoms of psychological distress. This distress leads to increased perceptions of threat, which in turn lead to political extremism and lack of support for political compromise.[88] The distress and perception of threat is particularly strong in cases of terrorism where random civilians are victims. This explains why peace negotiations often fail in conflicts that involve frequent episodes of terrorism.[89] The research also finds that the effect can be ameliorated by reassuring messages or aggravated by alarming messages.[90] The stress-based model contradicts the rationalist view, held by many negotiators, that the conflicting parties will be more motivated to accept negotiations and compromises when the violence becomes intolerable. On the contrary, they will be less compromising the worse the violence they experience.[91] This model is not based on evolutionary theory but on psychological theories of coping. Yet the findings are in perfect agreement with regality theory: people become more regal and uncompromising the more threatening they perceive their situation to be.

3.7. Contributions from social values theories

A network of social scientists have created the World Values Survey, which is a research project aimed at measuring people's values and beliefs around the world. Ronald Inglehart and Christian Welzel have analyzed these data and found that most of the variance in cultural values can be expressed by two factors:[92]

1. *Traditional versus secular–rational values,* reflecting the contrast between the relatively religious and traditional values that generally prevail in agrarian societies, and the relatively secular, bureaucratic, and rational values that generally prevail in urban, industrialized societies.

---

88   Canetti et al. (2013, 2015)
89   Canetti et al. (2015), Hirsch-Hoefler, Canetti, Rapaport and Hobfoll (2014)
90   Hirsch-Hoefler, Canetti, Rapaport and Hobfoll (2014), Canetti-Nisim, Halperin, Sharvit and Hobfoll (2009)
91   Canetti et al. (2013)
92   Inglehart and Welzel (2005, p. 48), Inglehart (2007)

2. *Survival versus self-expression values,* reflecting an inter-
   generational shift from emphasis on economic and physical
   security in the industrial society towards increasing emphasis
   on self-expression and subjective well-being, and the focus on
   rights above duties of a democratic, post-industrial society.

Figure 6 shows the average position of citizens in different countries
on these two dimensions in 2011 to 2012. The traditional values reflect
the importance of religion, deference to authority, traditional families,
moral standards, desire for a large number of children, and resistance
to abortion. The survival values reflect traditional gender roles, hard
work, confidence in government, and intolerance of deviants. Inglehart
and Welzel explain the traditional values on the first dimension by the
need for collective action in complex agrarian societies. The shift from
traditional to secular–rational values is connected with the change from
religious authorities to secular authorities that follows industrialization.
The survival values on the second dimension express the need for
physical security in terms of economy and health. The shift from survival
values to self-expression values reflect a relief from the immediate
threats of the industrial society to the higher economic security that
comes with a higher level of education and a modern welfare state. The
self-expression values also reflect an emancipation from authority and a
criticism of the risks of technology.[93]

Inglehart and Welzel's study does not focus on war or bellicosity, but
on how people react to changing perceptions of existential security.[94] As
existential security is essential in regality theory, we will expect both
factors to correlate with regality. The traditional versus secular–rational
values focus on how to deal with collective survival, while the second
dimension, survival versus self-expression values, has more focus
on individual threats. But the self-expression values also reflect an
emancipation from authority that we can relate to a kungic development.

Some studies described in the next chapter seem to support these
expected correlations with regality,[95] but we should not draw wide-
ranging conclusions because the studies are partially based on the
same data.

---

93   Inglehart and Welzel (2005, p. 48ff), Inglehart (2007)
94   Inglehart and Welzel (2005, p. 52)
95   See table 2

A further study of religious values shows that the transition from agrarian to industrial to postindustrial societies is followed by a decrease in religiosity.[96] This is explained by increasing security. Religion is more important to people when they feel insecure. The study shows that basic religious beliefs are not easily changed, but the importance of religion in people's lives is higher in societies with poverty, poor education, poor health, and high population growth. For example, the importance of religion is high in countries with high economic inequality such as the USA, Ireland, and Italy. The data do not support other theories of religiosity, such as secularization theory and religious market theory. Interestingly, both individual threats and collective threats seem to increase people's need to seek comfort in their religion.[97]

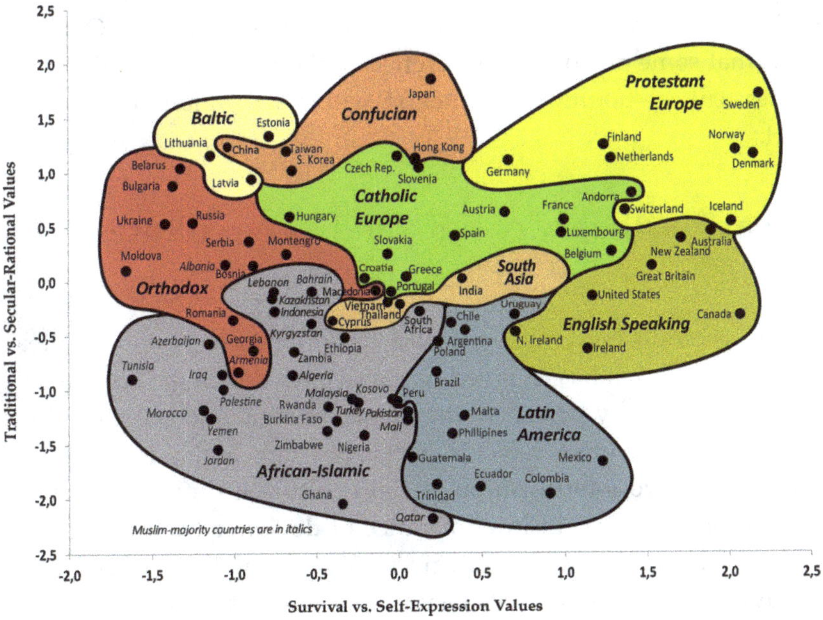

Figure 6. Inglehart and Welzel's Cultural Map (wave 6). World Values Survey Database.[98]

---

96   Norris and Inglehart (2011, chapter 4)
97   Norris and Inglehart (2011, chapter 11)
98   Public domain, http://www.worldvaluessurvey.org/WVSContents.jsp?CMSID=Findings

Another study found that religiosity increased after an earthquake in New Zealand.[99] Religiosity also increased after a financial crisis in Indonesia in 1997,[100] but not after a financial crisis in Europe in 2008.[101] Regality theory may give a clue why the latter financial crisis did not lead to increased religiosity. Regality is predicted to increase when a crisis is blamed on external enemies or uncontrollable forces of nature, but it is predicted to decrease when a crisis is blamed on one's own despotic leaders. The 2008 financial crisis was frequently blamed on greedy bankers and corrupt or incompetent politicians rather than on some nebulous external power or economic laws of nature. The prediction from regality theory is that this should cause a revolt, and indeed there were large protests in the streets in much of Southern Europe as a reaction to this crisis.

Whether personal values reflect shared norms of the surrounding society or just personal differences is a matter of debate. One study finds that some personal values reflect social norms while other values are less culture-bound.[102] Chapter 3.8 will look at values at the cultural level.

3.8. The theory of tight and loose cultures and other culture theories

The distinction between tight and loose cultures is an old idea in anthropology that has received renewed attention in the last decade.[103] This theory is very similar to the theory of regal and kungic cultures, but the two theories have in fact been developed independently without any cross-fertilization. A tight culture is defined as a culture with strong norms and low tolerance of deviance from these norms. A loose culture has weak social norms and a high tolerance of deviant behavior. The tightness of a culture is increased by ecological or external threats such as high population density, resource scarcity,

---

99   Sibley and Bulbulia (2012)
100 Chen (2010)
101 Healy and Breen (2014)
102 Minkov, Blagoev and Hofstede (2012)
103 Gelfand et al. (2011), Uz (2015), Harrington and Gelfand (2014)

disease, natural disasters, and external conflicts. These threats lead to the development of strong norms for enhancing the social coordination needed to effectively deal with such threats, according to the theory. The tightness is also provided by and reflected in social and political institutions such as the government, media, education system, legal system, and religion. People adapt to the strength of norms and intolerance of deviance by the psychology of self-regulation.[104] The older definitions of tightness have more focus on political organization,[105] while later definitions focus more on norms and deviance. Both sets of definitions are relevant to regality theory.

The tightness/looseness theory does not go into detail with any mechanisms beyond a general functionalist explanation and the assumption that cultural evolution somehow brings about the necessary norms and institutions. There is agreement between regality theory and tightness/looseness theory on the basic claim that collective threats lead to strict norms and psychological restriction, but the two theories disagree on the chain of causality. The basic model of regality theory is:

Collective threats → Psychological adaptation ↔ Cultural norms

whereas tightness/looseness theory says:

Collective threats → Cultural norms ↔ Psychological adaptation

The available data cannot distinguish between these two models because the direction of causality cannot be deduced from statistical correlations.

Tightness/looseness theory focuses mainly on ecological threats, while the threat of war plays only a minor role and bellicosity is not among the psychological reactions mentioned by this theory. While tightness and regality are very similar constructs, they are not the same. It would be more appropriate to say that tightness is one of several indicators of regality and in fact a very important indicator. Virtually everything that can be predicted from tightness/looseness theory can also be predicted from regality theory, while the opposite is

---

104 Gelfand et al. (2011)
105 Pelto (1968)

not true. We will now look at a number of statistical studies that have been made to test tightness/looseness theory and compare them with the predictions of regality theory.

In a pioneering study, Michele Gelfand et al. asked people in thirty-three countries how much they agreed with statements such as 'People in this country almost always comply with social norms'. This was not a direct measurement of cultural tightness but rather a measurement of perceived tightness. The results were correlated with a number of social and cultural variables and the results are shown in table 3.[106] Critics have pointed out that the perceived tightness may not be cross-culturally comparable, because people may have different kinds of norms in mind when answering the questions or they may be judging by different standards or frames of reference.[107] Later studies have used more direct criteria to gauge cultural tightness, and this turns out to give better correlations.

Psychologist Irem Uz has calculated three different measures of cultural tightness for sixty-eight countries: (1) a domain-specific index based on people's tolerance for various morally debatable behaviors, such as prostitution, abortion, divorce, euthanasia, and suicide in the 2006 World Values Survey; (2) a domain-general index based on people's endorsement of 124 different sets of values and behavioral practices in the World Values Survey. The index is based not on the mean but on the standard deviation of the responses. A high standard deviation of the responses is taken to indicate a high diversity of opinions, while a low standard deviation indicates that opinions comply with social norms; (3) a combination index based on a factor analysis of values in the domains of work, family, and religion in the World Values Survey.[108] A strong correlation was found between these three measures of cultural tightness, as shown in table 2. The correlation with Gelfand's perceived tightness was not statistically significant, but all three measures of tightness correlated significantly

---

106 Gelfand et al. (2011)
107 Minkov, Blagoev and Hofstede (2012), Uz (2015), Heine, Lehman and Greenholtz (2002)
108 Uz (2015)

with the secular–rational values and the self-expression values described in chapter 3.7. The measures of tightness showed highly significant correlations with various measures of collective threat and with various measures of tolerance.

A study by Jesse Harrington and Gelfand calculated an index of cultural tightness of the fifty US states based on objective indicators such as corporal punishment in schools, death penalty execution, punishments for marijuana law violations, alcohol restrictions, same sex civil unions, religiosity, and immigrant population. This index showed high correlations with indicators of collective threat such as natural disasters, health, and infant mortality, but not with population density. The index also showed strong correlations with various measures of law enforcement and attitudes toward regulation and deviant behavior.[109] The results are shown in table 4.

The findings of these three studies are in perfect agreement with the predictions of regality theory indicated in tables 2, 3 and 4. The indication > 0 means that a positive correlation is predicted. < 0 means that a negative correlation is predicted. 0 means that no correlation is predicted or that the prediction is ambiguous. The directions of the correlations found in the tightness/looseness studies are in agreement with the predictions of regality theory in all cases where the correlations are statistically significant ($p \leq 0.05$) as well as in most cases where the correlations are not significant ($p > 0.05$). These results are very interesting because they demonstrate the predictive power of regality theory. The methodology used in these studies can be adapted for future studies aimed specifically at testing the predictions of regality theory.

Another study by Lazar Stankov, Jihyun Lee and Fons van de Vijver found similar results for two factors that they called 'conservatism versus liberalism' and 'harshness versus softness', but their study did not investigate the influence of threats.[110]

---

109 Harrington and Gelfand (2014)
110 Stankov, Lee and van de Vijver (2014)

Variable	Cultural tightness			
	Domain specific (1)	Domain general (2)	Combination (3)	Expect from regality theory
Historical threat	.636***	.476***	.685***	>0
Current threat	.520***	.326***	.514***	>0
Traditional/industrialized	.549***	.405***	.630***	>0
Institutional repression	.626***	.402***	.639***	>0
Subjective well-being	-.279*	-.196	-.376**	<0
Feelings of freedom of choice and control	-.277	-.230	-.314*	<0
Willingness to live near dis-similar others	-.426***	-.383***	-.675***	<0
Tolerance for personal-sexual deviations	-.814***	-.502***	-.723***	<0
Tolerance for violation of legal rules	-.236	-.447***	-.219	<0
Behavior inhibition	.406***	.369*	.590***	>0
Inglehart and Welzel factors:				
Secular-rational values	-.577**	-.302*	-.448**	<0
Self-expression values	-.533**	-.284*	-.557**	<0
Hofstede and Minkov factors:				
Power distance	.242	.038	.332*	>0
Collectivism	.411**	.316*	.417**	>0
Masculinity	-.043	-.141	.028	>0
Uncertainty avoidance	.003	-.171	-.061	>0
Indulgence versus self-restraint	-.263*	-.303*	-.382**	<0
Stankov, Lee, and Vijver factors:				
Conservatism versus liberalism	.39	.33	.40	>0
Harshness versus softness	.20	.41	.46*	>0

Table 2. Correlation of the three tightness measures of Uz with various variables, compared with the predictions of regality theory.[111] Levels of significance: * $p < 0.05$, ** $p < 0.01$, *** $p < 0.001$. The signs of the coefficients have been adjusted for clarity.

Variable	Perceived tightness	Expect from regality theory
Percentage of population using left hand	-.61*	<0
Accuracy of clocks in major cities	.60**	>0
Justifiability of morally relevant behavior	-.48**	<0
Unrestricted sociosexuality orientation	-.44*	<0
Alcohol consumption	-.46**	<0
Preferences of political systems that have a strong leader or are ruled by the army	.38*	>0

111 Uz (2015), Stankov, Lee and van de Vijver (2014)

Most important responsibility of government is to maintain order of society	.61**	>0
Agreement on ways of life need to be protected from foreign influence	.57*	>0
Would not want to have immigrants as neighbors	.43*	>0
Percentage of population of international migrants (log)	-.32	?
Agreement on one's culture is superior	.60**	>0
Individualism	-.47**	<0
Power distance	.42*	>0
Uncertainty avoidance	-.27	>0
Masculinity index	-.08	>0
Long-term orientation index	-.05	0
Harmony	-.26	<0
Conservatism	.43*	>0
Hierarchy	.47*	>0
Mastery	.18	0
Affective autonomy	-.23	<0
Intellectual autonomy	-.28	<0
Egalitarian commitment	-.41	<0
Family collectivism	.49**	>0
Institutional collectivism	.43*	>0
Performance orientation	.35	0
Power distance	.32	>0
Gender egalitarianism	-.35	<0
Assertiveness	-.29	0
Uncertainty avoidance	.32	>0
Future orientation	.47*	0
Humane orientation	.30	0
Loyalty versus utilitarian involvement	.45*	>0
Traditional versus secular rational values	-.11	>0
Self-expression values	-.13	<0
Fate control	.44*	>0
Spirituality	.52**	>0
Reward for application	.60**	>0
Cynicism	.14	>0
Flexibility	-.20	<0
Vertical sources of guidance	.40*	>0
Guidance from widespread beliefs	.54**	>0
Guidance from unwritten rules	.18	>0
Guidance from specialists	-.18	>0
Guidance from coworkers	-.16	<0
Gross national product	.05	0
Global growth competitiveness	-.08	0

Table 3. Correlation of the perceived tightness measure of Gelfand[112] with various variables, compared with the predictions of regality theory. Levels of significance: * $p < 0.05$, ** $p < 0.01$.

112 Gelfand et al. (2011)

Variable	Tightness	Expect from regality theory
Natural disaster vulnerability	.84***	>0
Food insecurity	.46***	>0
Parasite stress index	.55***	>0
Infant mortality rate	.76***	>0
Environmental health: green index	-.77***	<0
Percentage of slave-owning families, 1860	.78***	>0
Population density (log)	.05	>0
Law enforcement employees	.29*	>0
Civil liberties	-.63***	<0
Attitude towards government control of media	.68***	>0
Behavioral constraint index	.81***	>0
Relativistic attitude to right and wrong	-.38***	<0
Attitude against immoral actions	.52***	>0
Desire for strict law enforcement	.49***	>0
Support for police use of force	.65***	>0
Support for buying US products	.78***	>0
Support for import restrictions	.51***	>0
Interest in foreign cultures	-.58***	<0
Collectivism	.37**	>0
Gender equality index	-.77***	<0
Discrimination charges per capita	.61***	0
Alcohol binge drinking	-.29*	<0
Illicit drug use	-.52***	<0
Murder rate	.19	0
Burglary rate	.22	0

Table 4. Correlation of tightness of US states with various variables,[113] compared with the predictions of regality theory. Levels of significance: * $p < 0.05$, ** $p < 0.01$, *** $p < 0.001$.

Most studies of culture types refer to modern industrial cultures, but a theory of Marc Howard Ross about cultures of conflict explicitly refers to non-industrial societies.[114] His hypothesis is that psycho-cultural dispositions for conflict and violence are formed through harsh childhood socialization and male gender identity conflict, and the targets for the aggressive tendencies can be either group-internal or group-external depending on structural factors, which he calls 'cross-cutting ties'. The correlation he finds between harsh childrearing practices and violent conflict is actually in agreement with the findings of regality

---

113 Harrington and Gelfand (2014)
114 Ross (1993)

theory, but there is disagreement about the direction of causality, which cannot be determined from the available statistical data. The idea that cross-cutting ties can mitigate conflicts is quite reasonable, but the hypothesis that internal and external conflicts form equivalent targets for an aggressive disposition is not in accordance with regality theory, and Sigmund Freud's drive-discharge theory of violence has often been criticized.[115]

Theories of culture types have also been developed for organizational culture. Geert Hofstede, Gert Jan Hofstede and Michael Minkov have defined a number of dimensions based on studies of organizational culture. The most important dimensions are:[116]

- Power distance. This is a measure of hierarchy and the degree of inequality between leaders and followers or between teachers and students.

- Collectivism versus individualism. This is a measure of the degree to which people identify with their organization or social network rather than rely on themselves and pursue their individual self-interests.

- Masculinity versus femininity. The masculine values include strength, competition, care for money and material things, and prioritization of economic growth. The feminine values include warm relationships, empathy, care for the weak, and preservation of the environment.

- Uncertainty avoidance. This is a reflection of fear and stress. Uncertainty avoidance includes conservatism, fear of change, fear of ambiguity, repression of deviance, need for precise and formal rules, nationalism, and xenophobia.

- Indulgence versus self-restraint. Indulgence includes personal life control, optimism, happiness, leisure, and individual freedom. Self-restraint includes moral discipline, order, and pessimism.

- Long-term orientation. This is a focus on perseverance, thrift, self-discipline, humility, learning, pragmatism, and adaptiveness. The opposite is short-term orientation, which

115 Sipes (1975)
116 Hofstede, Hofstede and Minkov (2010)

includes a focus on quick results, pride, saving one's face, and respect for tradition.

These dimensions include many elements that relate to regality theory, and we will expect the dimensions of power distance, collectivism, masculinity, and uncertainty avoidance to be positively correlated with regality, while indulgence should be negatively correlated with regality. We can test these predictions by using cultural tightness as an approximation for regality. The results in table 2 confirm the predicted correlations for power distance, collectivism, and indulgence, while there is no significant correlation for masculinity and uncertainty avoidance. The basic principle of long-term orientation is not related in any obvious way to regality, although both long-term and short-term orientation have elements that are typical of regality, namely self-discipline and respect for tradition, respectively. No significant correlation is found for this dimension.

We have now seen that psychological preferences or values can be described with many different measures or dimensions, and these can be studied both at the individual level as described in chapter 3.7, and at the collective level as described in this chapter. The regal-kungic dimension seems to be involved in or to interfere with many of the other dimensions that have been defined, both at the individual level and at the collective level, and the effects are in good agreement with the predictions of regality theory. More research is needed to figure out which of the many proposed dimensions are most useful, how they interact or overlap, and which factors they are influenced by. Regality theory is useful here because it identifies a mechanism by which many psychological preferences are influenced.

3.9. Contributions from human empowerment theory

Christian Welzel has developed a new theory of human empowerment based on the findings of the World Values Survey.[117] Welzel's theory posits that, as humans get more resources and freedom of action, they will give higher priority to principles of freedom and this will lead

---

117 Welzel (2013)

to democracy and other civic institutions that guarantee individual freedom. Welzel's model can be summarized as follows:

Action resources → Emancipative values → Civic entitlements

The concept of action resources defines the resources necessary for existential security, including material resources (equipment, tools, and income), connection resources (networks of exchange and contact), and intellectual resources (knowledge and skills). Under a shortage of action resources, people will give priority to survival values, as explained in chapter 3.7. When people have plenty of action resources so that existential security is no longer a concern, they will give higher priority to *emancipative values*. Welzel has replaced the factor named 'self-expression values' in the previous study with a very similar index named 'emancipative values'. The latter index is composed in a way that puts more emphasis on theoretical relations and less on statistical coherence (Cronbach's α is lower). Emancipative values give people the impetus to organize collective action against tyrannical overlords, replace an authoritarian government with a more democratic rule, and implement civic institutions that guarantee individual freedom. This model explains the growing democratization in the world during the last several centuries, according to Welzel's theory.[118]

Welzel's theory aligns neatly with regality theory, although it is seen from the opposite perspective. Regality theory is based on the model:

Collective danger → Psychological preference for a strong leader → Authoritarian political structure

Welzel's theory is expressed in the opposite terms:

Existential security → Psychological preference for emancipative values → Democratic political structure

The two theories are not perfectly equivalent, though: Welzel does little to explore the effects of danger, which is central to regality theory; Welzel's theory does not distinguish between individual and collective danger; and war plays hardly any role in Welzel's theory. Studying a phenomenon from a different perspective is likely to lead to different

---

118 Welzel (2013)

discoveries, and Welzel's theory throws new light on the processes during peaceful development.

A very important outcome of Welzel's study is that he finds evidence for the direction of causality by means of statistical analysis of time series over ten years or more. His analysis shows that the direction of causality is as indicated by the arrows above, whereas there is much less causal influence in the opposite direction. A similar study has not yet been carried out for regality theory because it would require the collection of large amounts of data over a prolonged period of time.

We may notice that democratic institutions come at the end of the causal chain rather than the beginning. This explains why many political attempts at forced democratization have failed. Political initiatives that aim at spreading democracy should focus on existential security before doing anything else.

The first link in Welzel's model, existential security, depends on external geographic factors. The most important factors can be summarized in what Welzel calls the *cool-water condition*.[119] This includes:

- A relatively cold climate, which makes work less physically exhausting, reduces the danger of infectious diseases, and reduces soil depletion.

- Permanently navigable waterways, which facilitate trade and democratic market access.

- Continuous rainfall all year round. This makes irrigation unnecessary so that farmers are autonomous and independent of a centrally controlled irrigation system.

These conditions are found in particular in Western Europe and Japan. Historically, farmers living under cool-water conditions had high productivity, autonomy, and existential security. This made it rewarding to switch from a strategy of maximizing fertility to a strategy of improving skills (from an *r*-strategy to a *K*-strategy). This demographic transition was a necessary condition for technological progress and economic growth, according to *unified growth theory*.[120] Artisans and urban markets came into being. The workforce was small and its quality high. The high

---

119 Welzel (2013, p. 15)
120 Galor (2011)

cost of labor was an incentive to search for technologies that save labor. Technological advances and higher productivity contributed further to existential security. Democratization and individual economic freedom made it profitable to invest in technological inventions. This explains why technological innovation was particularly strong in Western Europe, according to Welzel.[121]

If we want to compare the predictions of Welzel's theory of human empowerment with regality theory, we have to take war into account. Environmental factors that facilitate high productivity and easy transport are also factors that facilitate war. According to regality theory, we will expect to see wars, empires, and tyrannical kings in such an area rather than human emancipation. An authoritarian rule is not conducive to individual inventiveness, but the ruler may command the development of a large and efficient infrastructure, including cities, roads, ships, and harbors, which is a precondition for technological progress. Empires wax and wane for reasons explained in chapters 4.1 and 4.2, and there may be a window of opportunity for inventiveness and technological progress when a government is too weak to confiscate the profits of individual enterprise but not too weak to protect the private property of inventors.[122] Inventiveness and technological progress is also possible in remote areas that are less regal or when war is prevented by factors such as economic interdependence, alliances, or deterrence. It is no coincidence that the industrial revolution started in England, which had fewer enemies to fear than central Europe at the time. Technological progress brought prosperity and democracy, in accordance with Welzel's theory.

3.10. Moral panics: Contributions from the sociology of deviance

The human mind is notoriously bad at making rational decisions in the face of uncertainty and low probabilities.[123] Some people buy lottery tickets even though the average losses are higher than the average

---

121 Welzel (2013)
122 Acemoglu and Robinson (2012, chapter 3)
123 Tversky and Kahneman (1992)

gains, and some people have an extreme fear of terrorism even though the risks of dying in a traffic accident or from a lifestyle disease are many thousand times higher. Sometimes we ignore serious risks, and sometimes we overreact to even the smallest risks. Often, a certain group of people are blamed for causing danger and they are labeled as deviants, such as criminals or terrorists.

As already mentioned, the regal reaction to collective dangers depends not on the objective risk but on the perceived risk. An exaggerated perception of a minor or unlikely danger can have a strong regalizing effect. A highly emotional and exaggerated collective fear is called a 'witch hunt' or a 'moral panic'. The fear is likely to lead to strong reactions against a group of deviants who are seen as threatening the social order.[124]

The longest witch hunt in history—and the one that gave the phenomenon its name—took place in Europe around the period of the Renaissance. The Catholic Church needed scapegoats in order to consolidate its dwindling power, and the scapegoats were first heretics and later witches. The witches were accused of worshipping the devil and causing all kinds of evil. People who were accused of witchcraft were tortured until they confessed, and they were forced to inform against other witches so that the process could continue.[125] The dangers that the Church was fighting against may have been completely imaginary, but the social effects of the witch hunts were real. The regal effect of the witch hunts strengthened the authority of the Church and enabled it to defend its power against the threat of secularization.

Invisible dangers are particularly effective in witch hunts and moral panics because it is easier to exaggerate a danger when it cannot be seen and its objective magnitude cannot easily be estimated. This is also the case today. In modern society, the fears of many people have changed from a concern about how to get food to a concern about technological risks. Ulrich Beck calls this the 'risk society'.[126] Many risks are more or less invisible, such as pollution, nuclear radiation, and genetically modified foods. These risks—or rather the perception of these risks—can be manipulated up or down to serve the interests of various actors.

---

124 Cohen (2011), Goode and Ben-Yehuda (2009)
125 Ben-Yehuda (1980)
126 Beck (1992)

Environmental protection organizations claim that the risks of pollution are high, while the polluting industries claim that the risks are small. This contest over defining the risks is also a power game where the winner will gain more control of the situation.[127]

Governments and powerful elites often use witch hunts in order to consolidate their power. Well-known modern examples include McCarthyism, the 'war on drugs', and the 'war on terror'.[128] Take the 'war on drugs' in the United States as an example. Opinions are divided over whether this campaign has actually reduced the consumption of illegal drugs, but the intense focus on the dangers of drugs has fueled a 'tough on crime' attitude. The pursuit of drug criminals has led to many secondary crimes, such as economic crimes to pay for drugs, gang violence, weapons proliferation, corruption, vigilantism, and social deterioration. These secondary menaces have led to still more punitive sentiments and erosion of the standards of justice.[129] This self-amplifying process has driven the whole society in a more regal and punitive direction, which is part of the explanation why the USA has the highest incarceration rate in the world.

There is no clear distinction between a witch hunt and a moral panic. The term 'witch hunt' is used mainly when the fear and persecution of deviants is controlled by a powerful elite, while the term 'moral panic' usually describes a more spontaneous collective fear generated by gossip and mass media without any central control. An example of a moral panic is the child abuse panic, which began in the USA in the early 1980's and spread to large parts of the world. This panic was started by social workers, physicians, psychologists, and child protection organizations. The initial focus was on physical abuse and neglect of children,[130] but the focus gradually changed to sexual abuse, which was even more emotionally touching.[131] While child abuse is certainly a real problem, the reactions were highly emotional and many innocent people have gone to jail because of overreactions and disregard

---

127 Altheide (2014, p. 118)
128 Schack (2009)
129 Duke and Gross (1994)
130 Nelson (1984)
131 Best (1990)

for common standards of justice.[132] This is typical of a moral panic. The most characteristic symptoms of a moral panic include:

- There is a highly emotional collective reaction to a problem that has hitherto been largely ignored. The issue itself has high emotional appeal.

- Exaggerated claims are made about the incidence and effects of the problem.

- A group of scapegoats is stigmatized as responsible for the problem.

- The phenomenon is considered so dangerous that common standards of evidence and due process are violated.

- The definition of the deviance is unclear and expanding. The limit between normal and deviant shifts so that more and more people or behaviors can be defined as deviant and dangerous.

- If any experts contradict the claims, they are accused of being deviants themselves or allied with the deviants and therefore persecuted. Soon, there is hardly anybody left who dares to contradict the claims, and the new consensus allows even stronger claims to be made.

- A group of often self-appointed experts appear who claim to understand the problem and get resources to fight it.

The regal effect of a moral panic is obvious, but it is not always obvious who benefits from it. The experts who are allowed to define the problem and devise a strategy to fight it benefit in several ways. They get money and resources; they gain prestige and respect for their profession; and they gain power by being allowed to redefine the norms for the society and ultimately change social structure and laws in their own interest.[133] The problem is sometimes fought with means that have little connection with the problem.[134] Often, there is a competition between different professions over who 'owns' the issue, in other words, over which people are recognized as the

---

132   Eberle and Eberle (1986)
133   Ben-Yehuda (1990, p. 97), Foucault (1980, chapter 6)
134   Keen (2006, p. 98)

experts allowed to define the problem.[135] The 'issue owners' may be priests, psychiatrists, psychologists, social workers, police, legal experts, economists, military figures, industry figures, grassroots movements, politicians, etc. The social structure and power balance of a society may change profoundly when the ownership of social problems is transferred from one group of experts to another. Such societal changes are not necessarily the result of a planned strategy on the part of the persons involved, but a witch hunt or moral panic provides a convenient perpetual source of enemies to keep the hysteria going.[136] Those who gain power and prestige from fighting a social problem are indeed likely to keep promoting and expanding the fight.

Moral panics are characterized by exaggerated claims and exaggerated reactions. Many sociologists have discussed whether it is possible to determine that a claim is exaggerated if we have no objective standards to judge it by. Typically, we have to involve a different branch of science that deals with the subject matter of the problem. Erich Goode and Nachman Ben-Yehuda have proposed a list of simple criteria for judging exaggeration,[137] and Michael Schetsche has proposed an analytical framework that tries to separate the subject matter of the problem from its social construction.[138]

Conflicts about the definition of a social problem occur all the time, for example when environment protection organizations and industry lobbyists disagree over how dangerous a certain technology is. These conflicts rarely escalate into genuine moral panics with all the symptoms listed above. The less intense conflicts have little or no regalizing effect, but they are still important for defining social boundaries.

The kinds of deviance that cause strong emotional reactions are often of a religious or sexual nature.[139] Attempts to suppress sexual deviance are typically seen in connection with regal reactions. We may speculate why. It is hard to find a rational reason why anybody

---

135 Gusfield (1989)
136 Schack (2009)
137 Goode and Ben-Yehuda (2009, p. 101)
138 Schetsche (2000, p. 29)
139 Goode and Ben-Yehuda (2009, p. 18)

is concerned about, for example, homosexual activities carried out in private. The following hypotheses are possible explanations why concern over sexual deviance is so common:

- Controlling the sexual behavior of others may have important consequences for biological fitness. It is possible that a psychological tendency for controlling the sexual behavior of others has been shaped in our prehistory by selection mechanisms that are not yet fully understood.

- A social group under threat needs to produce many children. It is therefore likely that the regal reaction includes an impulse to suppress non-reproductive sexual behavior (see chapter 3.7).

- Elite members may control people by suppressing their sexuality.[140]

- Stories about sexuality have pornographic value regardless of whether the story is positive (about pleasure) or negative (about danger). A discourse about sexual deviance may be titillating because it appeals to the person's hidden, and perhaps repressed, desire to do the same.[141]

- Titillating stories are more likely to be retold. In other words, they have memetic fitness.[142]

- The mass media have an economic interest in telling titillating stories.[143] This is further discussed in chapter 5.1.

- Stories about sexual deviance have political consequences and may be promoted by groups with a certain agenda, such as feminists or religious groups.[144]

More research is needed to find out which of these effects are most important.

---

140 Foucault (1978)
141 Adams, Wright and Lohr (1996)
142 Brodie (1996, p. 103)
143 Greer (2012, chapter 5)
144 Plummer (1995, chapter 2)

4. Different Kinds of War in Human History

4.1. The rise of empires: Contributions from cultural selection theory

Regality is a self-amplifying phenomenon. Regal groups tend to grow bigger by conquering the territory of neighboring more kungic groups and become city states, kingdoms, and finally large empires. After a long period of growth, the empire becomes unable to grow any further and finally collapses. Reasons for the collapse are discussed in chapter 4.2. For now, we will concentrate on explaining the growth.

Figure 7. Charge of the Light Cavalry Brigade, 25th October 1854, under Major General the Earl of Cardigan. Painting by William Simpson, 1855.[1]

---

1   Public domain, https://commons.wikimedia.org/wiki/File:William_Simpson_-_Charge_
    of_the_light_cavalry_brigade,_25th_Oct._1854,_under_Major_General_the_Earl_of_
    Cardigan.jpg

© 2017 Agner Fog, CC BY 4.0   https://doi.org/10.11647/OBP.0128.04

The growth of a regal society is a cultural process, and we may seek explanations for this in the theory of cultural evolution and other cultural theories (see chapter 3.2).

One widely known version of cultural evolution theory is *memetics*. Cultural traits, called *memes*, are seen as heritable units analogous to genes.[2] Memetics can be explained by the example of a new food recipe. People will share a recipe with their friends if they find the new dish tasty. The friends will share it with other friends, and so on. The best recipes will be the ones that are shared the most. Such recipes may 'mutate' or be combined with other recipes to produce new, even better recipes.

A big problem with the theory of memetics is that it does not always make sense to divide culture into discrete and inherited units of information.[3] Some theories avoid these problems by allowing more fuzzy cultural phenomena without a single identifiable cultural parent.[4] Another variant includes the selection of quantitative traits. For example, big companies often have a competitive advantage that allows them to grow still bigger. This is an example of a selection process, but the selected trait is quantitative rather than qualitative.[5] The latter theory has the main focus on the dynamic structure of the whole social system rather than on individual actors. It is related to systems theory[6] and to the mathematics of feedback systems.[7]

Positive feedback can make a system unstable, while negative feedback can make a system approach equilibrium under certain conditions. An example of negative feedback in a social system is the economic law of supply and demand, which tends to drive a market towards a stable equilibrium. Positive feedback in social systems is often seen in the so-called Matthew effect: whoever has the most power and money can use their influence to manipulate the system in such a way as to allow themselves to gain still more power and money.[8] This

---

2   Tyler (2011)
3   Richerson and Boyd (2008), Edmonds (2005)
4   Sperber (1996, chapter 5)
5   Fog (1999, p. 64)
6   Laszlo and Laszlo (1997), Luhmann (1995)
7   Richardson, G. (1991)
8   Rigney (2010), Pierson (2015)

phenomenon is often seen in weak democracies where the president makes new laws that give more power to himself.

Both versions of cultural selection theory can explain some important aspects of the growth of regal societies. If a group of people invents a more efficient method for food production—for example, a form of agriculture—then they can produce more food per unit area. This allows them to feed more children and increase the population density. A bigger population can make a bigger army, and this enables them to win wars and incorporate neighbor groups under their rule. The new food production technology is now adopted by the newly assimilated population. This allows the population to grow still bigger and make still bigger armies, and so on. There is plenty of evidence that the growth of large civilizations and empires through history has been connected with the spread of agriculture.[9] Efficient weapons technology and transport technology may have been spread by a similar mechanism: the groups with the strongest weapons and highest mobility have been likely to win wars over other groups with inferior weapons. DNA studies show that the most prolific male lineages originate from areas with agriculture or pastoralism and domesticated horses.[10]

As the groups grow bigger and stronger, they also become more warlike in accordance with regality theory. A large-scale war with firearms is probably much more deadly than anything our ancestors saw in the evolutionary prehistory that shaped the regal response pattern. We can regard such a war as a supernormal stimulus leading to an extremely regal response. This allowed populations to accept the very hierarchical political structure and tyrannical leaders of the historical empires.

Certain religious memes may have spread in a similar way. A systematic study of Austronesian cultures shows that beliefs in supernatural punishment precede the evolution of larger, more complex, societies. People who believe that they are somehow punished for clandestine misdeeds or rewarded for bravery are more likely to obey their government and social norms, thereby making their society stronger.[11]

---

9   Eckhardt (1992, p. 178)
10   Balaresque et al. (2015)
11   Watts et al. (2015)

We can conclude that the growth of warlike societies and empires in history has involved the selection of qualitative cultural traits, such as agricultural technology, weapons technology, and religious beliefs, as well as quantitative parameters, such as food production, population size, and territory size. These factors have all supported and amplified each other in a positive feedback process.[12] A systematic study of civilizations through history shows a strong correlation between agriculture, territorial expansion, weapons technology, political centralization, social and economic inequality, slavery, strict discipline, authoritarianism, and warlikeness.[13]

A recent DNA study shows evidence of an extreme variance in male reproductive success around 5,000 to 7,000 years ago. In other words, a few men had many children and many men had none, while no such variance was seen for women. This is evidence of a wave of extreme polygamy, occurring first in Asia and the near East and later in Europe and Africa. This observation is explained by cultural factors connected to the spread of agriculture.[14] It is likely that agriculture and other cultural innovations allowed kings and emperors to rule over larger numbers of people and to have more wives. Cultural and religious norms later limited the amount of polygamy.

It has been proposed that the distinction between $r-$ and K-strategies in biological evolution (see chapter 3.3) can be applied somewhat analogously to cultural selection, though the analogy is far from perfect. A selfish meme—or rather a selfish complex of memes—can use different strategies for utilizing the resources of its hosts (that is, the persons holding the memes) to produce either a great quantity or a high quality of cultural offspring. A cultural r-strategy is a strategy where the meme complex makes its hosts spend a lot of resources on propagating their culture and beliefs to others. This is seen, for example, where a warring group impose their culture on the people they conquer, and also where a religious sect spends a lot of resources on winning new converts or competing with other religious groups. The opposite is a cultural k-strategy (written with a small k), which allocates few resources to winning new hosts and more resources to making its hosts

---

12   Gat (2006, chapter 12)
13   Eckhardt (1992)
14   Karmin et al. (2015)

satisfied so that they will not choose competing memes. This is typically seen in times of peace, when ideals of individual freedom become more popular than ideals of patriotism and unity.[15]

This theory predicts that the cultural r-strategy will be most efficient in times of intergroup conflict, while the cultural k-strategy will be most efficient where there is no culturally different neighbor group to compete with. Interestingly, the predictions of this cultural r/k theory are so similar to the predictions of the theory of a psychological desire for a strong leader in times of intergroup conflict that any of these two theories would provide a logical explanation for the same cultural effects. It can be argued that psychological responses to intergroup conflict are likely to be much faster than cultural r/k selection and therefore more effective, but the cultural r/k mechanism probably has at least some effect as well. For now, it is assumed that there is synergy between the two mechanisms, perhaps as the result of some kind of gene/culture coevolution[16] and we will use the terms *regal* and *kungic*, regardless of which of the two mechanisms is implied.

4.2. The fall of empires: Contributions from historical dynamics theory

Ethnic groups that are incorporated one by one into an empire as it grows often preserve some of their culture, ethnicity, religion, and language. Empires that are created in this way have an inhomogeneous population. It requires a great deal of power and discipline to prevent internal conflicts and to keep such an empire together. In terms of regality theory, an empire must be regal in order to stay together. The regality is built up and maintained mostly through territorial wars at the borders. We can predict that an empire will fall apart and split into smaller ethnic groups when it becomes unable to expand further and unable to make wars with its neighbors, and therefore unable to maintain the high regality. This can happen when the empire has become so big that communication between the center and the warzone at the border

---

15   Fog (1997)
16   Lumsden and Wilson (1981)

is difficult. Inhabitants at the center lose interest in what happens at the faraway border and lose the will to defend their country.[17]

Peter Turchin has given a very similar explanation of the rise and fall of empires based on the theory he calls historical dynamics.[18] Turchin explains the rise of empires by the historical observation that group solidarity, loyalty, and military strength grow in a conflict zone between culturally different peoples, such as a frontier between farmers and nomadic pastoralists. Such a conflict zone can form the nucleus of a growing empire. Turchin uses the word *asabiya*, borrowed from the fourteenth century Arab historian Ibn Khaldun, to denote these forces of social cohesion. The cohesive force, or *asabiya*, decreases when the borders are far removed from the political center and the state or empire stops growing due to logistic problems, internal competition, and economic collapse.

The main weakness of Turchin's theory is that his concept of *asabiya* is poorly defined and it is not very clear how it is generated. Obviously, it is very similar to the concept of regality in the theory of the present book. Regality theory offers the causal link that is missing in Turchin's historical dynamics. Turchin's explanation of the rise and fall of empires has so much in common with the explanation based on regality theory that a synthesis of the two theories seems natural.

Turchin has made a comprehensive study of empires through history and found several typical factors that are involved in the disintegration of empires.[19] The economy becomes unstable when the territorial expansion stops while the population keeps growing. The overpopulation and shortage of land in an agrarian economy leads to increases in the prices of land and food while the price of labor decreases. Impoverished farmers are forced to sell their land and go into unpayable debt. Elite landowners profit from this situation as they become able to extract more surplus from the peasants. The elite not only become richer at the expense of everybody else, they also become more numerous. The luxurious lifestyle of the growing elite is paid for through heavy taxation or debt.[20] There is growing inequality, not only

---

17   Fog (1999, p. 111)
18   Turchin (2006), Turchin and Gavrilets (2009)
19   Turchin and Nefedov (2009, chapter 10)
20   Gat (2006, p. 371)

between the elite and commoners, but also within the elite, where strong competition for the most attractive positions leads to conflict.

The starving population makes uprisings and food riots against the extravagant elite. The elite is split between competing factions, some of which may form alliances with the rebellious commoners and start a revolution or civil war. Suppressing the rebellion is so expensive that the government, which is already spending most of its surplus on external wars, goes into debt and starts to lose control. Banditry spreads, and farmland far from the cities is abandoned because it can no longer be protected against raids. The economy collapses and the state goes bankrupt. The weakened state becomes unable to defend its borders and is attacked, perhaps by nomadic pastoralists, perhaps by another growing empire.

This kind of downturn is sometimes ended by dramatic events such as a major war or pestilence. The increasing poverty, famine, and urbanization make breeding ground for disease epidemics, and such epidemics have wiped out large parts of the population several times through history. The survivors inherit the property of deceased family members, and a new cycle of prosperity and growth begins. An empire may go through several such cycles of growth and decline before it completely disintegrates. Each cycle may take one or more centuries.

Combining regality theory with cultural selection theory and historical dynamics, we may summarize the main factors leading to the rise and fall of empires. The growth of an empire can be explained by a positive feedback process involving war, growing regality, increasing food production, population growth, improved weapons technology, military strength, and expanding territory. The decline may involve the following factors:

- loss of regality when the empire has reached the limits to its growth

- overpopulation, poverty, and famine

- a growing elite, growing inequality, and conflicts between elite and commoners as well as between different factions within the elite

- collapse of the economy and weakening of the state.

4.3. General theories of war and peace

Studies of conflict and peace have produced many different theories of the causes of war. This chapter will review some of the most influential theories.

Political realism theory is based on the premise that the international system is anarchic. The international situation is determined by the distribution of power, where the stronger states are able to conquer territory from weaker states.

Every state needs to have a military power or build alliances in order to protect itself against attacks. A state that sees its neighbor increase its military power is often unable to know whether the intentions are offensive or defensive, so it must increase its own military power to at least the same level in order to assure its security. Such an arms race can leave both states in a less secure situation than before. This paradox is known as the *security dilemma*. Deterrence does not always work, and a war may result if one of the states launches a preemptive strike in order to remove the perceived threat from its neighbor.[21] The probability of war is increased when there is a change in the power balance, for example caused by an economic crisis or a revolution.[22] The same states often wage war against each other repeatedly in *enduring rivalries*.[23]

The *steps to war* model describes the process of conflict escalation that may lead to war. The first step is a disagreement over some issue such as a disputed territory. Both parties in the conflict may use various strategies, such as coercive threats, military build-ups, and forming alliances. Each step in this conflict spiral increases the probability of war. Leaders who plan for war are likely to build up military power, mobilize nationalism, and reduce democratic rights before initiating a war. The observed nationalism, etc. is thus not the ultimate cause of the war but part of the steps to war.[24]

The *bargaining theory* assumes that it is possible in most cases to negotiate a peaceful settlement that is less costly to both parties than

---

21   Gat (2006, p. 97), Levy and Thompson (2010, p. 30), Cashman (2013, chapter 7)
22   Cashman (2013, chapter 11)
23   Levy and Thompson (2010, p. 56)
24   Levy and Thompson (2010, p. 60)

a violent war. The conflict will only escalate into war if some factor prevents the parties from reaching a negotiated settlement. Such factors can include inaccurate information about the opponent's strength, lack of trust that the opponent is truly committed to accepting the negotiated agreement, or inability to reach a compromise over an indivisible issue.[25]

The theory of *liberal peace* argues that democratic structures, free international trade, and international institutions can promote cooperation, mitigate the effects of international anarchy, and prevent war.[26] Democracies are widely believed to be inherently more peaceful than non-democratic states. Democratic states almost never attack other democratic states, but peace researchers cannot agree why. Every theory of *democratic peace* has been met with counterarguments or contradicting examples, and statistical studies fail to confirm that democracies are less likely to wage war than autocracies.[27] Azar Gat argues that it is not democracy itself that leads to peace but other aspects of modernization that are typically seen in democratic states, such as economic prosperity, welfare, absence of hard work, non-violent domestic culture, sexual liberation, and lower population growth.[28]

Historically, peace has almost always come before democracy. Recent studies have found that states do not develop democracy until after all border disputes have been settled. In other words, there is strong evidence that peace causes democracy but little evidence that democracy causes peace.[29] Regality theory can easily explain this observation. Contested borders or other external threats to a state will activate regal reactions in the population, who will be likely to support a strong, autocratic leader. Peace and collective security, on the other hand, will make the population prefer democracy.

Politicians are well aware that the population shows more support for their government in cases of war or national insecurity. The *diversionary war hypothesis* supposes that state leaders sometimes take advantage of this effect and wage war in order to divert attention from internal

---

25 Levy and Thompson (2010, p. 63)
26 Levy and Thompson (2010, p. 70)
27 Cashman (2013, chapter 5)
28 Gat (2017, chapter 6)
29 Owsiak (2013), Gibler and Owsiak (2017)

problems and to make their population *rally around the flag* and support their leader. This is not a general cause of war, but it appears that leaders in some cases have escalated a threatening situation in order to rally support for themselves.[30] This phenomenon, which obviously relates to regality, is further discussed in chapter 6.3.

The interests of leaders and external actors are often decisive for the onset of conflicts. The *instrumental model of conflict* assumes that leaders manipulate their populations by emphasizing and amplifying public perceptions of ethnic or religious differences in order to rally support for a fight that mainly serves the interests of the leaders.[31]

Many wars are fought mainly for economic reasons. The *liberal economic theories of war* assume that state leaders want to maximize economic utility for their country. They will start a war only if the expected gains exceed the expected costs. A high level of international trade will make war unprofitable because the costs of disrupting the trade will be too high. Thus, this theory predicts that economic interdependence between countries will deter war.[32]

A different line of thought, called the *economic realist theory*, makes the opposite prediction. A high dependence on international trade makes a state vulnerable. State leaders are concerned that their access to export markets and foreign investments might be reduced by adversaries and that crucial imported goods, such as oil and other raw materials, could be cut off. They want to reduce their vulnerability and to control what they depend on. This will increase the probability of militarized conflict as states that depend on external raw materials will scramble to secure access to these raw materials and try to control or colonize countries that can supply these raw materials.[33]

The international relations researcher Dale Copeland has developed these theories further. His *trade expectations theory* confirms that economic interdependence between states can either increase or decrease the probability of war, depending on their economic prospects. Great powers need a strong and growing economy in order to sustain

---

30   Cashman (2013, chapter 6)
31   DeRouen (2014, chapter 4), McCauley and Posner (2017)
32   Cashman (2013, chapter 5), Copeland (2015, chapter 1)
33   Copeland (2015, chapter 2)

their positions in the international system. This requires a high level of trade. They need to buy many different raw materials and they need to sell products to many different countries. This makes them vulnerable to cutoff of markets and investments or supplies of raw materials. Their rivals may explore this vulnerability in coercive diplomacy.

Great powers are primarily driven by a desire to maximize their economic security and not by other goals such as welfare maximization, social cohesion, glory, or the spread of their ideologies, according to Copeland's research. Their foreign policy is driven mainly by the desire to reduce economic vulnerabilities and secure access to critical raw materials and markets. They may use military means to secure their access to vital raw materials, such as oil or minerals, if they see a risk that this access may be cut off in the future.

Copeland found that regime type and other domestic factors have much less influence than previously thought for the decision of great powers to make war or maintain peace. By studying the actual decision-making processes of various governments, he found that trade expectations often drove the patterns of peace and conflict even in cases that, on the surface, appeared to be about something else, such as ideology or religion. The conflicts over capitalism versus communism, or one religion versus another, may just be psychological rallying foci that hide more important underlying economic motives.[34]

Copeland also found that the economic security dilemma was just as dangerous as the military security dilemma. Attempts to improve the economic security of one state will decrease the economic security of its rivals, who are likely to respond with similar countermeasures. State leaders are aware of this danger of escalation. They need to signal their commitment to trade agreements in order to avoid escalation of economic conflicts.

The theories mentioned here are mostly concerned with 'conventional' wars between two sovereign states. This kind of symmetrical war has become rare since World War II. Instead, we are now seeing insurgencies, asymmetric wars, proxy wars, and other kinds of conflicts. This is explained in chapter 4.4.

---

34   Copeland (2015, chapter 8)

4.4. Changing patterns of war

There have always been violent conflicts between humans, but patterns of war have changed throughout history. Prehistoric hunters, gatherers, and nomads fought over territory, domestic animals, and other material possessions—and over women as well. The level of conflict was determined mainly by geographic and environmental factors. This will be explained further in chapter 7.

The introduction of agriculture and the settlement into cities led to larger concentrations of people and a more centralized political organization. The cities and agricultural fields had to be protected, and it became possible to assemble larger armies of warriors. Horses, wheeled vehicles, and large boats made it possible to transport the armies to distant battlefields where they could conquer more territory. The growing territories and growing wealth made it possible to assemble still larger armies and conquer still more territory. Improvements in weapons technology, transport, communication, and food storage contributed to this process. Cities grew into states, kingdoms, and finally large empires by means of this self-amplifying process. An ever more regal political organization was necessary for keeping large numbers of troops disciplined and keeping the empire together. An empire could continue growing until problems of transport, communication, logistics, and economics prevented further growth. The stagnation of growth started a kungic development and the empire began to disintegrate, until a new empire started to grow from a new center. The cyclic process of rise and fall of empires continued for thousands of years, as explained in chapters 4.1 and 4.2. The type and intensity of warfare in large parts of the world was ultimately determined by this dynamic process of the growth and collapse of empires.

Throughout history we have seen different kinds of political units that strived to expand their territories, such as dynasties, city states, nation states, and superpowers. Likewise, we have seen different motives or justifications for the expansive urge, such as prestige, religion, ideology, or alliances.[35]

---

35   Luard (1986, p. 133)

The end of World War II was a turning point in the history of warfare. Large-scale interstate wars became rare, while intrastate (civil) wars became more frequent. This development is shown in figure 8.

There are several reasons for this sharp decline in the frequency and scale of interstate wars. The improvement in weapons technology is a major factor in this development. The ever more advanced weapons and defense systems also became extremely expensive. The arms race was a large economic burden for even the biggest superpowers. The destructive capacity of the new nuclear weapons and other advanced weapons grew to alarming heights. The prospect of another large interstate war had become so frightening that it could deter any potential aggressor. Not only would such a war be so expensive that it would ruin the economy of both parties, it would also destroy large parts of the infrastructure and kill large numbers of people, not only in the attacked state, but in the aggressor state as well.[36]

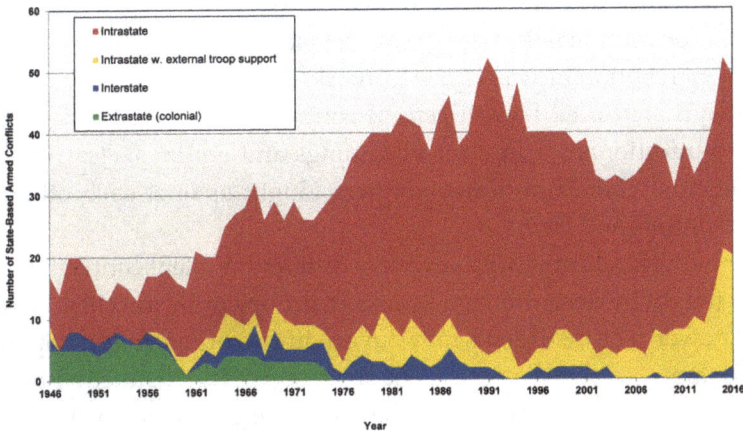

Figure 8. Historical development in different types of conflicts since World War II. Data from Uppsala Conflict Data Program. By Agner Fog, 2017.[37]

---

36 Mumford (2013, p. 2)
37 Data from Uppsala Conflict Data Program (2017), Pettersson and Wallensteen (2015)

The decline in interstate wars is also driven by the modern lifestyle, welfare, urbanization, lower population growth, non-violent attitudes, democracy, and other aspects of development.[38] Increasingly globalized trade has made most countries so dependent on both import and export that they cannot bear the disruptive effects of large wars on their trade and industry. The development of international law is also an important factor. The norms against war of aggression, promoted by the United Nations and other international organizations, will damage the reputation of any violator.[39] The Gulf War of 1990–1991 following Iraq's invasion of Kuwait was a crucial signal to potential aggressors that the international community will punish territorial aggression.[40]

While analysts have no problem describing the decline of international wars, there is more uncertainty when it comes to describing the kinds of conflicts that have replaced them. The new kinds of conflicts have variously been called irregular warfare, unconventional warfare, low-intensity warfare, fourth-generation warfare, asymmetric wars, new wars, and various other designations.[41]

The conventional theories of war developed over centuries are based on the perception of war as a symmetric conflict between two or more sovereign states, each with central control and a monopoly of force. Battles traditionally had clear beginnings and endings, clearly located battlefields where uniformed soldiers fought against each other, and clear winners and losers.

The new kinds of conflicts are very different. Violent conflicts are now often fought by decentralized insurgent groups or by guerillas against a state power. The insurgents may have a diffuse organization with many autonomous or semi-independent subgroups. The distinction between combatants and civilians is often blurred. The guerillas mostly use light or relatively cheap weapons. This kind of conflict has more similarities with the old forms of warfare that were common before the strong nation states were formed than with the kind of interstate war that we now call conventional warfare. In fact, insurgence and guerilla warfare

---

38   Gat (2017, chapter 6)
39   Münkler (2005)
40   Snow (2015, p. 203)
41   St. Marie and Naghshpour (2011), Rid and Hecker (2009), Münkler (2005), Hughes (2012)

is more common in weak states that lack a strong central government and a strong centralized military power. Some historians regard this as a return to old forms of warfare, while the modern kind of warfare based on heavy weapons has become too costly in terms of both money, casualties, and destruction of infrastructure.[42]

But modern conflicts are not only fought with old-fashioned guns. On the contrary, sophisticated methods, such as cyberattacks, propaganda, public relations smears, and economic undermining have become increasingly important in both local and global contests of power. The distinctions between civilians and combatants, between permissible and off-limits targets, and between economic and military targets seem to vanish when conventional warfare is replaced by such new tactics. These power games have been characterized as low-intensity conflicts because they never mobilize sufficient amounts of military power to win territory. Instead, we have seen prolonged conflicts that ebb and flow without ever reaching a definite resolution. The political and economic costs of increasing the level of conflict to the point where a clear victory is possible are simply too high.[43]

Insurgents and guerillas rely on many different sources of finance to sustain their operations. Robbery, plundering, and extortion are frequent sources of finance. Production and trade of psychotropic drugs is a very profitable source of finance for many insurgent groups. Other groups rely on the extraction of oil, minerals, and other natural resources. Even though involvement with organized crime initially seems unattractive for many insurgent groups, they quickly become dependent on it. Organized criminals involved in illicit trade depend on armed guerillas and local warlords for protection, while the guerillas rely on the illicit trade for finance. Once this symbiosis is established, it is difficult to break, and many of the actors involved have an interest in continuing the conflict.[44] The reliance on criminal sources of funding has a demoralizing effect on the guerillas. The high ideological principles that prompted their resistance in the first place are undermined by the

---

42   Münkler (2005, p. 24)
43   St. Marie and Naghshpour (2011, p. 3)
44   Münkler (2005, p. 15), Adams (1986)

necessary involvement with crime, and many groups have deteriorated into marauding gangs of bandits.[45]

Support from external sources is often crucial. Many insurgent groups rely on external sources of funding, either from a diaspora of emigrants and refugees or from foreign states.[46] Support from foreign states is very common but mostly secret. The clandestine support is typically provided by a secret service and channeled through some seemingly independent organizations in order to hide the original source. This principle is commonly known as *plausible deniability*: the donor can deny any involvement or responsibility when the support is channeled through an untraceable chain of intermediaries. Foreign support for insurgent groups often involves money, weapons, training, logistics, and intelligence.[47]

Support from foreign states and secret services never comes for free—there are always strings attached. Foreign donors use insurgents as proxies for promoting their own political and military goals in pursuit of their own geopolitical interests or trade interests. Insurgents often become so dependent on foreign sponsors that they lose their ideological purity. Foreign powers have often supported small and insignificant opposition groups and promoted them into powerful insurgents or even fabricated opposition groups where none existed.[48]

Such proxy wars often have two external stakeholders. This may be two rival superpowers or two coalitions of powerful nations competing to expand or defend their zones of geopolitical influence in the area. For example, one party may sponsor insurgents or coup makers while the other party supports the incumbent government. Instead of battling against each other directly—which would be too costly—the superpowers support their proxies in a war on a distant battlefield, conveniently avoiding any destruction in their own homelands.[49]

The party that supports the government of the troubled country may openly supply troops or air forces to support the government in the battle against what they consider illegitimate rebels.[50] This kind

---

45 Münkler (2005, p. 14)
46 Mumford (2013, p. 65), Hughes (2012, p. 38)
47 Hughes (2012, p. 15)
48 Bale (2012)
49 Mumford (2013, p. 3)
50 Mumford (2013, p. 14)

of interventionism has become much more frequent in recent years, as figure 8 shows. The yellow curve labeled 'Intrastate with external troop support' covers only those conflicts where external intervention or support includes official foreign troops. Less visible forms of external support on one or both sides is no doubt more frequent, so that in fact the majority of violent conflicts today can be characterized as proxy wars.

External sponsors can have many reasons for supporting a conflict in a faraway country. First of all, it is very cheap compared to traditional warfare. Proxy wars are fought mostly in poor countries where the living costs for the native fighters are low and the fighters mostly use small light weapons. The sponsor states can avoid large casualties from their own forces and reduce the risk of escalation by using proxy fighters, drones, mercenaries, or private military contractors.[51] The politicians of the sponsor countries seek to avoid political resistance from their own constituencies and also to avoid international opprobrium by hiding their support behind plausible deniability. In many cases, they have been able to escape accountability to their own jurisdiction as well as to international law by leaving the dirty work to irregular groups or mercenaries in a more or less lawless country and by making their own influence on these groups difficult to trace.[52]

A sponsor state can harm a militarily strong enemy, without engaging in direct conflict, by supporting insurgents in the enemy state or in a third state that is allied with their enemy. This enables the sponsor to divert their enemy's resources toward the intrastate conflict and away from external conflict.[53] It is often a big problem for the sponsor state, however, that its proxy fighters may be weak, incompetent, uncontrollable, and disloyal, and may have their own political or ideological agendas.[54]

Proxy wars can be very long-lasting because the external sponsors are never completely defeated, they do not easily run out of resources,

---

51 Mumford (2013, p. 4), O'Brien (2012)
52 Mumford (2013, p. 42), Carter (2014), Hughes (2012)
53 Hughes (2012, p. 38)
54 Hughes (2012, p. 47)

and the devastating consequences of the war mostly affects a distant country rather than themselves.[55]

Today's conflicts are mostly asymmetric conflicts between unequal parties—typically insurgents or separatists against a state government or an external occupying force. Peace negotiations are often tried but rarely successful. Obviously, the two parties in an asymmetric conflict have very different positions in the situation of a peace negotiation. A negotiation between unequal parties with unequal negotiating powers will naturally lead to a result that is more favorable to the stronger party than to the weaker party. The insurgents will almost unavoidably perceive the negotiation result as unfair. A negotiation result that addresses the grievances of the weaker party in a way that they perceive as fair can hardly be achieved without strong external pressure from a third party.[56]

The leaders or negotiators for the insurgents may be committed to accepting an unfair negotiation result because they see no alternative, but the negotiated peace remains fragile as long as the grievances are not resolved. The insurgents typically have no strong central organization that can enforce the negotiated peace agreement. Even small frustrated subgroups with uncompromising attitudes can easily break the peace and make the conflict flare up again.[57]

Growing regality is another reason why intrastate conflicts and asymmetric wars are difficult to end. The insurgents become regalized and radicalized because of the prolonged conflict. Insurgent leaders also deliberately promote a radical ideology in order to justify their cause and to unify their supporters. Leaders on both sides of the conflict will often manipulate religious feelings and identities in order to mobilize their population.[58]

Insurgent groups often use radical religion in their fight against corrupt and kleptocratic regimes. They see corruption as a result of poor morality, and the most obvious remedy against amorality is religion.[59] A process of selection occurs in countries where protesting is difficult: the

---

55   Mumford (2013, p. 107), Cummins (2010)
56   Cummins (2010, p. 6), Rouhana (2011)
57   Münkler (2005, p. 13), Rid and Hecker (2009)
58   DeRouen (2014, chapter 4), McCauley and Posner (2017), Toft (2007)
59   Chayes (2015, p. 64)

less radical protest groups are easily crushed by a suppressive regime, while the most regal groups are more likely to have enough strength and courage to survive. Some governments have even deliberately suppressed moderate protest groups while allowing more radical groups to continue. The existence of radical insurgent groups gives the incumbent government a convenient justification for their hardline policy.[60]

The leaders of insurgent groups face a tactical dilemma here. They can promote a very radical ideology in order to maximize the zeal of a small group of supporters, or they can express a less radical ideology in order to get support from the broader population.[61] The most radical insurgent groups are often out of touch with the sentiments of the majority of the population they want to represent, because the perceived level of danger is not high enough to justify their high level of regality. For this reason, the regality of an insurgent movement is likely to fizzle out after a victory. For example, the Iranian Revolution of 1979 led to a strict theocratic rule, but even though the subsequent war with Iraq also contributed to the regality, more kungic sentiments soon began to spread in the population, as evidenced by the fact that the Arab Spring gained considerable support in Iran in 2011.

Insurgents and guerillas have a tactical advantage because of their flexibility and their ability to hide. The state military often lacks this flexibility because it has a big, hierarchical organization that cannot adapt as quickly to changing situations. Traditional militaries have a problem of balancing hierarchy against innovation, or discipline against creativity.[62] The costs to the counterinsurgent force of maintaining order and legitimacy are much higher than the costs to the insurgents of undermining it.[63]

Guerillas and paramilitary forces are mostly found in weak states without a strong central military. While a strong military under central governmental control is taken for granted in modern first world countries, such an organization of power is less common in third world countries. A strong, centralized military organization has historically

---

60   Chayes (2015, p. 69)
61   Rid and Hecker (2009, p. 44)
62   Mumford and Reis (2014, pp. 4–17)
63   Rid and Hecker (2009, p. 3)

been formed in European countries because it was the format most suited for traditional interstate war. States that failed to build a strong military have long since been defeated by stronger and more militant neighbors.[64]

Third world countries have different organizations of their forces due to their different histories. Many of these countries have been formed not by interstate wars but by colonial wars. States that were decolonized through negotiation have inherited bureaucratic military organization from the departing colonial powers, but states that were formed through violent rebellion have tended to approbate the local guerillas that were active in the process of decolonization and to convert them into pro-state militias. These states do not need a strong central military if they are not threatened by their neighbor countries or if they are protected by the international community. Relying on paramilitary forces can be an advantage for a state government if they can use the local militias as proxies to quell insurgents and rebels while denying responsibility for any brutality and human rights violation in the counterinsurgency campaign.[65]

Weak states without sufficient centralized power are more susceptible to intrastate conflicts—and intrastate conflicts tend to weaken the state. This is a vicious circle. It is necessary to strengthen the state if you want peace, but international initiatives to support state building have often had little or no success.

The opinion of the population is essential to the outcome of an asymmetric conflict. Insurgent movements have an advantage over a foreign occupying force in this regard, because they often have the support of the local population while it is difficult or impossible for an external force to win the hearts and minds of the people. The insurgents have an enduring commitment to the cause and they can rely on steady recruitment from the local population, whereas an external occupying force often lacks the patience and political commitment to a long-lasting and expensive conflict.[66]

The importance of mass communication has increased dramatically as the types of conflicts have changed and new electronic media have

---

64   Ahram (2011, p. 8)
65   Ahram (2011, p. 4)
66   Rid and Hecker (2009, p. 3)

made mass communication accessible and affordable to even small insurgent groups. These groups have gradually learned to use electronic communication to their advantage. Insurgent groups can publish news, images, and political statements immediately after a violent event due to their decentralized structure, while a bureaucratically structured state force typically responds more slowly because they need verification, declassification, and political approval before they can issue a press release. This asymmetry in information handling can give insurgents a leading edge towards the highly competitive news media, hungry for breaking news.[67]

All in all, there are many obstacles that prevent weak states from obtaining internal peace and from establishing a stable central government and economic autonomy. The decisive factor for the outcome of asymmetric conflicts is often psychology rather than violent victories and losses on a diffuse battlefield. Both parties in the conflict need support from their constituency or home base. They cannot continue the battle if they become unpopular with the people they represent or claim to represent. Both parties will try to disseminate information favorable to their cause. Obviously, disinformation and selective information abound in modern conflicts. It has become common for military forces to invite selected journalists to travel with them and report from the battlefield. These embedded journalists live with the troops and depend on them for protection. Obviously, their reports will reflect a one-sided view of the situation. Insurgents and guerillas may have less access to journalists and rely more on electronic communication and social media. This information war is important for consolidating their support. International opinion is particularly important. If one party wins the undivided sympathy of the international community, then they have almost won the conflict. The best chance that insurgents have of winning an asymmetric conflict is actually to sway international opinion in their favor and call for sanctions against their opponent.[68]

Democracy cannot prevent symmetric wars, but it may reduce the probability of civil war and insurgencies because it gives minorities a forum where their grievances can be heard.[69]

---

67 Rid and Hecker (2009, p. 131)
68 Münkler (2005, p. 25), Kellner (2011)
69 DeRouen (2014, chapter 4)

Terrorism and accusations of terrorism are often significant elements in psychological warfare. The weaker party in an asymmetric conflict may turn their violence against civilians if there are no military targets that they can meaningfully attack, or if they consider that this is the only tactic that has any effect on their enemy. However, terrorism can be a very counterproductive strategy, as explained in chapter 6.1. The definition of terrorism is unclear and contested. It is easy to accuse someone of terrorism in a situation where the distinction between combatants and non-combatants is blurred or where there are no obvious military targets that can be attacked. The concept of terrorism is so vaguely defined, yet so highly laden with psychological, political, and legal significance today, that it has become common practice to accuse one's enemy of terrorism — with more or less justification.

Asymmetric wars and guerilla wars often generate many refugees because civilians are more directly affected than in conventional interstate wars. A massive flow of refugees can have significant effects — intended or not — on psychological and strategic warfare for several reasons: (1) the flow of refugees can attract the attention of the international community; (2) massive migration can destabilize the region that receives the refugees; (3) refugee camps can be hotbeds for recruiting new fighters; and (4) humanitarian aid to refugee camps often finds its way to the fighting insurgents.[70]

If we want to examine the regal effect of insurgencies and asymmetric conflicts we should look not only at the casualties but also at the psychological effects. These low-intensity conflicts typically have fewer casualties than interstate wars, but they pose more dangers to civilians and they can be quite long-lasting. Therefore, they have a very significant regalizing effect and they often lead to political or religious radicalization. Even the smallest ethnic or religious differences are amplified in this process. Ethnicity is a result of conflict as much as a cause of it.[71]

The regal effect is of course concentrated in the conflict zone, while the distant sponsors of a proxy war may be unaffected if their population pays no attention or has no knowledge of the clandestine sponsor activities of their governments. The sponsor countries may be affected,

---

70   Münkler (2005, p. 18)
71   Turton (2003), Francis (2009)

however, if they are hit by terrorist attacks or other forms of retribution or if they receive a massive influx of refugees from the conflict zone. The violence may spill over the boundaries of the initial conflict, and we may see blowback effects when radicalized proxy fighters become enemies of their former sponsors, as in Afghanistan, Syria, and several other countries (see chapter 5.5).[72]

A proxy war may be cheap for the external sponsors but devastating for the country that serves as battleground for the proxy war. The infrastructure is often severely damaged; powerful crime organizations do not disappear when the conflict ends, and a proxy may continue to be dependent on its sponsor.[73] The country cannot easily escape the vicious circle of lawlessness, corruption, instability, poverty, and economic chaos that follows after a prolonged conflict. Such a failed state can be very difficult to recover. It can continue for years to be an international center for drug production, organized crime, and a training ground for insurgents from other countries.[74]

The international community has been very slow in coming to terms with the new forms of conflict. The traditional understanding of civil war does not fit the new pattern of low-intensity warfare.[75] Analysts are using a variety of names for these conflicts suggesting that it is something exceptional. Research on asymmetric conflicts has mostly relied on the paradigms of counterinsurgency and counterterrorism based on the strategic interests of the stronger party in the conflict. This research is often sponsored by state actors who are parties to the conflict,[76] and there has been less research by neutral observers analyzing this as the 'new normal' kind of conflict.[77] International relations theory and international law and norms have not yet adapted to the new pattern of non-state actors and decentralized violence.[78]

The poor understanding of asymmetric conflicts has also led to ineffective responses. Many strategies aimed at ending the conflicts have

---

72 Mumford (2013, p. 78), Hughes (2012, p. 47ff), Williams (2012), Carment and Samy (2014), Anderson (2016, chapters 2 and 11)
73 Mumford (2013, p. 105), Münkler (2005, p. 75)
74 Williams (2012), Carment and Samy (2014), Münkler (2005, p. 128)
75 Münkler (2005, p. 135)
76 Speckhard (2012, p. 467), Richardson, L. (2006, p. 11), Bale (2012)
77 Arreguin-Toft (2012)
78 Keen (2006, chapter 1), Geiss (2006)

even been counterproductive. A highly militaristic approach is likely to escalate the conflict and make the population more regal and less inclined to accept a peaceful compromise. Prompted by the proclivity of the mass media to personalize conflicts and focus on individual leaders, politicians, and military strategists have often put the blame on tyrannical heads of state and individual terrorist leaders as responsible for the violence. These media phenomena are explained in chapter 5.1. The strategies have quite often had a narrow focus on removing 'evil dictators' or hunting down terrorist leaders.[79] Such a strategy ignores the regal psychology that brought these militant leaders to power in the first place. A strategy that removes a militant leader by violent means is likely to make the affected population still more regal and make sure that his successor will be even more militant. Attempts to create democracy by violent means have failed time and again.[80]

The international community will not be able to deal appropriately with asymmetric conflicts until it develops a better understanding of the root causes and grievances behind these conflicts and develops norms to regulate external sponsorship.

Attempts to establish peace, stability, prosperity, and democracy in third world countries have often been sabotaged by wealthy elites and foreign actors, especially in resource-rich countries. This will be explained in chapter 5.4.

4.5. Theories of revolution

Regality theory predicts that rebellion or revolution is likely to happen when there is an overly repressive regime and no external conflict or other dangers that can justify such a strong regime. In other words, a regal regime with a kungic population will be unstable. We will now look at existing theories of revolution to see if this is a likely path from a regal to a kungic society.

Early theories have often interpreted revolutions as manifestations of a class war where peasants and workers fought against a capitalist elite. However, this view had to be revised when historical studies

---

79   Keen (2006, chapters 2 and 5)
80   Keen (2006, chapter 2)

found that members of a divided elite play a leading role in revolutions, and that some revolutions were not based on class issues at all but on other political or religious issues.[81]

The sociologist and political scientist Jack Goldstone has studied revolutions through history and found that they have a number of factors in common. The most important condition for a revolution to be possible is that the government is weakened or unstable. There is a remarkable similarity between the factors that make a revolution possible and the factors that make an empire collapse, as discussed in chapter 4.2, even though a revolution is not necessary for an empire to disintegrate. Economic crises and conflicts between members of the elite are typical factors that make a state weak and vulnerable to revolution.[82]

Elite members who lose their privileges, see their wealth being undermined, or are excluded from influence will not support a weak regime with economic problems. A regime can also lose the allegiance of the elite if it squanders its wealth, wastes human lives in wars that it cannot win, or fails to respect the culture and religion of the people. A government must provide effectiveness and justice in order to maintain popular support.[83]

Revolutions happen more often in periods of high population growth. Overpopulation may lead to falling wages and rising prices for food and land. Inequality grows and the economy becomes unstable. There is growing competition for elite positions and increasing conflict between different factions of the elite. A critical event such as international pressure, military defeat, state bankruptcy, or famine—which a healthy state would be able to withstand—may trigger a revolution.[84]

Another consequence of high population growth is the youth bulge effect, discussed in chapter 3.4. A combination of high population growth and a growing and divided elite can produce a surplus of well-educated and resourceful, but frustrated, young people who have both the skills and the motivation to lead a revolutionary movement.[85] A large popular mobilization around a common ideology, and an alliance

---

81   Skocpol (1994, p. 273), Goldstone (2001)
82   Goldstone (2001)
83   Goldstone (2001)
84   Goldstone (2003)
85   Heinsohn (2003)

between commoners and frustrated elite members is necessary for a
revolution to succeed. A dissatisfied and impoverished population is
not enough to start a revolution. A strong and stable government with a
loyal elite can always crush a rebellion.[86]

Sociologist John Foran has studied revolutions in third world
dependent states where the interplay of internal and external factors is
decisive. Colonial rule or a government that depends heavily on foreign
support and foreign interests will always cause structural problems,
inequalities, political exclusion, and repression. A political culture
and ideology of resistance may develop, but this is not sufficient for a
successful revolution. A revolutionary movement can only succeed if
the state is weakened by economic problems, lack of loyalty from parts
of the elite, and changes in the international situation.[87]

A revolutionary movement can often be seen as a kungic reaction
against a regal and tyrannical government, but this is not always the
case. Revolutions can happen for other reasons as well. An autonomous
ruler who excludes the country's elite from influence is vulnerable to
rebellion because he lacks the support and loyalty of the elite. This
was the case in the Iranian Revolution of 1979, for example, which
happened because the Shah failed to respect the religion and culture of
the population.[88]

Democracy is often the goal of a revolution, but rarely the result. It
is difficult for a new revolutionary regime to establish the political and
economic stability and safe living conditions that the old regime lacked.
The failure of the new regime to create stability and welfare may lead
to continued conflicts where the moderate reformists are replaced by
more radical forces. The youth bulge adds fuel to internal conflicts that
can last for years. In order to deal with these conflicts, the new regime
may end up being more centralized, bureaucratic, and repressive than
the one it replaces.[89]

The more violent a revolution is, the less likely it is to produce
democracy. Democracy is more likely to emerge by gradual reform
or by non-violent 'velvet revolutions' or 'color revolutions' than by

---

86   Goldstone (2003)
87   Foran (2005, chapter 6)
88   Skocpol (1994, chapter 10)
89   Cincotta and Doces (2012), Goldstone (2001, 2003)

violent overthrow of a despotic ruler.[90] The outcome of a revolution also depends on whether important resources can be monopolized. Where industrialization has produced great concentrations of economic power in the form of large factories, mines, oil fields, and heavy infrastructure, a revolutionary regime is likely to seize these resources in its drive for power and create an autocratic regime. But when the sources of wealth are widely dispersed in the form of farmland, small shops, and artisan businesses, the revolutionary regime cannot seize control of the economy. In such cases, the likely result is a less centralized regime.[91] These historical findings confirm the theory of contest competition versus scramble competition discussed in chapter 3.1. Concentrated resources that can be monopolized and defended by a group are more likely to lead to a centralized government and a regal society than widely dispersed resources.

90 Pop-Eleches and Robertson (2014)
91 Skocpol (1994)

5. Economic Determinants of Conflict and Fear

5.1. Fear is profitable: The economy of the mass media

We have learned that the regality of a society depends on the perceived level of collective danger. Now we will look deeper into the factors that shape people's perception of their surrounding world as either safe or full of dangers. If we can identify the most important factors that make people perceive their world as dangerous or safe then we may be able to predict whether a society will move in the regal or kungic direction.

A very obvious factor is the mass media. People can directly observe only what happens in their immediate vicinity. They have to rely mainly on the mass media for information about the political processes in their country, social problems, international conflicts, and almost everything else that happens in society at large.

Before we look at the factors that influence the contents of the mass media, we will look at the theories of how the mass media influence people's opinions. We need to incorporate the scientific disciplines of *media effects theory* and *social cognition theory* in order to understand how the mass media affect people's perception of their social world. The mass media affect their audience in several different ways, including:[1]

---

1   Bryant and Oliver (2009), Iyengar and Kinder (2010)

© 2017 Agner Fog, CC BY 4.0   https://doi.org/10.11647/OBP.0128.05

- Selection: for example, selective reporting of good or bad news about a particular subject, or preferential treatment of opinions for or against a controversial issue.

- Agenda setting: telling people what topics are important and what issues to think or be concerned about.

- Framing: telling people how to look at a particular issue. For example, nuclear power may be framed either in terms of (1) technological progress, (2) environmental hazards, or (3) a strategy for making the energy supply independent of oil from unstable foreign countries.

- Priming: making particular aspects of a phenomenon salient, for example, the criteria by which a politician is evaluated.

- Cultivation: the long-term effect of repeated exposure to similar stories.

The effects of framing and priming can be further explained in terms of social cognition theory: in modern society, people are bombarded with such a large amount of complex information that it is impossible to digest it all. We have to economize the use of our cognitive brain capacity. We do this by fitting new information that we receive into existing mental structures or models. Such a mental structure or model is variously called a cognitive schema,[2] interpretive frame,[3] social script,[4] or paradigm.[5] For example, if news media report a home robbery, readers or viewers will immediately activate their cognitive schema for a home robbery. This schema includes a catalogue of objects that are typically involved, such as a house, stolen goods, weapons, vehicle, etc. and the roles of people involved, such as robber, victim, police, and witness. The cognitive schema also includes information about the supposed actions, motives, and emotions associated with each of these roles. The reader will easily understand what happened, even if some information is missing in the news report. Information that does not fit into the schema is likely to be ignored. The priming effect may influence

---

2   Dimaggio (1997), Graber (1993, chapter 8)
3   Schetsche (2000, p. 109)
4   Abelson (1981)
5   Nersessian (2003)

which schema is activated for a particular story when more than one fitting schema is available.

If people read or hear a story about a new phenomenon for which they have no appropriate cognitive schema, they may either: (1) lose interest and ignore the story; (2) apply a less appropriate schema and thereby misunderstand the story; or (3) build a new cognitive schema. The last reaction requires a lot of mental effort and is therefore less common. We will expect the framing effect to be particularly strong and possibly long-lasting in this situation, but this is an under-researched area.[6]

A person's world view is based on cognitive schemas (also called schemata), shaped by the mass media and by other information sources. People with different world views and different cognitive schemas are very likely to disagree on political issues. Cognitive schemas also influence the decision of political leaders about war and peace.[7]

With these theories in place, we will now look at the factors that influence the contents of the mass media. Scholars of media studies have listed a number of influential factors that determine the contents of the media:[8]

- journalists
- editors
- media owners
- information sources
- audience preferences
- advertisers
- technology
- economic constraints
- economic market forces

We need not go into detail about all these factors here, but the economic factors are particularly important. Commercial mass media in our

---

6   Tewksbury et al. (2000)
7   Rosen (2005, chapter 2)
8   Shoemaker and Reese (2014), Wahl-Jorgensen and Hanitzsch (2009, p. 59), Ericson, Baranek and Chan (1989)

modern society operate in a very competitive market. The media compete not only for readers or viewers but also for advertisers. In the view of many economists, the primary business of a commercial TV station is not selling news to its viewers, but selling the eyeballs of the viewers to the advertisers.[9] The competition in the advertising market forces TV stations to produce entertaining programmes that appeal to the broadest possible audience in order to get as many viewers as possible to watch the commercials. The media appeal to our emotions to such a degree that the competition between the media has been called an emotional arms race.[10] The result is trivialization and a blurring of the boundaries between news, entertainment, and advertisement. Controversial and complicated issues are avoided, while emotionally touching topics such as sex, violence, and danger get maximum coverage.[11]

Most economists believe that free competition in the media market is a guarantee that the most diverse range of topics, interests, and opinions is covered.[12] However, many studies contradict this. The media shape the cognitive schemas of their audience. Once these cognitive schemas are formed, the audience will be less receptive to messages that do not fit into the already formed schemas. People prefer to hear uniform opinions rather than a diversity of opinions.[13]

The unrestrained competition has another unintended consequence: It gives the media fewer resources to produce quality content. In a monopoly situation where a country has only a single TV station, this station will have plenty of resources for investigative journalism and for producing a diversity of programmes of high quality. But if there are many commercial TV stations competing for the same advertising money, then there will be less money for each TV station to spend on the production of programmes, and consequently it will be unable to produce the same level of quality as in a less competitive market. This is a consequence of the fact that the fixed costs are high while the variable costs are virtually zero. The costs of producing a TV programme are the same whether there is one viewer or a million viewers.[14] With few

---

9   Meyer (2004)

10   Fuller (2010, p. 71)

11   McManus (2009)

12   McManus (2009)

13   McManus (2009), Gunther and Mugham (2000, chapter 12)

14   Berry and Waldfogel (1999, 2001)

resources for investigative journalism, political news broadcasts are often reduced to an uncritical relaying of statements from leading politicians.[15]

The consequences of the fierce competition in the media market include: trivialization and tabloidization; more entertainment and less information; more gossip, celebrity scandals, sex, violence, crime, and disaster; less controversy, and less detailed analysis of complicated problems; and more focus on persons and less focus on principles and issues.[16]

This trivialization leads to a dumbing-down of the audience. People who have been fed this kind of news all their lives will lack the cognitive schemas necessary for understanding complicated issues, and they will be less likely to pay attention to alternative information sources that provide detailed discussion of complicated issues.[17]

Humans have an innate tendency to pay attention to danger because this has always been important for survival. This tendency makes us pay attention to all kinds of media stories about violence, crime, deviance, and disaster.[18] The news media in a competitive market take advantage of this human tendency by presenting an extraordinary amount of bad news. Fear is profitable. Dramatic stories about crime and disaster make people buy the newspaper or watch the TV news. If nothing bad is happening then the media may even report people's fear of bad things that might happen. The intense media focus on fear-provoking stories has created a 'culture of fear', as Barry Glassner calls it.[19] All too often, the media make us afraid of the wrong things. Minor dangers are hysterically blown out of proportions, while much more serious dangers in society go largely unnoticed.[20] The exaggerated fears often lead to moral panics, unnecessary precautions, poor legislation, and 'gonzo justice'.[21]

---

15   McManus (1995), Picard (2004)
16   Fuller (2010, chapter 7), Picard (2004), Esser (1999)
17   Baek and Wojcieszak (2009), Bennett and Iyengar (2008)
18   Fuller (2010, chapter 6), Shoemaker (1996)
19   Glassner (1999)
20   Altheide (2014), Glassner (1999)
21   Goode and Ben-Yehuda (2009), Altheide (1995, p. 97; 2002, p. 196)

The effect of receiving a constantly high dose of crime and violence from television has been called the *mean world syndrome*.[22] Heavy TV viewers come to perceive the world as a gloomy and dangerous place. Fearful people take drastic measures to protect themselves and their children—measures that are out of touch with the objective risk. As predicted by regality theory, fearful people become authoritarian and easy victims of political manipulation.[23]

The person-centered framing of issues makes people blame social problems on individuals rather than on political principles or social and economic structures.[24] We see the effects of person-centered framing in several areas:

- Crime stories focus on perpetrators and victims. They focus more on violent crime and less on property crime. They report individual cases rather than statistics. Crime is blamed on moral defects in the perpetrator rather than on structural causes. The political consequence of this type of crime reporting is more prisons and fewer preventive measures,[25] even though most experts agree that crime prevention is more effective than punishment.[26]

- Political news focuses more on the personalities and strategies of politicians and less on issues and ideologies. This makes voters cynical and disinterested.[27]

- Social problems are blamed on individuals who are seen as responsible rather than on the underlying social or economic structures. This lack of focus on root causes undermines society's problem-solving capacity.[28]

- International conflicts are blamed on the 'evil' leaders of enemy states rather than on political and economic structures. This has led to many attempts to solve international conflicts by removing foreign leaders and governments by force. Such

22 Signorielli (1990)
23 Signorielli (1990), Hetherington and Suhay (2011), Reith (1999)
24 Altheide (2014, p. 122), Iyengar (1996)
25 Altheide (2002; 2014, p. 144)
26 Welsh and Pfeffer (2013)
27 Capella and Jamieson (1997)
28 McManus (2009), Fuller (2010, chapter 7)

coups and assassinations have often exacerbated the problems by leading to chaos and by making the conflict zones even more regal than they already were.

All in all, the implacable dependence of the commercial mass media on economic market forces has led to a framing and public perception of social problems that is superficial and out of touch with the root causes. At the national level, this has in many cases led to a very punitive and inefficient strategy against crime. At the international level, it has led to chaos, escalation of conflicts, and general regalization. Among the unintended consequences of market forces is that the mass media make people perceive the world as more dangerous than it is. This 'mean world syndrome' makes the whole culture more regal. Mass media competition is thus a root cause of one of the most important—and also most neglected—forces of regalization in our modern age. These media effects are seen most strongly in countries that have a very competitive media market and where non-commercial news media have a very low market share.[29]

This situation may change as the internet is radically changing the media landscape these years. It has become so cheap to disseminate information that even small political groups and grassroots organizations can afford to have their own website, blog, Facebook page, or Twitter account. Several states and governments are financing non-commercial news media with a worldwide target audience. Some of these media purvey partisan views and propaganda, while others are dedicated to presenting a diversity of opinions. However, in many countries it is still not possible to receive non-commercial local news of high quality.

The alternative or non-commercial news media are slow to penetrate the market. People do not easily switch to watching a new and unfamiliar kind of information source. They have to get used to different formats and learn new cognitive schemas before they can appreciate the less sensation-seeking alternative news media.

But social media also make a contribution here. A message from even the most obscure source can be spread all over the world in a matter of days through Facebook and other social media if people find it sufficiently interesting. This mechanism makes it very difficult for

---

29 Fog (2013)

governments and powerful elites to cover up serious wrongdoing and scandals.

Messages are disseminated through social media according to the laws of memetics. The study of memetics tells us that messages are particularly likely to be spread through social media if they are interesting, important, easy to understand, and emotionally touching.[30] The social media can be regarded as a new 'marketplace of ideas' no less competitive than the commercial mass media, but controlled by different mechanisms of competition with different criteria. The theory of memetics tells us that a message can spread equally fast whether it is true or false, as long as no proof or disproof is immediately available. People soon realize that messages on social media are not always reliable unless it can be verified that they come from a trustworthy source. This weakens the role of social media as an information source.

The media landscape is changing fast and we have yet to see which way it is pushing our culture and world views. Different people follow different media, and the internet has niches for virtually any kind of special interest. The diversity of media also means a diversity of cognitive schemas and opinions. People who follow different media will have different cognitive schemas and therefore different interpretations of political events. We can predict that this will lead to a wider diversity of political opinions and ideas.[31]

The commercial media are shaped by anonymous market forces, and social media are shaped by the equally uncontrollable forces of memetic competition. The whole media landscape can be seen as a headless monster that nobody is able to control. Which way is this headless monster taking our culture and our world view? The answer is anybody's guess.

5.2. Economic booms and busts

A stable economy is important for people's existential security. This makes us predict that economic stability and security will stimulate a kungic climate, while an unstable economy or crisis will have a

---

30   Tyler (2011), Berger (2013)
31   Bennett and Iyengar (2008)

regal psychological effect. In fact, conflict studies have found a strong connection between affluence, peace, and democracy.[32]

People want to strengthen their nation or their social group when they experience a collective threat, and several studies show that economic crises can have such a psychological effect, just like other collective threats.[33] The regal effect is strongest if the crisis is perceived as a collective threat. There is little or no regal effect if a bad economy is perceived as an individual problem. This does not mean that individual insecurity is inconsequential, but the consequences of individual insecurity are better explained by biological r/K theory (see chapter 3.3) than by regality theory (see chapter 2.2).

Economic insecurity and poverty mean poor living conditions and higher mortality. The r/K theory predicts that this will lead to higher fertility, which in turn may lead to overpopulation, epidemics, and war. Economic insecurity also means poor education, which may lead to youth crime, conflicts, and political radicalization. It is well-known that economic inequality can lead to political instability.[34]

A good, stable economy, on the other hand, will foster a more K-like strategy involving good education, which promotes the demographic transition (see chapter 3.4). These phenomena have obvious effects on regality. We can therefore expect that individual economic insecurity leads to a number of social problems which may lead to regality, while collective economic insecurity also leads to regality directly without these intermediate steps.

Here, we will first look at some of the mechanisms that cause economic instability. After this, we will look at historical examples of economic booms and busts and discuss their psychological consequences.

Some theories of economic instability

Economic activity in society goes up and down due to many factors, both endogenous and exogenous. High economic activity involves high employment, consumer optimism, and high consumer spending, while an economic downturn involves unemployment and less consumption.

---

32 Gat (2006, p. 587)
33 Fritsche, Jonas and Kessler (2011), Inglehart and Welzel (2005, p. 161), Rickert (1998)
34 Galbraith (2012), Turchin (2012)

Fluctuations in wages, public spending, interest rates, inflation, credit, savings, investment, and speculative bubbles are all factors that complicate the equation. Consumers and investors react to changes in the market, but the effects of these reactions are delayed, and the delays can cause fluctuations. Markets are also influenced by many external factors, such as varying crop yields, changes in international markets, shortage of natural resources, technological changes, national and international political changes, and violent conflicts.

These fluctuations in economic activity are traditionally called business cycles, but in reality they are random fluctuations with no predictable time periods. It may be argued that, if the fluctuations were predictable, then people would speculate against them and the effect of such speculation would cancel out any predictable changes. There are many different theories of business cycles—too many to review here—where the different theories have their main focus on different factors.

While these short-term fluctuations appear to be relatively small and random, there are various mechanisms that may explain major long-term effects and more dramatic changes. In chapter 4.2, we saw how the dynamics of empires in agrarian societies can lead to state bankruptcy and political collapse with periods of a hundred years or more.

A further explanation of the rise of despotic empires and kleptocratic regimes is provided in the theory developed by economist Daron Acemoglu and political scientist James Robinson.[35] Political and economic institutions can be either *inclusive* or *extractive*, according to their theory. Inclusive institutions are institutions that give rights and protection to all members of a society, which is typical for a democracy. Extractive institutions are institutions that extract income and wealth from common people to serve the interests of a small elite. Extractive political institutions can facilitate the consolidation of extractive economic institutions, and vice versa. This positive feedback creates a Matthew effect that tends to stabilize such institutions and make them quite durable, as discussed in chapter 4.1. The opposite situation is seen when inclusive political and economic institutions mutually consolidate each other to create a democratic society that makes individual entrepreneurship attractive and generates a thriving economy.[36]

---

35   Acemoglu and Robinson (2012, p. 80)
36   Acemoglu and Robinson (2012, p. 81)

The situation in most Western democracies today is that economic institutions and market forces are gaining more and more *de facto* power and influence, while the options available to politicians are limited by economic constraints. The nation state has lost much of its autonomy and territorial sovereignty to globalized economic market forces. Economic power is also becoming increasingly impersonal and anonymous.[37] There is no longer any identifiable king or government who controls the economy. Instead, economic influence is concentrated around a group of major international banks who own shares in each other.[38] The banks are formally controlled by their shareholders, but when these shareholders are other banks, and so on, it is hard to claim that any identifiable persons are in control at the top of the economic pyramid. It looks like the economy is controlled, to a large degree, by the impersonal market forces. The banks are required to maximize the returns on investment for their shareholders. This means that they are extractive economic institutions. We are facing the paradoxical situation today that Western society is based on inclusive political institutions (democracy) but extractive economic institutions (banks). The gradual shift in power from inclusive political institutions to extractive economic institutions is part of the reason why economic inequality is now growing fast nationally as well as internationally despite a political desire for more equality. A few extremely rich people now own as much as the poorest half of the world's population.[39] This growing inequality is likely to cause both political and economic instability.

The inequality and instability is driven to a large degree by globalization. Free trade across borders has led to widespread competition between countries for attracting industries and cheap labor. Many countries use low corporate tax rates, poor working conditions, and poor environmental regulation as methods for attracting profitable industries in this so-called race to the bottom.[40]

Another important reason why severe economic crises and collapses continue to happen is the inherent instability of the money system. Most countries have a fractional reserve banking system that allows

---

37   Bauman and Bordoni (2014, chapter 1)
38   Vitali, Glattfelder and Battiston (2011)
39   Credit Suisse (2015), Oxfam (2017)
40   Buzbee (2003), Zodrow (2010), Duanmu (2014), Verschueren (2015), OECD (1998)

banks to provide loans for more than they hold in reserves. This process sets electronic money in circulation that is not backed by any material wealth. Most of the money in circulation in the world today originates from debt to private banks.[41] As this debt accumulates interest, there is much more debt in the world than there is money in circulation, as shown in table 5.

Money	US$ trillion
Coins and banknotes (M0)	5
Narrow money (M1 = M0 + checkable deposits)	29
Broad money (M2 = M1 + savings, time deposits, and money market accounts)	81
All debt (government and private)	199
Derivatives (estimated)	630

Table 5. World money supply and debt in 2014.[42]

As this debt is repaid with money originating from new debt, the total debt keeps spiraling and there is no realistic chance that it can ever be repaid with the available money. The inevitable consequence is that somebody will be unable to pay their debt, even if nobody is culpable. It may be individuals who default on their debts and lose everything they own, or it may be companies, banks, or entire countries that go bankrupt. This unfortunate flaw in our money system has remained in obscurity for centuries,[43] but today it is widely discussed among economists and political organizations.[44]

Economists have proposed many different explanations of long economic cycles (or Kondratiev cycles). A particularly useful explanation is Carlota Perez's theory of technological revolutions.[45] Important technological breakthroughs such as the industrial revolution or electronic information technology can start a boom of investment,

---

41   Werner (2014), McLeay, Radia and Thomas (2014)
42   Data sources: Desjardins (2015), Dobbs, Lund, Woetzel and Matafchieva (2015), CIA, "The World Factbook" (2016), https://www.cia.gov/library/publications/the-world-factbook
43   Zarlenga (2002)
44   Benes and Kumhof (2012), Mellor (2010)
45   Perez (2002, 2010)

development, and economic growth. As new technology matures and the market for new products begins to saturate, capital investments begin to move from material wealth to paper wealth in the form of complex financial products. Those products generate a virtual growth that becomes more and more decoupled from the real economy. Such a decoupling of paper wealth from the real economy is evident in the amount of derivatives in table 5.

The boom ends when one or more asset bubbles collapse and the political need for regulating the wayward financial market becomes obvious. Some of the excess capital that was accumulated during the period of rapid growth is likely to be invested in new inventions that may start the next technological revolution. The whole cycle may take half a century or more.[46]

Most of the money in circulation originates from bank credit, as explained above. Today, only a small part of this bank credit is invested in new production equipment or other things that could benefit the real economy. Instead, most of the credit is used for changing the ownership of existing assets, such as real estate, factories, and intellectual rights. This process inflates asset prices. Housing costs are increased by inflated real estate prices, while the costs of production are increased by the inflated prices of the means of production. The consequence is that an increasing amount of money is extracted from the real economy and into the financial economy, where it serves to enrich a small financial elite. The result is rising inequality, instability, and insecurity. We have seen many examples of severe austerity, poverty, and destitution when asset bubbles burst and the spiraling debt can no longer be repaid.[47]

We have now looked at some—perhaps unorthodox—theories that seek to explain economic instability. It must be emphasized that there are many other theories of economic booms and busts. Whether one subscribes to one theory or another, it is a historical fact that economic crises, crashes, and collapses occur from time to time, and that economists and politicians have been remarkably bad at predicting them.[48]

---

46   Perez (2002, 2010)
47   Bezemer and Hudson (2016)
48   Reinhart and Rogoff (2009)

Booms and busts in the twentieth century

World War II was a characteristic example of the connection between economic crisis and regality. The Great Depression of the 1930s was one of the worst economic crises in modern history, and many historians believe that this crisis played an important role in the outbreak of the war. Other factors that contributed to the war are discussed in chapter 6.4.

However, an economic crisis does not necessarily lead to regality and conflict. It depends on whether it is perceived as a collective threat or an individual threat. It also depends on whether the threat is perceived as coming from an external enemy. It is remarkable that the 2008 financial crisis had little or no regal effect, while the Great Depression had a dramatic effect. The difference between these two crises requires an explanation. The earlier crisis was worse, but this is not the whole explanation for their remarkably different effects. How people perceive a crisis is important for its psychological effect. Many banks were bailed out by national governments in the 2008 crisis. Even though this was a controversial move, it gave the population an impression that the governments were in control of the situation. The bank runs in the 1930s, on the other hand, gave an impression of chaos with nobody in control. Many people lost their savings. A large surplus of poor and unemployed young people served as fuel for the war (see chapter 6.4).

An important question in relation to regality theory is who is blamed for a crisis. If a crisis is blamed on foreign enemies or on general misfortune, then people will seek protection in collective action and desire to strengthen the government and the state. In other words, a regal reaction. But a crisis that is blamed on one's own leaders may have the opposite effect, namely a kungic rebellion against leaders perceived as incompetent and despotic.

In Germany in the 1930s, the massive propaganda of the Nazi regime blamed the crisis on Jews and other foreigners and there was hardly space for any counter-propaganda. The 2008 crisis, in contrast, was mostly blamed on 'greedy bankers' and inept political regulation. In other words, the latter crisis was blamed on an internal elite. A kungic reaction against the political and economic elite was seen in 2011 in the form of street protests in southern Europe and in the Occupy Wall Street movement, which was a protest against plutocracy.[49]

---

49   Bauman and Bordoni (2014, chapter 1)

Between the two busts there was a boom. In the 1960s, we experienced extraordinary prosperity and economic growth. The climate of existential safety led to fast cultural changes in the kungic direction, epitomized by the student protests in 1968 and the 'Flower Power' movement. The kungic trends were particularly strong in Northern Europe. The Nordic countries had made it through the war in a relatively benign way compared to other European countries. Their strong traditions of social welfare were further developed into what is known as the Nordic welfare state with a strong social safety net, free education, and free universal healthcare. The existential security that the welfare state provided to all members of society had a strong kungic effect. It is no coincidence that the Nordic countries have a strong reputation as peace brokers. Despite their small size, the Scandinavian countries have been among the avant-gardes with respect to social security, sexual liberation, women's liberation, and environment protection. However, the Nordic welfare state is now being undermined by globalization and the race to the bottom.

5.3. Greed or grievance: Economic theories of civil war

In the late 1990's, a group of economists at the World Bank and the University of Oxford set out to investigate whether economic factors could explain civil wars. They found that the risk of civil war was correlated with various economic variables, while the relationship between the onset of civil war and ethnic diversity was non-significant or non-monotonic.[50] The economists interpreted these findings as an indication that insurgent leaders were motivated by economic profit rather than by grievances such as ethnic discrimination.

A later modification of this theory focused more on opportunities and less on greed. The revised theory claims that predictions whether there will be conflict or peace cannot be based on the presence of grievances because grievances can be found (or invented) everywhere. The distinguishing factors that determine whether an insurgency breaks out or not are the economic and other factors that make insurgency financially and militarily feasible. The study portrays rebel leaders as

---

50   Collier and Hoeffler (1998, 2004)

sadistic predators and their followers as irrational.[51] A group of political scientists at Stanford University made similar findings, supporting the theory that civil wars depend on economic factors rather than on grievances.[52]

These publications sparked a long debate about whether insurgents were motivated by greed or by grievance. Critics argued that the economic studies had no relevant measures of grievances other than some very broad measures of ethnic diversity and income distribution.[53] In the initial study, the economists actually found a correlation between civil war and ethno-linguistic fractionalization, but they used this variable as an indicator of transaction costs rather than possible ethnic discrimination.[54] Similarly, the economists used the size of an ethnic diaspora as a measure of external economic support but ignored the alternative explanation that the size of the diaspora might indicate the number of people who fled because of ethnic discrimination or other grievances.[55]

The distinction between greed and grievance may be an over-simplification, and some studies have found mixed results for the influence of resource wealth on civil wars.[56] It appears that lucrative economic resources have more influence on the duration of armed conflicts than on their onset.[57] Case studies of a number of individual conflicts have revealed more complex causal mechanisms that defy the distinction between political and economic motives for conflict.[58]

The publications that explain insurgencies in terms of economic factors rather than grievances are widely cited and quite influential despite their admitted methodological weaknesses and poor data. Perhaps the popularity of the 'greed' explanation is due to the fact that it is politically convenient. It legitimizes the continued repression of ethnic

---

51 Collier, Hoeffler and Sambanis (2005), Collier, Hoeffler, and Rohner (2009)
52 Fearon and Laitin (2003)
53 Cramer (2002), Cederman, Gleditch and Buhaug (2013, chapter 2), Keen (2012, chapter 2)
54 Collier and Hoeffler (1998)
55 Collier and Hoeffler (2004)
56 De Soysa and Neumayer (2007), Weinstein (2005)
57 Ballentine and Sherman (2003, p. 267)
58 Ballentine and Sherman (2003, p. 259), Collier and Sambanis (2005)

minorities by making their grievances irrelevant and by portraying insurgents as greedy, irresponsible, and irrational.[59]

A different group of conflict researchers collected more detailed and reliable data related to the grievances of ethnic groups. They found strong evidence that the onset of civil war is related to grievances such as economic inequality and political exclusion of ethnic groups.[60] Such conflicts are particularly frequent in sub-Saharan Africa, where the boundaries of former colonies are not aligned with cultural and ethnic boundaries.[61] Another group of political scientists found that the regime type and degree of democracy is the most important determinant of conflict, while economic conditions, geography, and demography are less important.[62]

It is useful to distinguish between factors that influence the onset of civil wars and factors that influence their duration. Conflict researchers have found that strong rebel groups who can pose a credible challenge to the government tend to fight shorter wars and are more likely to see an outcome favorable to them, while weak and peripheral groups with strong grievances but little power are likely to see long and intractable conflicts. Negotiated settlements are unlikely in the highly asymmetric conflicts between a state power and an excluded minority group.[63] There is general agreement among scientists that external support or intervention tends to lengthen intranational conflicts.[64]

Part of the disagreement in the greed-versus-grievance debate stems from different interpretations of the data. For example, conflicts are more frequent in areas rich in oil, minerals, or other valuable resources. This may be interpreted as greed when the insurgency is financed by these resources, or as grievance when the insurgents protest against the government's exploitation of resources in a territory that they perceive as theirs.

Just as it is misguided to ignore grievances as a cause of insurgencies, it would be equally misguided to ignore the economic and other factors

---

59   Keen (2012)
60   Cederman, Gleditch and Buhaug (2013, p. 206), Cederman, Wimmer and Min (2010), Buhaug (2006)
61   Cederman, Gleditch and Buhaug (2013, chapter 4)
62   Goldstone et al. (2010)
63   Cederman, Gleditch and Buhaug (2013, chapter 8)
64   Collier and Hoeffler (2004), Cederman, Gleditch and Buhaug (2013, p. 189)

that make insurgency possible and feasible. Violent insurgencies cannot take place if the potential rebels have insufficient funding sources or insufficient access to weapons, no matter how strong their grievances.[65] On the other side of the conflict, the interest of a powerful elite to suppress democratic rebellion or to support anti-democratic coups is often based on strong economic motivations.[66]

Civil wars are more likely to break out when the state is weak, when the insurgents are strong and well financed, and when they have an ample supply of young recruits. A mountainous terrain also seems to be favorable to insurgents.[67]

There are inconsistent findings on the influence of education on conflicts. One study finds that ethnic conflicts are more likely when people have money but no education, while class conflicts and revolutions are more likely when people are educated but poor.[68] Education is a resource that can help rebels organize, build alliances, analyze political dynamics, and make propaganda, while lack of access to education can contribute to the grievances that motivate people for conflict. Education is associated with advancement of the demographic transition, lower population growth, and higher welfare—all factors that promote peace.[69]

Finally, we should not ignore the influence of third parties who profit from a conflict[70] or depend on the export of critical resources from the conflict area.[71] In chapter 4.3, we saw that economic factors and access to critical resources are often the root causes of symmetric wars. The same factors may be decisive in asymmetric wars, especially as motivating factors for the strong part in the conflict. This is further discussed in chapter 5.4.

---

65   Regan and Norton (2005)
66   Acemoglu and Robinson (2005)
67   DeRouen (2014, chapter 4), Collier, Hoeffler and Rohner (2009), Fearon and Laitin (2003), Hegre and Sambanis (2006), Buhaug and Lujala (2005)
68   Besançon (2005)
69   Lechler (2015), Collier and Hoeffler (2004), Cincotta, Engelman and Anastasion (2003, chapter 2), Hegre et al. (2013), Sambanis (2005)
70   Keen (2012)
71   Copeland (2015, chapter 8)

5.4. The resource curse

Natural resources are becoming increasingly scarce and expensive as a consequence of the insatiable demand for growth in the modern economic system. Many third world countries have rich occurrences of natural resources in high demand such as oil, minerals, and fertile land. Paradoxically, these valuable resources have often turned out to be a curse rather than a blessing because they give rise to conflicts and corruption rather than enriching the country.

In chapter 3.1, we saw how a concentration of resources in small defendable patches leads to contest competition and a higher level of conflict. This still applies in the modern world even though the conflicts are no longer over food patches but over oil wells and mineral mines.

An important economic study at the Peace Research Institute Oslo has found that many resource-rich countries are poorly developed because of the way the rents from the resources are spent. Entrepreneurs often use the profit from resource extraction in a way that economists call 'unproductive rent extraction' or 'grabbing', rather than productive activities that benefit the country. Countries with weak government institutions are more 'grabber friendly' as these economists express it. The grabbing entrepreneurs spend a significant part of their profit on supporting a corrupt and grabber-friendly government that allows them to export most of the profit out of the country rather than benefiting local industry. Countries with strong and transparent government institutions, on the other hand, are able to avert the resource curse and use their resources in a way that benefits local industry and the country's economy.[72]

Oil is no doubt the number one cause of resource curse problems today, but other important resources such as food, water, land, timber, and minerals are also frequent sources of conflict.[73] International investors actively buy land and mining rights, for example, in order to secure their future access to resources that are expected to become scarce.[74]

---

72   Mehlum, Moene and Torvik (2006)
73   Ross (2012), Carmody (2011)
74   Jacur, Bonfati and Seatzu (2015)

Extractive industries have little or no interest in establishing democracy. The voters in a democratic country will want to distribute the wealth produced by extraction of natural resources to purposes that benefit the country's population, while a dictatorial government can be manipulated or bribed by transnational companies to allow large amounts of wealth to be exported. There is plenty of evidence that foreign actors have secretly supported or promoted undemocratic forces in resource-rich countries although they purportedly support democracy.[75] For example, the coup in Iran in 1953 that brought the Shah to power was led by US and British secret services in order to reverse Prime Minister Mohammad Mosaddegh's nationalization of oil.[76] Even today, the USA is stepping up military operations in almost all African countries in order to defend its security interests.[77] By doing this, the USA may inadvertently be impeding progress and supporting kleptocratic and grabber-friendly regimes.

The resource curse can be a serious impediment to economic development. For example, the Republic of Niger currently has the lowest development index of all countries in the world despite—or rather because of—a rich export of uranium and oil. Frequent conflicts over the valuable resources have destabilized the country, which has seen four coups and several coup attempts since its independence in 1960.[78] A similar fate has befallen the Democratic Republic of the Congo, which holds more than half of the world's reserves of the mineral coltan that is important for the electronics industry.[79] The poor development of resource-rich countries can be explained by a combination of vicious circles involving both political, social and economic effects. The rent from oil extraction, mining, or other extractive industries benefits a small elite but produces few jobs for the local population. The elite have an interest in shaping policy to secure their own advantageous position by investing in security rather than in the development of other productive sectors. A surplus of labor and subsidized unproductive employment leads to poor development and delayed demographic transition.[80]

---

75   Hiatt (2007, chapter 1), Looney (2012, chapters 3, 6, 12, 14 and 24)
76   Ross (2012, chapter 2)
77   Turse (2015)
78   Carmody (2011, chapter 6)
79   Carmody (2011, chapter 6), Keen (2012, chapter 2)
80   Auty (2012)

Rich resources can lead to conflicts between national and foreign actors, and between different foreign actors such as the USA, Europe, and China, who compete for access to the resources of third world countries. There are also frequent conflicts between insurgents and a government that squanders wealth rather than spending it for the benefit of the people. Secession conflicts can arise when different groups compete for resource wealth that is concentrated in a particular part of a country. For example, a conflict over oil resources in Sudan led to a split of the country in 2011, and the conflict continues.[81]

Rich resources can often fuel and escalate conflicts that may have started for reasons unrelated to the resources. Governments of resource-rich countries are likely to use their riches to finance military operations in internal as well as external conflicts, but resource wealth may also end up in the hands of insurgents, so that both parties in a civil war are financing their operations from the same resource.[82] There are several ways in which insurgents can use resources as a means of financing their operations. Resources that are easy to loot and smuggle are particularly useful for insurgents, as we have seen with diamonds in Sierra Leone.[83] The lawlessness and chaos in a civil war makes it possible for guerillas to profit from the production of drugs and other illegal trade. Finally, insurgents can profit from an existing industry by stealing the product and by extortion. Oil extraction and mining in remote areas are particularly vulnerable to extortion because of the large sunk costs that are invested in the industry and because the industry cannot easily be moved to another place.[84]

After World War II, the global oil market was dominated by a cartel of oil companies, commonly known as the Seven Sisters. The behavior of the Seven Sisters became so intolerable that, in the 1970s, most oil-rich countries nationalized their oil production. However, government institutions in many of these countries were not strong enough to handle the windfall of new wealth accountably. The lack of oversight enabled autocrats to consolidate their power by increasing spending, lowering taxes, buying the loyalty of the armed forces, and concealing their own

---

81 Ross (2012, chapter 5), Peck and Chayes (2015, p. 6), le Billon (2004)
82 Ross (2012, chapter 1)
83 Le Billon (2004), Wilson, S. (2013)
84 Ross (2012, chapter 1), Clarke (2015)

corruption and incompetence.[85] Many of the oil-rich countries also used their new wealth to finance wars and conflicts.[86] Oil prices are quite volatile because the market is inelastic, and extreme price fluctuations have caused political instability.[87]

When the economy of a country is dominated by a single sector, other sectors are likely to suffer. This is known as the *Dutch disease*. The high income from oil production leads to an increase in the real exchange rate of the oil-rich countries. This makes other sectors such as agriculture and manufacturing less profitable. The export of other products goes down and the import of food and other products outcompetes much of the domestic production. The result is lack of investment in other sectors, lack of development, unemployment, economic inequality, and further political instability.[88]

The low demand for labor in these countries has consequences for the role of women in society. While men may find jobs in the construction industry, there are fewer jobs in export-oriented industries, such as textile production, which typically hire more women. Statistically, women who work at home have more children than women who make a career outside of the home. This leads to higher population growth and less economic development, whereby the demographic transition is thwarted. The patriarchal values of countries in the Middle East have often been explained with reference to Islam, but statistical studies show that oil production explains the low status of women better than does religion.[89]

Resource conflicts are not always fought by violent means. International actors often use more subtle economic weapons to gain access to valuable resources. A quite common strategy used by the international extractive industry is to control resource-rich countries by imposing unpayable debt on them. The countries are offered large infrastructure projects to be financed through loans from international banks, the World Bank, or the International Monetary Fund (IMF). Most of the borrowed money goes to international contractors rather than to local businesses. The profits from the finished projects are invariably

---

85   Ross (2012, chapter 1)
86   Peck and Chayes (2015)
87   Ross (2012, chapter 2)
88   Ross (2012, chapter 2)
89   Ross (2012, chapter 4)

less than estimated and insufficient to service the debt. Unable to pay back the debt, the countries are required to implement so-called *structural adjustment reforms* as a condition for debt relief. The structural adjustment reforms include privatization and deregulation of industry as well as lowering of taxes and trade barriers. These reforms are supposed to improve the economy of the country by attracting foreign investment, but often the paradoxical consequence of these reforms has been to allow international companies to expand the extractive industry that benefits mostly foreign investors and suppresses or outcompetes local industry.[90]

The economic mechanisms that lead developing countries into growing debt are poorly understood by contemporary politicians. The common system of fractional reserve banking creates more debt than money, as explained in chapter 5.2. Private banks earn rents simply because most of the money in circulation originates from debt. The privileged status of the US dollar as a reserve currency and the role of the dollar in international trade increase the demand for dollars and allow American banks to extract rent from the international circulation of dollars. Of particular importance is the fact that the international oil trade is based on US dollars. This so-called petrodollar system gives the dollar a privileged status and generates rent for US banks at the cost of oil-producing countries and oil consumers.

Some analysts believe that the American desire to uphold the petrodollar system has played an important role in conflicts with oil-producing countries. For example, the seemingly groundless war against Iraq and its leader Saddam Hussein makes more sense when we know that Hussein decided prior to the war to switch to euros as the currency for the country's oil exports.[91] In the same vein, analysts suspect that the US-led wars against Libya and Syria, and the tensions with Iran and Russia, are connected with attempts to control the oil market and the desire of these countries to trade oil in currencies other than the dollar.[92] It must be emphasized, though, that this is an unproven theory and that both sides in a conflict are likely to spread deceptive propaganda.

---

90   Hiatt (2007, chapters 1, 9, 11 and 12), Shiva (2013, p. 31), Babb (2005), Hudes (2009), Münkler (2005, p. 89)
91   The background for the war in Iraq is further discussed in chapter 6.3
92   Clark (2005), Klare (2012), Scott (2003)

5.5. Example: Proxy war in Afghanistan

Now we will look at the history of Afghanistan and discuss how a long series of conflicts and proxy wars has made this one of the most regal countries in the world.

The landlocked country of Afghanistan is rugged with high mountains and sparse vegetation. It has a typically continental climate with extreme variations in temperature. People have lived for several thousand years as herding nomads and farmers in this area. Many villages are concentrated at plateaus and along rivers where irrigation of the fields makes high agricultural production possible.[93]

The geography of the country makes violent conflicts likely. The fertile, irrigable spots of land are likely to be coveted, and warriors are able to travel far with few obstacles. Until the introduction of modern technology, however, the sizes of traveling armies were limited by the poor roads and the sparsity of food. The influences of environmental factors on the level of conflicts in a geographic area are discussed in chapter 3.1 and chapter 7.

Afghanistan has a strategically important position in Central Asia. Historically, the Silk Road and other important trade routes have gone through this area. Afghanistan has often been called 'the graveyard of empires'. Throughout more than a thousand years, the area has been a border zone of one empire after another that tried to conquer the land from all directions. The shifting empires were unable to fully control the land, however, with its many autonomous villages separated by large areas of infertile land.[94]

In the mid-eighteenth century, the area gave rise to its own empire, the Durrani Empire, with its capital in Kandahar. The Durrani dynasty ruled for almost a century until a series of wars with Great Britain that started in 1839.[95] The present-day Afghanistan became independent in 1919 after the Third Anglo-Afghan war.[96]

Starting in 1956, Afghanistan developed increasingly strong ties with the Soviet Union. The country received large amounts of economic and

---

93　Barfield (2010, chapter 1)
94　Barfield (2010, chapters 2 and 4), Rashid (2002, p. 7)
95　Abbas (2014, p. 23)
96　Barber (2008, p. 44)

military aid from the Soviets. A communist government was established in 1978 after two violent coups. The anti-religious ideology of the communists was very far from the traditions of the deeply religious Islamic population, and soon a revolt against the government started in the town of Herat. Soviet troops invaded Afghanistan in 1979 to crush the rebellion. This became the start of a series of civil wars. Millions of Afghans fled the country during the following years. A large part of them ended up in refugee camps in Pakistan.[97]

A growing resistance movement of Islamic warriors, known as the Mujahideen, received billions of dollars in aid as well as training and weapons from the USA, China, Saudi Arabia, and several other states. Most of this aid was channeled through Pakistan's intelligence service, the ISI, which played a major role in the conflict.[98]

Thousands of radical Muslims from forty-three Islamic countries joined the Mujahideen guerillas to support the fight against the ungodly communists. The US Central Intelligence Agency (CIA) actively supported the recruitment of Muslims for the Mujahideen because they did not want to see the Soviet empire expanding. The war lasted until 1989, when Soviet president Mikhail Gorbachev ordered a withdrawal of Soviet forces. A million Afghans had been killed and six million had become refugees.[99]

The Afghan infrastructure was totally demolished by the long war. Roads, fields, livestock, and irrigation canals had been bombed by the Russians to destroy the villages where the Mujahideen guerillas were hiding. The Mujahideen were poorly educated and hardly able to govern a state. Chaos, lawlessness, and banditry prevailed. Large parts of the country were controlled by local warlords who had got hold of the many weapons left from the war.[100]

A group of leaders in local religious schools—the so-called madrassas—took action against the lack of security and started to punish warlords and bandits for stealing, raping, and bullying the population. This became the start of the Taliban movement. The Taliban quickly became popular because they established order and security

---

97 Rashid (2002, p. 13), Abbas (2014, p. 32), Nojumi (2002, p. 197)
98 Abbas (2014, p. 54), Nojumi (2002, p. 83), Murshed (2006, p. 33), Abbas (2005, p. 110)
99 Rashid (2002, p. 19, 129), Nojumi (2002, p. 95), Abbas (2005)
100 Rashid (2002, p. 21), Abbas (2014, p. 55), Murshed (2006, p. 39)

in the lawless country. Local military commandos switched sides and brought weapons with them to the Taliban. The movement grew as returning refugees joined them. The Taliban interpreted their success as a sign that it was God's will that they should prevail, and within a few years they controlled large parts of the country. The Taliban finally captured the capital Kabul in 1996 and established the Islamic Emirate of Afghanistan, which ruled almost ninety percent of the country. They received support from Pakistan, and from Saudi Arabia and other Arab countries.[101] The northern part of Afghanistan was still under Russian influence and against the Taliban. The Northern Alliance was formed in 1997 to fight against the Taliban. It received support from Russia, Iran, and India.[102]

The Taliban rule was based on their own extremely strict interpretation of Sharia law. Women were not allowed to work outside the home and were required to cover their faces in public. Girls were not allowed to go to school. Music, dancing, and games were banned. Music tapes, videos, and televisions were destroyed. The strict religious discipline that had enabled the Taliban to establish order in the lawless country came at a high price. The Taliban were vigilante militias with a religious education, but they were not educated for governing a country and they made many mistakes. In 1998, they killed ten Iranian diplomats, and this forced Iran to go into conflict with the Taliban.[103]

Many of the people who joined the Taliban were young Afghan refugees who had grown up in refugee camps in Pakistan. The only education they had been able to get was in the madrassas. Pakistan's ruler Muhammad Zia-ul-Haq, who had come to power by a military coup in 1977, supported the expansion of the network of conservative madrassas in his country. Zia was not a religious extremist himself, but he supported the conservative religious forces for strategic reasons. He probably understood that a deeply religious population would be less likely to demand democracy than a more secular population.[104] The madrassa students were recruited from deeply religious people, poor

---

101 Rashid (2002, p. 21), Abbas (2014, p. 62), Murshed (2006, p. 42), Richardson, L. (2006, p. 89)
102 Murshed (2006, p. 49)
103 Rashid (2002, p. 204), Abbas (2014), Nojumi (2002, p. 168), Murshed (2006)
104 Abbas (2005, p. 100)

refugees, and orphans. The education in the madrassas was aimed at religious work rather than trade or other occupations. Whether the madrassa leaders realized it or not, they were educating loyal cannon fodder for the civil wars in Afghanistan, Kashmir, and elsewhere.[105]

The Taliban attracted Islamic extremists from many countries, and the areas under Taliban control became a sanctuary for radical international Islamist groups. One of those Islamists who sought protection under Taliban was Osama bin Laden. In 1989, bin Laden became the leader of al-Qaeda—an Islamist network, mainly Salafist, which was fighting for a caliphate and against Western interference in the Islamic countries. Al-Qaeda fought along with the Taliban in the Afghan civil war, but they also had a more international focus. Allegations that the CIA supported bin Laden cannot be confirmed, but the Pakistan Inter-Services Intelligence (ISI) had recruited bin Laden, and the CIA supported the ISI as well as the Taliban.[106]

Al-Qaeda was linked to the first terror bombing of the World Trade Center in New York in 1993 and was actively involved in the bombing of US embassies in Kenya and Tanzania in 1998. The USA switched sides when they discovered that the anti-Russian forces they had so generously supported and armed were also anti-American. US forces bombed targets in Afghanistan and Sudan in retaliation and started a hunt for bin Laden.[107]

The terror attacks against New York and Washington on 11 September 2001 were a turning point for Afghanistan. The US administration immediately named al-Qaeda and bin Laden as the culprits behind these dramatic terror attacks. The USA demanded that the Taliban expel al-Qaeda and extradite bin Laden. The Taliban refused to comply with these demands until they had received evidence that al-Qaeda was responsible for the terror attacks. They were not satisfied with the weak evidence provided. Less than a month after the terror attacks, the USA went to war with Afghanistan, with the aim of removing the Taliban from power, destroying al-Qaeda, and hunting down bin Laden. Rebuilding the nation had low priority. A coalition led by the USA and NATO fought for thirteen years in Afghanistan. The coalition forces

---

105 Butt (2012), Bird and Marshall (2011, p. 167)
106 Rashid (2002, p. 131), Abbas (2014, p. 73)
107 Rashid (2002, p. 134)

withdrew in 2014, but conflicts continued. The chaos and insecurity that had led to the formation of the Taliban twenty years earlier plagued the country once more, and the Taliban has grown again, though they do not rule the country.[108]

The introduction of drones has changed the way of fighting. The USA focuses more on drone-killing leading members of Taliban and al-Qaeda and less on the reasons why these organizations exist. The death toll may be smaller in a drone war than in other kinds of war, but there are still many civilian casualties. Drone strikes are frightening to the general population because they are seen as an uncontrollable external force that kills randomly without warning. The US drone strikes have led to increased recruitment for the Taliban, and we can conclude that this new military technology has a regalizing effect comparable to other ways of fighting.[109]

Both the Northern Alliance and the Taliban have relied heavily on drug production as a funding source. There is a strong symbiosis between guerillas and drug producers. The guerillas need the drug money, and the producers need the guerillas for protection and smuggling. As a consequence of this self-sustaining cycle, Afghanistan has become the world's largest producer of heroin and hashish, with hundreds of drug laboratories.[110] It appears that the USA did nothing to stop this booming drug industry.[111]

Another factor in the conflict was the large unused resources of oil and minerals in Central Asia. There was intense competition between two oil companies, the US-based Unocal and the Argentinian Bridas, for permission to build a gas pipeline through Afghanistan. Access to the oil resources was an important focus for the US involvement in the conflict prior to 2001.[112]

Afghanistan has been the battleground for many simultaneous conflicts within this period: a contest between the two superpowers USA and the Soviet Union; a conflict over oil and other natural resources; conflicts between the many ethnic groups in the country; the illegal

---

108 Rashid (2002), Abbas (2014, p. 117), Murshed (2006, p. 294), Bird and Marshall (2011, p. 111)
109 Abbas (2014, p. 166, 201)
110 Scott (2003, p. 31), Peters, G. (2011)
111 Risen (2006, p. 154), Scott (2003)
112 Rashid (2002, p. 160)

drugs trade; conflicts between Muslims and anti-religious communists, Sunni and Shia Muslims, moderate Sufis and the more fundamentalist Salafists and Wahhabis; conflicts between Iran and Pakistan; and a central focus in the USA's global war on terror. At the same time, Afghanistan has been a training ground for fighters in the conflict between Pakistan and India over Kashmir as well as for al-Qaeda's fight against US dominance in the Middle East.

These conflicts have been fought to a large extent by local militias with poor education but with large amounts of external support from each of their sponsor. All warring parties used local militias as proxies because local knowledge of the rugged terrain was essential for military success. Arming local militias whose loyalty could change was a dangerous strategy.[113] The USA, Pakistan, Saudi Arabia, and several other countries have supported the Taliban with billions of dollars as well as with weapons, training, and logistics. Russia, Iran, and India have supported the Northern Alliance on the other side. Everyone supported their local proxies based on the dubious logic that my enemy's enemy is my friend. The external support has enabled these impoverished tribespeople to fight on a much larger scale then they would have been able to do otherwise. The casualties, the number of refugees, and the destruction of infrastructure have been much higher than any 'natural' conflict would bring without external support. This is why the regal effect of this conflict has been so high and why Afghanistan under the Taliban was one of the most regal societies we have seen in modern times.

Regal influences from neighbor countries also played a significant role. The political climate in Pakistan had been radicalized by the conflict with India over Kashmir and East Pakistan—now Bangladesh—while Zia-ul-Haq actively supported radical Islamic groups for strategic reasons.[114] At the same time, the Islamic revolution in Iran and the war with Iraq had given Iran strict religious rule. Pakistan and Iran both contributed to the conflict in Afghanistan by supporting their chosen sides.

The extreme regality of the Taliban showed itself in many ways. The Taliban gained momentum initially by punishing rape, homosexuality,

---

113 Giustozzi (2012)
114 Abbas (2005, p. 112)

and other sexual crimes. Culprits were publicly punished and often killed and hanged in the streets for everybody to see. The regality was also reflected in cultural life, as predicted by our theory. Music, dance, paintings, and most games were banned, but all this returned quickly when the Taliban was removed from power.[115] Ancient pieces of art from before the Islamic period were destroyed in the National Museum in Kabul. International condemnation was universal when the Taliban destroyed two giant Buddha statues from the sixth century.[116]

A very conspicuous sign of regality was the seclusion and complete veiling of women under the Taliban. Before the Taliban rule, there was a diversity of opinions about the liberation of women and whether they should be veiled. Women were free to go with or without the veil, and some even wore miniskirts when this was in fashion in the 1960s and 70s (compare figures 9 and 10).

The majority of women lived in villages where veiling was incompatible with the hard work. Rich women in the cities carried a veil as a status symbol to show that they did not have to work hard. Many women put on the veil during the war with the Russians as a symbol of their cultural difference from the Russians. But under the Taliban there was no choice. Women had to cover themselves completely, including the face.[117]

Many commentators have described the Taliban as a Frankenstein's monster. After US support had helped them come to power, the Taliban and its associated organization al-Qaeda became bitter enemies of the USA.[118] The US administration and the CIA have not learned from their mistakes in Afghanistan. They are doing the same thing in Syria right now. The USA and its allies have supported opposition groups in Syria in an attempt to topple President Bashar al-Assad, while Russia supports Assad.[119] Opposition groups became powerful with US support, only to grow into one of the most brutal and feared 'terrorist organizations', variously known as Islamic State, ISIS, ISIL, or Daesh.

---

115 Rashid (2002), Baily (2015)
116 Rashid (2002, p. 76), Abbas (2014, p. 71)
117 Rahimi (1986), Iversen and Stray (1985)
118 Abbas (2005, p. 13), Barfield (2010, chapter 5), Rashid (2002)
119 Mumford (2013, p. 98), Anderson (2016, chapters 2 and 11)

There are presently many rumors about who is supporting whom in Syria but little reliable evidence. It is too early to write the history books as events are still unfolding, but we can already see the signs of a new proxy war between the USA and Russia in Syria.

Figure 9. Students at the Higher Teachers' College, Kabul, Afghanistan, 1967. Before the war, the religious dress code was not always observed. Photo by William (Bill) F. Podlich, 1967.[120]

Figure 10. Afghan women wait outside a USAID-supported health care clinic, Afghanistan, 2003. The increased regality is clearly reflected in the women's dress. Photo by Nitin Madhav, 2003.[121]

---

120 All rights reserved, reproduced with permission, http://www.pbase.com/qleap/image/120404891

121 Public domain, https://no.wikipedia.org/wiki/Fil:Group_of_Women_Wearing_Burkas.jpg

6. Strategic Uses of Fear

6.1. Terrorism conflicts

Terrorism often leads to a significant regalization of the attacked society. This leads to further repression of the political interests that the terrorists are fighting for and therefore more terrorism. The current chapter explains this vicious circle.

There is no common agreement on how to define terrorism or on deciding who should be called terrorists. It is often said that one man's terrorist is another man's freedom fighter. Some definitions of terrorism focus on the terrorization of people through shocking acts of violence. Other definitions point at political violence against civilians. And some definitions regard terrorism as a form of political communication where the intended audience is broader than the immediate victims of the violent acts. The word 'terrorism' was originally used about tyrannical leaders who terrorized their population, but today the word is mostly used about rebels fighting against a government—many definitions explicitly exclude government actors as terrorists.[1]

All attempts to reach a common definition of terrorism have failed because the main use of the words 'terrorist' and 'terrorism' has been to delegitimize one's enemies. Politicians and professionals have generally been careful to tailor their definition of terrorism so that it fits their enemies but not their friends.[2] The power of words and definitions has

---

1   Richardson, L. (2006, p. 20), Ganor (2002), Reid (1997), Jackson, Breen-Smyth, Gunning and Jarvis (2011), Lutz (2008, p. 9)
2   Reid (1997), Jackson, Breen-Smyth, Gunning and Jarvis (2011)

© 2017 Agner Fog, CC BY 4.0   https://doi.org/10.11647/OBP.0128.06

long been recognized,[3] and there is no neutral ground in this war on words.[4] The one who succeeds in applying the label of 'terrorists' to his enemies has won the moral battle.

The present book will not contribute to the futile attempts to reach a precise definition of terrorism. Instead, the word will be used in a more interpretive and historical meaning to describe any kind of political violence that is considered terrorism by the surrounding society.

Today, terrorism is mostly seen in *asymmetric conflicts*. It makes no sense for a small group of rebels with few resources to use conventional weapons when fighting against an immensely more powerful state or government.[5] Terrorism is used as a last resort when other strategies fail. For example, Palestinian terrorists say that they have tried everything else.[6] This logic was explained with remarkable clarity by the French philosopher Jean Baudrillard after the terror attacks against the USA on September 11, 2001: 'It is the system itself that has created the objective conditions for this brutal retort. By taking all the cards to itself, it forces the other to change the rules of the game'.[7] The playing cards that Baudrillard metaphorically refers to include the superior military and economic power of the USA, as well as the discursive power. Most of the international news media are US-owned, and the USA dominates the academic discussions as well. The much weaker group of Islamists has strong grievances against the USA, but they would get nowhere if they chose to fight by more conventional means. However, terrorism is difficult to justify morally or ideologically, and some terrorists have left their organizations when they realized how much they had hurt their innocent victims.[8]

There is a strange relationship between terrorists and the mass media. Fear is profitable for the media, as explained in chapter 5.1. The more shocking and scary the event, the more profitable it is for the media to report about it. Insurgents complain that the only time the media want to talk with them is when they have carried out a terror

---

3   Foucault (1980, chapter 6)
4   Rid and Hecker (2009, 45)
5   Richardson, L. (2006, p. 28)
6   Speckhard (2012), Adams (1986, p. 48)
7   Baudrillard (2001)
8   Moghadam (2012)

operation.[9] Terrorism is a form of political communication, aided by the mass media, but terrorists pay a high price for this media attention.

The mass media prefer a simple picture of good guys versus bad guys because moral ambiguity is bad for the media.[10] The media tend to frame terrorism stories with a focus on current episodes of violence rather than on the underlying political conflicts; and terrorists do not always regard media coverage as beneficial to their cause.[11] The media often rely heavily on government sources when reporting terrorism incidents, and they tend to support the government's position on the conflict.[12] In this way, the mass media amplify the asymmetry of the conflict and thereby block the road to peace. Peace negotiations have been more successful in situations where the media have diminished the asymmetry by representing the negotiators of the two parties as equals.[13]

The media coverage of terrorism since 9/11 shows all the signs of a moral panic described in chapter 3.10: the talk is highly emotional; the reaction is exaggerated, in the sense that other dangers with much higher death tolls get much less attention; a group of scapegoats is stigmatized (such as Muslims after 9/11); common standards of justice are undermined; the definition of terrorism is unclear and ever expanding (eco-terrorism, narco-terrorism, cyber-terrorism, etc.); and a large number of experts claim to know the motives and *modus operandi* of the terrorists without ever meeting a terrorist.[14]

Social scientists have discovered that the threat of terrorism can have profound effects on public attitudes and sentiments towards a variety of issues.[15] There is a general 'rally around the flag' effect, with increased trust in the government and support for the president, and increased social identification, nationalism, and patriotism.[16] There is a

---

9   Speckhard (2012)
10   Weimann and Winn (1994)
11   Paletz and Schmid (1992, pp. 41, 67), Cooke (2003)
12   Brinson and Stohl (2009)
13   Shinar and Bratic (2010)
14   Jackson, Breen-Smyth, Gunning and Jarvis (2011), Rothe and Muzzatti (2004), Altheide (2009), Keen (2006, p. 98)
15   Woods (2011)
16   Chowanietz (2010), Nacos, Bloch-Elkon and Shapiro (2011, chapters 1 and 2), Schmid and Muldoon (2015), Huddy, Feldman and Weber (2007), Olivas-Luján, Harzing and McCoy (2004)

tendency towards a more hierarchical orientation and authoritarianism in political debates.[17] Democratic principles, civil liberties, and human rights are undermined,[18] freedom of the press is limited,[19] and there are more hate crimes and scapegoating.[20] There is more war, militarism, and armament, and more state crimes after terrorism campaigns.[21] All these effects are indicators of the strong regalizing effect of terrorism.

We may wonder why the psychological effect of terrorism is so strong compared to other dangers. A major reason is that terrorism lends itself to coverage in the mass media. It offers very dramatic and graphic images with opportunities for person-centered stories of victims, heroes, and villains. People pay attention to these stories because terrorism can hit anybody.[22] Political entrepreneurs have worked effectively in many cases to boost the rallying effect by skillfully using the media to whip up public emotions and by organizing big mourning ceremonies for the victims.[23] Terrorism events are often followed by tough legislation implemented hastily in the highly emotional atmosphere after the event, and this legislation is not always rolled back when the peaceful atmosphere returns.[24] Common standards of justice are typically eroded by this process, which spills over into other areas of justice unrelated to terrorism.[25]

There is a long-standing debate about how effective terrorism is as a weapon.[26] The political scientist Robert Pape has found that suicide terrorism may be successful in achieving at least some proximate goals,[27] while his colleague Max Abrahms maintains that terrorism is generally a counterproductive strategy.[28] The disagreement boils down to different definitions of terrorism and different criteria for measuring success.

---

17   Olivas-Luján, Harzing and McCoy (2004), Perrin (2005)
18   Rothe and Muzzatti (2004), Huddy, Feldman and Weber (2007), Welch (2006, chapter 9), Human Rights Watch (2014), Amnesty International (2004)
19   Simon (2002)
20   Welch (2006, chapter 5)
21   Huddy, Feldman and Weber (2007), Carnagey and Anderson (2007), Welch (2006, chapter 7), Mayer, J. (2008)
22   Hirsch-Hoefler, Canetti, Rapaport and Hobfoll (2014)
23   Oates, Kaid and Berry (2010)
24   Lynch (2012), Douglas (2014), Carlile and Owen (2015)
25   Donohue (2012)
26   Compare Dershowitz (2002) and Abrahms (2006)
27   Pape (2003)
28   Abrahms (2011)

Terrorist groups have in some cases had success in the short-term goals of strengthening their organization and their political position, but rarely in their ultimate strategic goals of gaining national independence or expelling a foreign regime.[29] Historically, terrorism has sometimes been successful in colonial wars.[30]

A considerable number of statistical studies have found that rebel groups who attack civilians practically never win a conflict, and only rarely are they able to obtain a negotiated settlement,[31] while guerilla groups that attack only military targets have success in some cases.[32] In the few cases where terrorists have had some success, this success may be due to factors other than their tactic of attacking civilians.[33] The Madrid bombings have often been mentioned as an example of a successful terrorism action because they influenced an election result and political decisions. However, this success was mainly due to the fact that the Spanish government had lied about the terror incident and that the lie was exposed on the day of the election (see chapter 6.3).

Asymmetric conflicts tend to be long-lasting,[34] and conflicts that involve terrorism often become intractable. Statistical studies show that conflicts where civilians are attacked can be very long-lasting.[35] There are several reasons why terrorism is such an unsuccessful strategy. Government negotiators are mostly unwilling to negotiate with terrorists for fear that their concessions might inspire more terrorism.[36] The rebels have no chance of winning by means of conventional warfare, yet their motive for fighting persists as long as their grievances are not resolved.

The political scientist Max Abrahms uses attribution theory to show that people very often misinterpret the intentions of terrorists by confusing the consequences of their actions with the motives behind them. The targeted population believes that the terrorists want to destroy their society and their values and create chaos, while the real goal of the terrorists may be, for example, to gain independence, or to

---

29 Krause (2013)
30 Münkler (2005, p. 103)
31 Abrahms (2006), Abrahms and Lula (2012), Abrahms and Gottfried (2014), Fortna (2015)
32 Abrahms and Lula (2012)
33 Abrahms and Lula (2012), Cronin (2009)
34 Thies (Thies 2001)
35 Fortna (2015)
36 Richardson, L. (2006)

end occupation of their land. This misinterpretation leads to hardline and uncompromising responses. The negotiators of the targeted country assume that concessions will lead to more terrorism not less.[37]

The rebels may use violence in order to get media attention, but the media coverage they get is almost completely negative, with a focus on the violence and not on the grievances of the terrorists. This does not lead to more sympathy for the terrorists or understanding of their grievances.[38] Instead, we see a spiral of hate, violence, and political extremism where the attacked country implements draconian countermeasures that lead to new violent reactions.[39] This spiral of violence makes both parties more regal and less willing to make concessions or to negotiate a compromise. Ethnic, religious, or cultural differences between the two conflicting parties are amplified by this process and used strategically by the leaders.[40] If the conflicting parties have no important cultural identities to hinge their conflict on, they will invent them. For example, both the Palestinians and the Israelis have strengthened their nationalism and sense of identity as a consequence of the conflict between them. Before the conflict, the inhabitants of present Palestine were likely to idenfity as Jordanians or Arabs, while 'Palestinian' was not a strong or important identity.[41]

We may ask why terrorism is used at all as a political strategy when it is so clearly counterproductive. We would expect rebel groups to learn from history and realize that the strategy of violence against random civilians only benefits their enemy. Analyzing the situation, we may suggest several hypotheses for explaining why terrorism exists at all:

- Some terrorists are simply bad strategists. Their minds are occupied with other things, such as political, ideological, or theological discussions, internal hierarchy, organizational matters, security, rallying ceremonies, fundraising, recruiting new members, and propaganda.[42]

---

37   Abrahms (2006)
38   Abrahms and Lula (2012)
39   Canetti et al. (2013), Chowanietz (2010)
40   Richardson, L. (2006, p. 188)
41   Brand (1995), Peretz (1996)
42   Kassimeris (2008)

- Terrorists are generally driven by strong grievances. They become radicalized by seeing great injustice being committed against a group that they identify with, and they want revenge. Influenced by a supportive group and a legitimizing ideology, they may resort to terrorism when other strategies seem useless. They see themselves as altruists fighting for a just cause, while their enemies see them as evil psychopaths.[43]

- Rebel groups are generally unable to cause any significant damage to their enemy if they fight with conventional weapons against a superior military power. Attacking civilian targets is perceived as a last resort.[44]

- Rebel groups are unable to get media attention unless they commit shocking acts of violence. They are encouraged by the intense media coverage they get when they commit acts of terrorism.[45]

- Rebel groups are inspired by similar groups in other countries. They imitate the groups that get the most coverage in the international news media rather than the groups that are most successful.

- Terrorists overestimate their chances of victory, because they draw a false analogy with guerrilla successes.[46]

- Usually, the ultimate goal of a rebel group is to win a strategic victory over their enemy, but at the same time they are pursuing the more proximate goals of strengthening their own organization, recruiting new members, gaining adherents and supporters, fundraising, and strengthening their position vis-à-vis rival groups within the same movement. While the strategy of terrorism does not help them reach the ultimate goal, it may be effective in reaching the other proximate goals. In several cases, more radical and violent groups have outcompeted relatively moderate groups fighting for the same cause.[47]

---

43 Speckhard (2012), Richardson, L. (2006, p. 63)
44 Speckhard (2012), Richardson, L. (2006)
45 Richardson, L. (2006)
46 Abrahms and Lula (2012), Richardson, L. (2006)
47 Richardson, L. (2006, p. 105), Krause (2013)

- The conflict may have side effects that are beneficial to certain stakeholders on one or both sides of the conflict. Leading members of the rebel groups, government officials of the attacked country, and, of course, weapons dealers may have little incentive to end the conflict because of the political, economic, and psychological payoffs it provides for them. These payoffs include cover for suppression and abuse of the population, exploitation of natural resources, protection for the drugs trade and other criminal economic activity, profit for the weapons industry, profit for the mass media, uniting a population around the psychological need for a strong leader, warding off democracy, and the opportunity for the military to interfere in politics.[48]

- In some cases, rebel groups are manipulated by their enemies into using counterproductive tactics. This is explained in the next chapter.

- Spiraling violence causes both parties in a conflict to become more regal. This causes people to act more emotionally, to use a more radical and uncompromising rhetoric, and to become more violent or to support the more violent groups.

The strong psychological impact of terrorism makes the attacked population more regal and more likely to react with draconian countermeasures. The rebel group feels that the counter-attacks against them are disproportionate and unfair, and this is ground for further radicalization. Their civil rights and access to a fair system of justice are often undermined. This gives the rebel group still more grievances and reasons to fight. For example, during the conflict in Northern Ireland, the republican rebels were often more focused on fighting against the lack of due process and other injustices against them than on fighting for their original cause.[49] In Palestine, many people feel that their life conditions in the conflict zone are so intolerable that they would rather die an honorable death as martyrs or suicide bombers than continue their hopeless lives.[50]

---

48   Keen (2006, chapter 3)
49   O'Day (1993), Soule (1989)
50   Speckhard (2012, chapter 8)

This vicious circle is further aggravated by the widespread policy of never negotiating with terrorists. The rebels may conclude that their only option is to use more violence when they experience that their enemies are unwilling to negotiate with them or make any concessions.[51]

Such a vicious circle is not easily ended. An escalating conflict may end with complete victory for the strongest party when no third party intervenes. This was the normal outcome of asymmetric conflicts in ancient times. Today, however, the international community will not accept the genocide that such an outcome would entail. The intervention of third parties or international organizations has often been strong enough to prevent a genocide but not strong enough to stop the conflict and enforce a negotiated solution. Conflicts that involve terrorism may end if the rebels lose their support and funding,[52] but the conflicts may flare up again unless the grievances are resolved.

The international community may facilitate a more lasting solution to an asymmetric conflict by putting enough pressure on the stronger party to deal with the grievances of the weaker party. This is what happened in South Africa when apartheid was abolished. However, many rebel groups lack sufficient resources to influence international opinion in their favor. The Irish Republican Army (IRA) in Northern Ireland did indeed have a strong propaganda strategy, but they had little luck in bypassing British censorship and the self-censorship of the British media.[53] News media in other countries simply relayed reports from the British news agencies without informing their audience that they were purveying censored news. The impetus for any kind of third party intervention that might have facilitated the negotiation of a compromise solution was therefore missing. The IRA eventually changed their strategy and ended the violence when they realized that they lacked public support and that the strategy of terrorism was futile.[54]

While the tactics of terrorism are mostly counterproductive, so are attempts to combat terrorism. Politicians often respond with extreme measures and rhetoric, fueled by media hysteria in the aftermath of a

---

51   Richardson, L. (2006)
52   Cronin (2009), Clarke (2015)
53   Kingston (1995), Cooke (2003)
54   Alonso (2001), Phayal (2011)

terrorism event.[55] The regal reaction to terrorism events often includes infringements on civil liberties and lowered standards of justice.[56] This is fuel for the ideology of the terrorist organizations and their claims that they are victims of oppression and injustice. In this way, counterterrorism measures work against their purpose; they contribute to radicalization of terrorist organizations and help them to recruit new members. The Dutch historian Beatrice de Graaf has shown that there is a positive correlation between the amount of dramatic counterterrorism measures and the radicalization of rebels.[57]

Those who fight against terrorism often ignore the more fundamental but less visible causes of terrorism and focus on something tangible such as specific persons and organizations.[58] Much of the counterterrorism literature ignores the importance of grievances. Rebels will still be motivated to fight as long as they have strong grievances. Any attempt to suppress them by brute force will only add to their grievances. The strong asymmetry of power makes it unlikely that negotiations can lead to a compromise that is acceptable to the rebels. The politicians of the attacked country face a serious dilemma. If they make concessions to the terrorists they may be encouraging more terrorism. But if, on the other hand, they employ a hard-line policy, they may exacerbate the grievances and contribute to further radicalization of the terrorists. Even the most dovish politician of a country struck by terrorists would refrain from making concessions to terrorists for fear that he or she might be accused of being 'soft on terrorism' and rewarding terrorists when the population is panicked by the attacks.

Attempts to combat terrorist organizations are often focused on eliminating specific leaders, who are portrayed as evil. Arresting the leader of a terrorist group may be an effective strategy only if the group depends on a single charismatic leader. In other cases, it will lead to further radicalization, especially if the leader is killed under dramatic circumstances.[59] Attempts to eliminate specific terrorist organizations with militaristic means have only contributed to escalation of the conflict

---

55   Altheide (2006), Mueller, J. (2006)
56   Amnesty International (2004), Simon (2002), Carlile and Owen (2015), Norris, J. (2016)
57   De Graaf (2011)
58   Keen (2006, chapter 2)
59   Inglehart and Welzel (2005, p. 297), Richardson, L. (2006, p. 12)

and radicalization of the terrorists, while the defeated terror cells are continually replaced by new ones.[60] It may be an efficient strategy to target the funding sources of the terrorists, but this does not remove the grievances.[61] In most cases, the grievances can be dealt with in a way that the weaker party perceives as fair only when a third party mediates in the conflict and puts pressure on the stronger party to make concessions.

While terrorism is not a winning strategy for the rebels, it may be quite useful for the government of the country that is being attacked. Terrorism often produces a strong sense of patriotism, rallying around the flag, support for the incumbent government, and tolerance of quite repressive measures of justice. These effects are so strong that the governments of terror-ridden countries may have an interest in letting an ongoing terrorism campaign continue. They may even provoke terror attacks against their own countries. It has happened many times in modern history that governments have instigated, or deliberately failed to prevent, terror attacks against their own countries. Whether this is intended or not, terror attacks give the government the advantage of a regal climate that strengthens the power of the leaders. The paradoxical phenomenon that political leaders may somehow be complicit in terror attacks against their own population is described in the next two chapters.

6.2. The strategy of tension in Italy and elsewhere

Italy was hit by a large wave of terror attacks in the period from the late 1960s to the early 1980s—the so-called Years of Lead (named so after bullets made of lead). The situation in Italy in those years was very complicated but also very interesting from a theoretical point of view. The political landscape was chaotic, and political violence was widespread. There were frequent clashes between neofascists and left-wing activists in the streets, and the violence escalated into political assassinations, kidnappings, terrorism, and attempted coups.[62]

---

60 Keen (2006, chapter 11)
61 Adams (1986), Clarke (2015)
62 Bull, M. (1992), Weinberg (1995)

A particularly notorious series of terror events occurred on 12 December 1969. A terror bomb exploded in front of a bank on Piazza Fontana in Milan, killing sixteen people and wounding eighty-eight. Three bombs exploded in other Italian cities on the same day, and one more bomb was left unexploded. The bombings were initially blamed on a small anarchist group, and twenty-seven left-wing activists were arrested under intense media coverage. After several years of investigation, it was found that among the approximately ten members of the anarchist group were two infiltrators: a neofascist activist and a police agent. They had produced false evidence against the group. New evidence pointed to the neofascist organization Ordine Nuovo, which was also responsible for several other terror bombings.[63]

There were hundreds of terror bombings committed by various neofascist groups in these years, and quite often they were blamed on left-wing groups. Sometimes, right-wing terrorists left false leads to implicate left-wing groups. Corruption at all levels of the state apparatus was evident. The police would often follow the so-called 'red track' (left-wing suspects) and ignore the more likely 'black track' (right-wing suspects). Police investigations were obstructed, evidence disappeared, false evidence was fabricated, witnesses were murdered, and processes at one court were interrupted and moved to another court with another judge. A general pattern was that suspects were convicted only to be later acquitted at a higher court. This happened to both right-wing and left-wing suspects. Thirty-three years of trials, re-trials, and appeals ended with acquittals for most of the suspects.[64]

Some left-wing groups became radicalized as a response to the neofascist violence, the political situation, and the illegitimate behavior of the state apparatus. A particularly militant left-wing group was Brigate Rosse (the Red Brigades), who committed a series of assassinations and political kidnappings. The former Prime Minister Aldo Moro was kidnapped by Brigate Rosse in 1978 and killed after fifty-five days in captivity. The kidnapping happened at a time when Moro, as president of the Christian Democrats, was negotiating a historical compromise to include the Italian Communist Party in a coalition government. Those plans died with Moro. Preventing the communists from entering the

---

63   Ferraresi (1996, p. 9)
64   Cento Bull (2007, p. 24), Willan (1991, chapter 7), Bale (1996), Ferraresi (1996)

government was perhaps not the most logical thing for a left-wing group to do, but it appears that they were against the revisionist, or compromising policy of the Communist Party.[65]

The population was confused over whether left-wing or right-wing groups were behind the many terror attacks, and there was much suspicion that the secret service, clandestine paramilitary groups, the CIA, and other foreign agencies were involved.[66] Research has confirmed many of these suspicions as more and more evidence has been uncovered through the following decades. It is now evident that the terrorism in Italy in these years was instrumental in the so-called strategy of tension, with the purpose of preventing a communist takeover of power in Italy.[67]

The strategy of tension can be traced back to a symposium on the topic of revolutionary warfare in 1965, organized by the obscure Alberto Pollio Institute of Military History. The Pollio Institute was set up by an agent of the secret service with money from companies interested in winning defense contracts. Present at the meeting were leading members of various neofascist organizations and future terrorists, as well as high-ranking members of the secret service and the armed forces.[68]

There are many different interpretations of the strategy of tension, and each of the involved groups and organizations had their own agenda. Italy had the strongest Communist party in Western Europe at the time, and many were afraid of a communist takeover of power in Italy. The supporters of the strategy of tension deliberately fueled political violence in order to create psychological tension in the population. They wanted to 'destabilize in order to stabilize'. By creating insecurity, they hoped that people would seek security in a strong government and that this would pave the way for a more authoritarian government or a neofascist coup.[69]

The groups that were most active in the strategy of tension were Ordine Nuovo, Avanguardia Nazionale, and the masonic lodge Propaganda Due, all with a neofascist ideology and with links to the secret service.

---

65 Amara (2006)
66 Hajek (2010)
67 Ferraresi (1996, p. 86), Cento Bull (2007, chapter 4)
68 Weinberg (1995), Ferraresi (1996, p. 71), Cento Bull (2007, p. 57), Willan (1991, p. 40)
69 Bull, M. (1992), Cento Bull (2007, chapter 4), Willan (1991), Bale (1996), De Lutiis (1998, chapter 4)

Propaganda Due included high-ranking and influential members of the police, military, and secret services, as well as industry leaders and politicians. The secret services were dominated by neofascists and anti-communists, and the connections between all these groups made cover-ups and the systematic obstruction of justice possible.[70] Secret documents from the Italian military intelligence service SIFAR reveal that they were very afraid that the communists would win elections and that the SIFAR was actively working against the socialists and communists.[71]

The Brigate Rosse and other radical left-wing groups in Europe were infiltrated by right-wing activists and secret service agents, who manipulated them and made them more radical and more violent in order to use them as tools in the strategy of tension.[72] The kidnapping of Moro was carried out with the highest degree of military professionalism — far beyond the capabilities of a small group of political activists. It is unknown whether they were helped by undercover agents from the secret services or by some paramilitary group.[73] However, we can be certain that the secret services could have stopped the Brigate Rosse at any time if they wanted to, given the information available to them from infiltrators.[74] During the period of Moro's captivity, the police messed up or failed to follow obvious leads that could have taken them to the hiding places of the Brigate Rosse.[75]

Historians have been debating for years about who were the masters pulling the strings in the strategy of tension. Were the various groups acting autonomously, were they controlled by some powerful elite in the country, and did they receive support from abroad? Some historians believe that the CIA and NATO played a major role[76] while other historians disagree.[77]

There is evidence of strong connections with Aginter Press in Portugal, a press agency that served as a cover for an anti-communist mercenary organization that trained its members in covert action

---

70   Ferraresi (1996, p. 89), Cento Bull (2007, chapters 3 and 4), Willan (1991, chapters 2 and 3), Amara (2006, chapter 3)
71   Ferraresi (1996, pp. 63, 77)
72   Weinberg (1995), Willan (1991, chapter 10), Bale (1996)
73   Amara (2006)
74   Willan (1991, chapter 10)
75   Pellegrino (1997), Amara (2006, chapter 1)
76   Feldbauer (2000), Rowse (1994), Ganser (2005)
77   Davies (2005), Nuti (2007)

techniques including infiltration and counterinsurgency. Members of the Aginter Press participated in the Pollio Institute meeting, and there are close similarities between the strategy promoted by Aginter Press and the Italian strategy of tension.[78]

The possible involvement of the CIA is difficult to prove because that organization routinely hides covert actions behind other organizations so that its involvement can plausibly be denied.[79] Nevertheless, there are many indications that the CIA and NATO supported the strategy of tension because they wanted to limit communist influence in Western Europe and because of the strategic importance of US and NATO military airbases in Italy.[80] It is well documented that the CIA had secretly been interfering in Italian elections since 1948 in order to keep the Italian Communist Party from power.[81]

The CIA, the National Security Council (NSC), and the US Joint Chiefs of Staff (JCS) were involved in a plan called Operation Demagnetize, whose purpose was to reduce the strength of the Communist Party, according to secret documents that have since been declassified.[82] The US Secretary of State Henry Kissinger strongly criticized Moro's policy, and a US intelligence official warned Moro of serious consequences if he continued his policy of dialogue with the communists. Moro was so shocked by this direct threat that he feared for his life.[83] However, there is no direct proof of active support from the CIA for the violent actions, and some historians argue that the terrorism was determined mainly by internal forces within Italy.[84] Still other theories suggest that KGB, Mossad, and other foreign secret services were also involved.[85]

If we want to analyze the motivations of the terrorists, it is clear that the right-wing and left-wing terrorists had very different motivations. The right-wing, or neofascist, terrorists followed a carefully planned strategy, the strategy of tension. The neofascist groups were anti-intellectual, and

---

78 De Jesus (2012, chapter 2), Ferraresi (1996, p. 61)
79 Agee (1975)
80 De Lutiis (1998, chapter 4), Cento Bull (2007, chapter 4), Willan (1991), Pellegrino (1997)
81 Corson (1977)
82 Cento Bull (2007, chapter 4), Willan (1991, p. 27), Ferraresi (1996, p. 76), De Lutiis (1998, p. 133)
83 Willan (1991, chapter 11)
84 Coco (2015)
85 Fasanella, Pellegrino and Sestieri (2000)

their members were not very concerned with strategy. They were more moved by feelings, group spirit, rituals, marches, group loyalty, and orders from their leaders.[86] The organizational culture and hierarchy of these neofascist groups clearly matched their regal ideology. We can assume that the actions of the low-ranking members were more determined by authoritarian or regal sentiments and orders from above than by individual rationality. The higher-ranking decision makers, on the other hand, had clearly thought out a strategy.

The strategy of tension definitely makes sense in the light of regality theory. Evidently, those who devised this strategy had a theory that people would support a more authoritarian government if random violence and terror created psychological tension. The plan was that this should pave the way for a neofascist coup. However, the strategy of tension was not very successful. The Italians still remembered the cruel fascist rule under Benito Mussolini and there was no general support for returning to a fascist dictatorship. The only success of the strategy of tension was that the murder of Aldo Moro thwarted plans to include the Communist Party in the government. Such a coalition government would probably have been quite unstable anyway.[87]

The failure of the strategy of tension can be explained by the fact that the chaos and political violence was blamed on internal rather than external factors. A threat from external powers would have increased the support for a more authoritarian government, according to our theory, but the main cause of the chaos and violence was the obvious corruption, which of course did not increase people's trust in the government.

The motivation for the left-wing activists was quite different. They had long discussions about theory and ideology, and their motivation for using violence was defense against the violent fascists and the corrupt state apparatus.[88] Their strategy was, of course, even more unsuccessful, as they were manipulated to serve interests opposite to their own. We cannot predict what Italy would have looked like without the political violence, but it appears that the political climate in the country was more

---

86   Ferraresi (1996, pp. 156, 194)
87   Willan (1991)
88   Ferraresi (1996, p. 193), Amara (2006)

influenced by economic progress and by the democratic developments in the neighboring countries than by the strategy of tension.

Figure 11. Aldo Moro in captivity, 1978. The kidnappers published this photo, which made sure that everybody associated the action with the left-wing group Brigate Rosse.[89]

The strategy of tension in Italy was extreme in its range and magnitude, but far from unique. In a classic study of counterrevolution, Arno Mayer describes the deliberate fabrication of violence and chaos as a common counterrevolutionary strategy.[90] The Aginter Press was not only active in Italy and Portugal, but also in Algeria, Congo, Biafra (part of current Nigeria), and other African countries. It is uncertain whether it used similar strategies in those countries.[91] A mysterious organization, the Hyperion language school in Paris, delivered weapons to Brigate Rosse, making it appear that the weapons originated from the Palestine Liberation Organization (PLO). There were claims that Hyperion also delivered weapons to the Red Army Faction (RAF) in Germany, to the Basque separatist group Euskadi Ta Askatasuna (ETA) in Spain,

89 Public domain, https://it.wikipedia.org/wiki/Caso_Moro#/media/File:Aldo_Moro_br.jpg
90 Mayer, A. (1971, chapter 3)
91 De Jesus (2012)

and to the IRA in Ireland,[92] but this has not been verified. The Italian investigators initially believed that the Hyperion school was a cover organization for the CIA, but it might have been more than that. Later investigations have led to the theory that Hyperion was a networking point for Eastern and Western secret service organizations, with the purpose of maintaining world stability by preserving the balance of powers that was established in the Yalta agreement of 1945. Therefore, the Hyperion people fought against anybody who threatened to sway this balance of powers, including the new left, Aldo Moro, Olof Palme, and many others.[93]

Heads of police from European countries decided at a meeting in Cologne in the early 1970s to implement a common strategy of infiltrating terrorist groups at the leadership level. Their undercover agents had to be the bravest and cruellest members of the groups.[94] In Belgium, undercover intelligence agents were heavily involved in a series of false flag terrorism attacks that claimed more than thirty lives in the 1980s.[95] Turkey had a particularly powerful secret service that infiltrated left-wing groups with agents provocateurs, massacred whole villages and blamed it on the Kurdistan Workers' Party (PKK), and created general chaos before a coup.[96]

The situation in West Germany (the Federal Republic of Germany) and the later united Germany was particularly complex. The undercover agent Peter Urbach infiltrated several left-wing groups. He delivered Molotov cocktails to an otherwise peaceful demonstration, he delivered bombs to a Marxist organization, and he supplied weapons to the infamous RAF.[97] Other undercover agents failed to infiltrate the RAF but delivered weapons to other radical groups and even attempted to create a new terrorist organization.[98]

The RAF had three generations of militant activists. The first generation received some support from the Stasi, the secret service of East Germany (the German Democratic Republic), but they did not

---

92   Willan (1991, chapter 10), Feldbauer (2000, p. 66)
93   Fasanella, Pellegrino and Sestieri (2000, part 3), Igel (2012, p. 136ff)
94   Amara (2006, p. 22)
95   Jenkins (1990)
96   Çelik (1999)
97   Rosenfeld (2014), Winkler (1997), Peters, B. (2004)
98   Gössner (1991, pp. 183–215)

want to be controlled by the Stasi. The second generation received paramilitary training from the Stasi, and many RAF members fled to East Germany, where they became involved in international terrorism as paid agents of the Stasi.[99] However, the relationship between the Stasi and the RAF appeared to be ambivalent, and western agents and double agents were also involved.[100]

The third generation of the RAF showed hardly any signs of being an autonomous organization. Some researchers claim that the third generation was nothing more than a phantom organization created by Western security forces,[101] while later research reveals a heavy involvement of the Stasi.[102] But if we assume that the third generation of the RAF was controlled by the Stasi, then we have a problem explaining why the RAF continued its activities for almost a decade after the fall of the Berlin Wall, when the Stasi no longer existed. Regardless of who was behind the later generations of the RAF, it was able to carry out bombings and assassinations with military professionalism, and most of its crimes remained unsolved.[103] The similarity with Italy is obvious, but there is so far no evidence of involvement of western secret services.

There is evidence of cooperation between the RAF in Germany and Brigate Rosse in Italy. The German industry leader Hanns Martin Schleyer was attacked by the second generation RAF. Schleyer and his safety escort were stopped on a road in September 1977. His chauffeur and three bodyguards were shot while he was abducted alive. He was killed after forty-three days in captivity. Six months later, exactly the same happened to Moro in Italy. There is evidence that some of the RAF militants traveled to Italy and assisted in Moro's abduction.[104] If the RAF was controlled by the Stasi, then why did they cooperate with Brigate Rosse, which was almost completely controlled by right-wing and Western security forces at that time? On the other hand, if the RAF was controlled by right-wing or Western forces in the same way as Brigate Rosse, then why did they kill Schleyer? There were obvious right-wing and Western motives for killing Moro but not for killing

---

99   Igel (2012, p. 214), Müller and Kanonenberg (1992), Schmeidel (1993)
100   Bale (2012), Igel (2012, p. 288), Müller and Kanonenberg (1992)
101   Wisnewski, Landgraeber and Sieker (2008)
102   Igel (2012, p. 159)
103   Schmeidel (1993), Wisnewski, Landgraeber and Sieker (2008)
104   Igel (2012, p. 173)

Schleyer. Was he killed in order to delegitimize left-wing movements in a strategy of tension?

There are many unanswered questions. What we know so far is that both Eastern and Western security services and paramilitary forces were heavily involved in terrorism in Western Europe in those years. It seems that the Western security forces relied on the theory that terrorism can incite authoritarianism and harm the apparent supporter, as evidenced by the strategy of tension. The Stasi, on the other hand, appeared to be more concerned with ideology and prestige.[105] Perhaps the Eastern security forces simply relied on the belief that terrorism would harm its victim.

6.3. Fabrication of threats and conflicts

The economist and conflict researcher David Keen has studied many violent conflicts around the world and made the surprising observation that, in many cases, the fighting parties appear to prolong the conflict rather than trying to win and end it. For example, they often use tactics known to be counterproductive, or they provoke an enemy to attack, and in some cases they even sell weapons to their enemy. Keen suggests that some participants deliberately engage in endless conflicts because it gives them certain advantages. Economic advantages include the weapons trade, but also the extraction of valuable resources such as oil in Iraq, minerals in the Democratic Republic of the Congo, or opium in Afghanistan. The advantages can also be of a political nature. A violent conflict can justify the suppression and exploitation of people, sabotage an emerging democracy, create national unity, and create the need for a strong leader. Keen suggests that such conflicts can be self-sustaining— intentionally or not—because certain key players benefit from the continued conflict.[106]

Regality theory can contribute to the explanation of the paradoxical behavior of deliberately protracting a violent conflict. Continuous fighting is not in the interests of the general population, of course, but it may be in the interests of the leaders because it helps them sustain

---

105 Schmeidel (1993)
106 Keen (2012; 2006, chapter 11)

a regal culture. We saw in chapters 2.4 and 2.5 how a regal culture increases the biological fitness of the leader. Let us look further at this phenomenon.

Throughout history, we have seen that many of the leaders of great empires had large numbers of children and provided luxury and advantages for themselves and for their relatives. In other words, the leaders of regal societies had a huge advantage in terms of reproductive fitness. The more regal the society, the greater the fitness of the leader at the expense of his followers. DNA evidence shows that a large fraction of the modern human population is descended from a few successful emperors. Genghis Khan is the record holder with millions of descendants.[107] It is therefore obvious that any behavior that makes a leader more powerful could be promoted by natural selection.

The diversionary theory of war, mentioned in chapter 4.3, suggests that leaders may wage unnecessary wars for the sake of creating a 'rally around the flag' effect that will consolidate their own status. The extreme fitness advantage of powerful leaders leads us to the prediction that leaders could be inclined to deceive their followers and make their society more regal than necessary, for example by fighting unnecessary wars, or by exaggerating or fabricating dangers to their own society. We can expect this strategy to be employed when leaders, or prospective leaders, see a chance to expand their power, but also when leaders see their power threatened by subversive or rebellious movements.

In this chapter, we will search for historical examples of such deceptions in order to find out how far leaders are willing to go in terms of spreading fear and fabricating unnecessary threats and conflicts against their own group.

In chapters 3.10 and 5.1 we saw how leaders and the mass media can benefit from witch hunts and moral panics. Terrorism threats are particularly efficacious for boosting regality, because, as we have seen, they produce strong emotional effects while causing much less damage than regular intergroup wars. There are many ways in which a leader and his government can consolidate their position by manipulating terrorism threats and other dangers, possibly in collusion with the mass media, including the following:

---

107 Balaresque et al. (2015)

- opportunistic exploitation of an unpredicted attack by creating maximum publicity around the event, showing leadership in fighting the attackers, showing empathy for the victims, and implementing stricter legislation

- exaggerating an existing threat by massive media coverage, by overstating the capabilities of potential enemies, by excessive security measures, or by setting national alert levels higher than necessary

- blaming actual terrorism attacks on the wrong culprits

- making false or exaggerated reports about terrorism plots that have been thwarted

- exaggerating the danger of a relatively weak adversary and fighting it with dramatic means

- using undercover operations to lure somebody into planning an attack and arresting them when they are about to carry out these plans (entrapment)[108]

- fabricating and staging a victory over an insignificant or imaginary adversary

- provoking a potential enemy to attack, or deliberately escalating a low-level conflict

- infiltrating peaceful protest groups with violent agents

- selectively eliminating moderate leaders of rebel groups while more radical leaders remain at large

- deliberately failing to avert known terrorism plots

- paving the way for known terrorists by removing obstacles to their plans, disabling surveillance, alarm, and rescue systems, or actively increasing the damage they are causing

- infiltrating potential terrorist groups with undercover agents who help them with advice, intelligence, logistics, training,

---

108 Altheide (2014, pp. 38, 68), Norris, J. (2016)

weapons, money, and so on, and induce them to attack a particular target or use particularly dramatic tactics

- fabricating a violent attack and blaming it on a convenient enemy (false flag attack).

Manipulations at the beginning of this list are very common, while the more serious deceptions at the end of the list are probably less common but also more difficult to document. Let us look at some notable historical examples.

Shelling of Mainila

On 26 November 1939, the Russian village of Mainila, near the Finnish border, was hit by seven explosions. The Russians claimed that Finland had attacked Russia with artillery, killing four and wounding nine. A Finnish border guard who had watched the incident from a distance of 800 meters reported that the shots were fired from the Russian side. He did not see any bodies being carried away from the scene.[109] Russian and Finnish historians later found documents confirming that there were no casualties and that the Russians had staged the incident as a pretext for attacking Finland.[110] The Soviet Union renounced the non-aggression pact with Finland and started the Winter War between the two nations four days later.

Gulf of Tonkin incident

On 2 August 1964, the US destroyer USS Maddox was on a secret intelligence gathering mission in the Gulf of Tonkin near the North Vietnamese coast when it was approached by three North Vietnamese torpedo boats. Fire was exchanged and there were losses on the North Vietnamese side but only minor damage on Maddox. The United States Congress was misinformed about the incident and told that Maddox was on a routine patrol in international waters when it was attacked by the torpedo boats. The truth is that Maddox fired first and pursued the torpedo boats. Two days later, two US destroyers misinterpreted radar

---

109 Edwards (2006, p. 105)
110 Sokolov (2000), Aptekar (2001), Leino (2009)

signals and reported that they were being attacked by torpedo boats, when in fact there were no other boats in the vicinity. Congress was informed that boats had been attacked again, which was not true. This led Congress to pass the Gulf of Tonkin Resolution, which allowed the US military to engage armed forces in Vietnam and escalate the Vietnam War. While some of the misinformation was due to human error rather than deliberate deception, the mistakes were covered up and the false information was used as a pretext for attacking North Vietnam.[111]

Sendero Luminoso

During the 1990s, President Alberto Fujimori of Peru gained popularity through his uncompromising fight against violent insurgent groups such as the Sendero Luminoso, or Shining Path. Even at a time when the insurgents were effectively defeated, Fujimori kept fighting them in dramatic military actions without completely eradicating them. These highly publicized actions helped Fujimori and his increasingly autocratic government to stay in power. When the opposition protested the inauguration of Fujimori for his third term in 2000, the regime infiltrated a group of peaceful protesters with violent agents who set fire to the National Bank. The attack, which took the life of four bank guards, was blamed on the protesters in order to discredit the opposition.[112]

Madrid bombing

The Spanish Partido Popular lost the election in 2004 when it was revealed that it had deliberately deceived the population and blamed a terror bombing on the separatist group ETA, when in fact the Islamicist group al-Qaeda was responsible. The Partido Popular would probably have won a landslide victory had the deception not been revealed on the early morning of the election day.[113]

---

111 Moïse (1996), Hanyok (2001)
112 Burt (2008)
113 Jordan and Horsburgh (2008)

Operation Mongoose and Operation Northwoods

During the Cold War, the US government tried to destabilize and overturn Fidel Castro's government in Cuba in many ways. A campaign of psychological warfare, known as Operation Mongoose, tried to build an opposition within Cuba.

In the 1960s, the CIA covertly funded private radio stations inside and outside of Cuba, including Radio Swan on the Swan Islands, which sent propaganda messages from anti-Castro Cubans into Cuba. A radio station in Mexico City promoted a press campaign about epidemics of hoof and mouth disease and smallpox in Cuba. A CIA radio station in Zambia was spreading misinformation about the conduct of Cuban troops in neighboring Angola in the 1970s. Among the false stories was a fictitious scene in which Cuban soldiers raped fifteen-year-old girls. The station even disseminated faked photographs of the trial and executions of the Cuban soldiers.

In 1961, CIA planes bombed targets in Cuba and pretended that the pilots were Cuban defectors. Many other operations of misinformation and sabotage were proposed or planned, some of them quite bizarre, but most of these plans were never carried out.[114]

One set of plans, known as Operation Northwoods, included false flag attacks on US and Cuban soil. For example, it was proposed to blow up a US ship in Guantanamo Bay and blame Cuba. Other plans were to blow up an unmanned vessel in Cuban waters, or a civil airplane, and to conduct funerals for the non-existing victims. False evidence would be fabricated to blame those terror attacks on Cuba. There were also plans of false flag terror attacks in Florida and Washington, or attacks on Cuban refugees blamed on Cuba. The intention was that such actions should strengthen the opposition in Cuba and serve as justification for a US military intervention there.[115]

---

114 Elliston (1999)
115 National Security Archive (2001)

The strategy of tension in Italy and elsewhere

The period from the late 1960s to the early 1980s brought a sharp increase in terrorism to Italy. Both right-wing and left-wing terrorists were involved. This story is told in chapter 6.2 above.

The Italian terrorists are classified as elite-sponsored terror groups, according to a typology developed by Jacob Ravndal.[116] However, if we try to identify the elite that controlled or masterminded the strategy of tension, we get a very confusing picture of freemasons, various neofascist and paramilitary groups that were supported or even created by the secret services, and high-ranking persons within the military, possibly with international support. These were powerful people within the state apparatus but not within the official government. A government commission to investigate the corruption even saw this as a dual government or dual loyalty.[117] Whoever orchestrated the strategy of tension, it certainly included the most serious kinds of deception on our list.

The strategy of tension, in our interpretation, might look like this: powerful actors wanted to create political violence, indiscriminate terror, and chaos, in the hope that this would create public support for a more authoritarian government and a neofascist coup. In particular, they wanted to make their opponents look more violent in order to delegitimize them. This was done by deliberately failing to stop left-wing violence, by infiltrating and manipulating left-wing groups to make them more radical and violent, and by staging false flag terror attacks. However, as discussed above, this plan did not really achieve its goal.

Russian apartment bombings and other terror attacks in Russia

A series of explosions hit four apartment buildings in the Russian towns of Moscow, Buynaksk, and Volgodonsk from 4 to 13 September 1999, killing 293 people and injuring 651. Quite remarkably, foreknowledge of these attacks was apparent in a document that was leaked in Moscow a few months before the incidents. This document talked about terror

116 Ravndal (2015, p. 21)
117 Pellegrino (1997)

attacks that would be blamed on the mafia and Chechen criminals.[118] In another leak of plans, a speaker at the Duma announced that a building in Volgodonsk had been blown up, three days before it actually happened.[119]

The attacks were in fact blamed on Chechen terrorists, but the evidence against the suspected terrorists was nebulous. Historians and independent investigators have presented a strong case supporting their theory that the Federal Security Service (FSB, formerly KGB) was behind the attacks.[120] A fifth attack against an apartment building in the town of Ryazan was averted by vigilant residents who reported suspicious behavior to the police. The police found a bomb consisting of 150 kg of the advanced explosive RDX (called hexogen in Russia), a detonator, and a timer. Two suspects were apprehended when they tried to leave the town. They were not Chechen terrorists but FSB agents. An intercepted telephone call was also traced to the FSB, and identical explosives were discovered in a military depot 30 km from the town.

The incident in Ryazan was also initially blamed on Chechen terrorists, but after the media had published the leads pointing to the FSB, the FSB claimed that it was an exercise and that the bomb was only a dummy containing sugar. This explanation has been convincingly refuted.[121] The police, who had analyzed the contents, maintained that it was RDX. The investigations of all five incidents were characterized by massive cover-ups, perversion of justice, coerced confessions, intimidation of witnesses and journalists, and assassination of independent investigators.[122] The FSB was also behind the other four apartment bombings, according to a confession by FSB major Vladimir Kondratiev as well as large amounts of circumstantial evidence.[123]

The terror attacks in September 1999 helped a previously unknown former KGB officer, Vladimir Putin, attain the presidency. The attacks also served to justify the invasion of Chechnya on 1 October 1999 in the Second Chechen War. Historians believe that an independent

---

118 Litvinenko and Felshtinsky (2007, chapters 5 and 6)
119 Dunlop (2012, p. 248)
120 Dunlop (2012)
121 Litvinenko and Felshtinsky (2007, chapter 5), Dunlop (2012)
122 Litvinenko and Felshtinsky (2007), Dunlop (2012, p. 184ff), Satter (2003, chapter 2)
123 Litvinenko and Felshtinsky (2007, chapter 6), Dunlop (2012)

investigation of the attacks will not take place as long as Putin is in power.[124]

The FSB is suspected of being involved in several other terror attacks that were blamed on Chechen terrorists, including a railway bombing in Moscow in 1994 before the First Chechen War, the Moscow theater hostage crisis in 2002, and the Beslan school siege in 2004.[125] One of the worst terrorist acts in modern Russian history was the siege of a school in Beslan in 2004, in which more than a thousand people were held hostage, most of them children. Many of the terrorists were Ingush criminals who had been released from prison a few months prior to the incident or who had mysteriously escaped imprisonment. At least some of these terrorists had to carry out certain operations for the FSB as a condition of their release. One of the terrorists was a double agent. Weapons had been brought into the school before the attack, and it appears that the FSB had prior knowledge of the attack but did not prevent it. Rather than negotiating with the hostage takers, the police stormed the school and killed most of the terrorists. Several hundred hostages died as well.[126] Some historians have argued that there was collusion between the Russian government and the Chechen rebels to stage the terrorism incidents and to keep the civil war going.[127]

Terror attacks of September 11, 2001

The topic of 9/11 is difficult to avoid in this context, even though we do not really know what happened. The official account says that Islamic terrorists led by Osama bin Laden in Afghanistan flew hijacked airplanes into the Pentagon and the two towers of the World Trade Center, and that the towers collapsed because of the ensuing fire.[128] Important parts of this explanation are contradicted by large amounts of technical evidence. Independent investigations have found evidence that the collapse of the towers was induced by explosives and thermite which was placed inside the buildings.[129] We do not know who placed it

---

124 Dunlop (2012), Satter (2003, p. 46)
125 Goldfarb (2007), Dunlop (2006)
126 Goldfarb (2007), Dunlop (2006, p. 29ff), Kesayeva (2008)
127 Keen (2006, p. 63), Dunlop (2006, p. 104)
128 National Institute of Standards and Technology (2005, 2008)
129 Harrit et al. (2009), Jones et al. (2008)

there or why. Theories range from insurance fraud to the involvement of the CIA or foreign intelligence agencies. A large number of publications of varying quality, too many to review here, have tried to prove or disprove various accounts of what happened on 9/11. In the present situation, neither the official account nor any of the alternative theories are supported by convincing evidence.

It may seem odd to write about an incident here when we do not know what happened. The point is, however, that the regalizing effect does not depend on what happened, only on what people believe happened.

Initially, most people believed the official account, but later many people doubted it. In one opinion poll in 2006, 16% of US respondents found it very likely and a further 20% found it somewhat likely that 'People in the federal government either assisted in the 9/11 attacks or took no action to prevent the attacks because they wanted the United States to go to war in the Middle East'.[130] Many other polls show widespread disbelief of the official account. In some polls, the majority of respondents believed that the Bush administration was not telling the whole truth about the event. (These poll publications are no longer available, but the results are summarized with archive references on Wikipedia.)[131]

There can be no doubt that the US government took advantage of the psychological tension after 9/11 to get support for the wars in Afghanistan and Iraq. Many people believe that the US government did more than that and used some of the more severe forms of deception further down the list on page 146 (chapter 6.3), but this remains a matter of speculation.

Weapons of mass destruction in Iraq

Prior to the Iraq War in 2003, the US government claimed that it had evidence that Iraq possessed weapons of mass destruction (that is, nuclear, chemical, and biological weapons) in violation of a UN resolution, and accused the Iraqi government, led by Saddam Hussein,

---

130 Stempel, Hargrove and Stempel (2007)
131 Wikipedia. "Opinion Polls about 9/11 Conspiracy Theories", https://en.wikipedia. org/wiki/Opinion_polls_about_9/11_conspiracy_theories

of involvement with the militant organization al-Qaeda and in the terror attacks of 9/11. The US administration had put strong pressure on the CIA to produce intelligence reports that supported these claims. Much of the purported evidence relied on misinterpretations, some was knowingly false, and some was even fabricated by an Italian agent.[132] The media and populations in the USA and its allied countries mostly believed the propaganda, and this was the basis for going to war in Iraq.[133] The former UN weapons inspector, Scott Ritter, had already asserted before the war that the weapons had been destroyed in 1991 and that the production facilities had also been destroyed.[134] Ritter later revealed that the UN-sanctioned weapons inspection programme was manipulated and compromised by the CIA to create the illusion that Iraq was resisting disarmament, when in fact Iraq had already disarmed.[135] The award-winning journalist James Risen claims that the CIA deliberately ignored reports from some thirty spies in Iraq who all said the same thing: that the nuclear, chemical, and biological weapons programmes had all been abandoned in 1991.[136]

Whistleblowers in three other countries later confirmed that there was no conclusive evidence of any banned weapons.[137] Iraq claimed that the weapons had been destroyed in 1991, and for all that we know following intensive searches before, during, and after the war, this claim appears to be basically true.[138] The purported alliance with al-Qaeda did not exist. On the contrary, the Iraqi government was an enemy of al-Qaeda.[139]

The US government supported coup attempts against Saddam Hussein in 1992 and 1996.[140] The decision to wage war against Iraq was made by US President George W. Bush and his staff in 2001.[141] This decision was not based on any rational evaluation of available

---

132 Blix (2004), Eisner and Royce (2007), Pillar (2011, chapter 2)
133 Pillar (2011, chapter 3), Wilkie, A. (2010, p. 95), Calabrese (2005)
134 Ritter and Pitt (2002)
135 Ritter (2005, p. 288)
136 Risen (2006, p. 563)
137 Wilkie, A. (2010), Aagaard (2005, p. 75, 90), "Profile: Dr David Kelly". *BBC News*, January 27, 2004, http://news.bbc.co.uk/2/hi/uk_news/politics/3076869.stm
138 Blix (2004)
139 Pillar (2011, p. 43), Ritter and Pitt (2002, p. 45)
140 Ritter (2005, p. 161), Woodward (2004)
141 Woodward (2004, p. 38)

information and options but, according to several analysts, on motives relating to oil supply, geopolitics, revenge for Hussein's attempt to assassinate Bush's father, and a naive belief that it was possible to spread democracy in this way.[142] The accusations of weapons of mass destruction and terrorist connections served mainly as pretexts for justifying the war. While the decision to attack Iraq lacked in rationality, the deception that was used to justify the war was definitely the result of deliberate planning, although an element of self-deception was also present.[143]

Conclusion

Fabrication of dangers and unnecessary conflicts has occurred repeatedly throughout modern history, and many leaders and contenders for leadership have deceived their own populations into believing that their society was more endangered than it actually was. The above accounts are just examples. These examples show that the fabrication of dangers towards one's own population has occurred primarily in the following situations:

- Before a war, in order to justify the war and create psychological support for the war. There are many examples of this, including World War II (see below), the Russian-Finnish Winter War, the Vietnam War, the Chechen Wars, the Iraq War, and possibly the Afghanistan War.

- Before an undemocratic coup or in general to facilitate a transition to a less democratic form of government. This strategy achieved its goal in Germany (under Hitler, see next chapter) and Russia (under Yeltsin and Putin), and was attempted without success in Italy.

- To strengthen a government when its power is dwindling. This strategy was used in Peru and attempted in Spain.

- To defame an enemy, as in the case of Cuba.

---

142 Pillar (2011, chapters 2 and 3), Wilkie, A. (2010, p. 62), Woodward (2004)
143 Pillar (2011, chapters 2 and 3), Wilkie, A. (2010)

Terrorism provokes strong psychological reactions because it makes civilians feel threatened, even though the casualties are much lower than in conventional war.[144] In other words, terrorism is a cheap way of creating regality. Even warnings about possible terror attacks have a regal effect. For example, it has been found that terror warnings have increased the approval ratings for the incumbent US president.[145]

Beliefs in 'conspiracy theories' are quite common.[146] After almost every major terror attack in recent years, especially since 9/11, there have been rumors and suspicions in the social media that it was a false flag attack. It is quite logical to entertain such suspicions when it is observed that terror attacks are almost always counterproductive. The suspicions and rumors of false flag attacks are quite often false, of course, but they can in fact be true, as we have seen in this chapter. The often-used term 'false flag attack' is actually somewhat misleading. The leaders do not want to use their own forces to make the deceptive attacks, but prefer to manipulate others to do it. In many cases, such as the Red Brigades in Italy, some of the terrorists involved were actually 'true flag' militants, who honestly believed that they were fighting against the political system, unaware that they were being manipulated by infiltrators to work against their own interests.

Deceptions like these are obviously difficult to study and document, and many cases may have gone undetected. Objective and reliable information is hard to get, and many information sources are tainted by deception, propaganda, and political bias. We must keep this caveat in mind and consider that there is often more than one interpretation of a situation, and that the above accounts may be inaccurate.

The deceptive fabrication of threats to one's own population is predicted by regality theory, as discussed at the beginning of this chapter. The leader simply gets a personal fitness advantage by boosting the regality of his society. Some of the great emperors in history have fathered hundreds of children, but such extreme levels of fecundity belong to the past. Powerful leaders today do not in general have many children, due to the widespread norm of monogamy. Many leaders have mistresses, but they rarely get their mistresses pregnant because

---

144 Hirsch-Hoefler, Canetti, Rapaport and Hobfoll (2014)
145 Willer (2004)
146 Stempel, Hargrove and Stempel (2007)

of the use of birth control. It is hard to demonstrate that the powerful leaders who live in modern society today have any fitness advantage. If there is any raise in the fertility of powerful leaders at all, it is probably offset by the risk of being killed in a violent overthrow. One of the most powerful evolutionary forces that has driven political leaders to strive for grandeur, majesty, and power has disappeared, but the psychological predisposition for this behavior remains. The behavior of strong leaders and warlords today is a strange mixture of rationality and irrationality. Military strategies, as well as strategies of psychological warfare, are often carefully planned with the use of the best available experts, while the ultimate motive behind the wars and imperialism is hidden in psychological predispositions that evolved in a distant past, driven by fitness advantages that no longer exist today.

6.4. Example: Why World War II started

A war generally depends on a spiral of increasing regality, and we have to look at the factors that contribute to this spiral of regality in order to explain a war. This chapter will not review or explain the whole history of World War II but will focus on only a few key factors that started the spiral of regality that drove the war.

Many psychologists have tried to explain World War II by analyzing the peculiar psychology of Adolf Hitler, and a long list of diagnoses have been postulated.[147] Such an approach gives only a very limited understanding of the causes of the war. I will venture the theory that all countries have a potential Hitler, and that—under different circumstances—Hitler could have remained a painter, as he once was, rather than a warmaker.

Regality theory allows us to go beyond the explanation of the war as a consequence of Hitler's postulated psychopathology and try to answer some more basic questions such as: why was Hitler bellicose? Why did the Germans vote for him? Why did they not replace him with someone more peaceful and agreeable? Why did they allow him to abandon democracy and make himself a dictator? How did he manipulate his people to make them support his war?

---

147 Redlich (1998, pp. 255, 333)

If we want to know why Hitler was bellicose, we may try to start with the same focus as psychologists habitually employ. Psychologists like to analyze people's childhoods, and it cannot be denied that childhood experiences have a strong influence on people's personalities. Hitler lost his father and a brother while he was a child, and later lost his mother. Three older siblings had died before Adolf was born.[148] These and other traumatic events may have had an influence, but other people with similarly traumatic childhoods have grown up to become peaceful adults. Hitler played war games when he was a child, but this is something that most normal boys do.

At the age of twenty-five, Hitler volunteered for a Bavarian regiment to fight in World War I. He was a brave soldier and was promoted to the rank of corporal. He was wounded twice, but survived many dangerous episodes. Thousands of men in his regiment were, however, killed.[149]

These experiences during World War I are sufficient to explain Hitler's regal disposition, although childhood experiences may have laid the foundation. This regal disposition can explain his passion for war and his disdain for democracy. It also explains why he was so receptive to anti-Semitic ideas, which were common in Germany at the time.

A study of Hitler's personality may explain the behavior of one person, but it does not answer the more important question of why the German people supported him and his ideas. We have to look at the general political, social, and psychological climate of the time.

The first half of the twentieth century was a period with many wars in Europe. The Ottoman Empire was in decline, and other countries with imperial ambitions were eager to fill the power vacuum. Shifting alliances, contests of power, and arms races led to World War I, which ended with total humiliation of Germany.

Europe was not a peaceful place after World War I. There were many wars and violent uprisings all over Europe that contributed to a general insecure and warlike climate.

The Great Depression of the 1930s hit Germany particularly hard, with unemployment rates over 30%. The population was relatively young due to a high rate of population growth. The combination of a

---

148 Redlich (1998, p. 5)
149 Redlich (1998, p. 36)

youth bulge and a high unemployment rate is a dangerous cocktail. It was easy to mobilize the large surplus of unemployed and frustrated young people by means of a violent ideology.[150] The psychological consequences of economic crises were discussed in chapter 5.2.

The aftermath of World War I and the economic crisis provided fertile breeding ground for the nationalist and racist ideology called National Socialism, or Nazism. This ideology was promoted by the National Socialist German Workers' Party, led by Hitler. There is general agreement among historians that the economic crisis contributed to the rise of Nazism. A study of Germany in the interwar years has found significant correlations between economic indicators and various measures of authoritarianism. The strongest correlation was between the unemployment rate and the votes for Hitler.[151]

The Nazi ideology was promoted by massive propaganda, organized by the Ministry of Public Enlightenment and Propaganda, headed by Joseph Goebbels. The most powerful mass media at the time were radio and cinema films. The Nazis promoted the distribution of radios and the sale of cheap radios so that, by 1938, 60% of all households owned a radio receiver. Radio broadcasting in Germany was initially intended to be an apolitical cultural institution with the purpose of educating the population. Nationalist sentiments were common on radio programmes in Weimar Germany, but this was not perceived as controversial. When the Nazis came to power, they used the radio for propaganda purposes and broadcast political speeches. The radio propaganda must have had a strong influence on the German population, since they were unaccustomed to organized propaganda at this scale, while counterpropaganda was only available from foreign media. The Nazi propaganda promoted an ideology of racial superiority and territorial expansion (Lebensraum) and blamed the economic crisis on Jews and other foreigners.[152]

Fabrication of threats, as discussed in chapter 6.3, was used both for suspending democracy and as a pretext for starting the war. In February 1933, the Parliament building (Reichstag) in Berlin was set on fire. A young, homeless, visually impaired Dutchman, Marinus van der Lubbe,

---

150 Heinsohn (2003, p. 23), Weber (2013), Moller (1968)
151 Padgett and Jorgenson (1982)
152 Von Saldern (2004), Zimmermann (2007)

was arrested inside the burning building. Van der Lubbe, a communist, readily confessed that he had set the fire for political reasons, and he was sentenced to death.[153]

Figure 12. Burning German Parliament, 1933. Bundesarchiv.[154]

The German government, led by Hitler, blamed the attack on the Communist Party and arrested a large number of communists including all of the communist members of the parliament. The Nazi party now had an absolute majority in the parliament, which allowed them to suspend democracy and civil liberties under the pretext of preventing a communist uprising. This key event made Germany a dictatorship until the end of the war and allowed Hitler to rule without opposition.

---

153 Hett (2014)
154 CC BY-SA 3.0, https://en.wikipedia.org/wiki/Reichstag_fire#/media/File:Reichstags brand.jpg

However, technical evidence indicates that van der Lubbe could not have started the fire alone, and Nazi leader Hermann Göring admitted that he was responsible for the fire, according to one witness testimony.[155]

In 1939, the Germans made a propaganda campaign to justify an invasion of Poland. This campaign, known as Operation Himmler, involved various false flag operations that simulated Polish attacks on Germany. On 31 August that year, German soldiers dressed in Polish uniforms attacked a number of buildings in Germany, including a radio station, and left behind dead bodies, most of them in Polish uniforms to make it appear that Polish attackers had been shot down. The dead bodies were in fact not Polish soldiers but prisoners from concentration camps that had been killed for this purpose. Hitler made a public speech the next day, claiming twenty-one alleged attacks as justification for military invasion of Poland. This was the start of World War II.[156]

---

155 Hett (2014), Shirer (1960, p. 191)
156 Whitehead (2008)

7. Regality Theory Applied to Ancient Cultures

In this chapter we will look at a number of ancient, non-industrial civilizations and see whether the predictions of regality theory fit these cultures. A statistical study of the results follows in chapter 8, but the discussions of each of these cultures can also be read as interesting and illustrative examples in themselves.

In chapter 3.1, we saw how the ecology and the environment can determine the level of violent conflict between groups or determine whether war is possible at all. Conflicts are likely if multiple groups are competing for the same ecological niche, according to the competitive exclusion principle, while conflicts are less likely for groups that have adapted to their own unique niche. Using the distinction between contest competition and scramble competition, we found that conflicts are more likely where food or other valuable resources are concentrated in patches that can be monopolized and defended, while conflicts will be rare where food is sparsely distributed. Logistic factors such as weapon technology, ease of travel, efficient food production, and storage and transportation of food must also be taken into account.

To summarize, we can predict that the level of violent intergroup conflict in a non-modern culture will be high when the following factors are present:

- An ecology and technology that enables a high food production per unit area and thus a high population density.

© 2017 Agner Fog, CC BY 4.0   https://doi.org/10.11647/OBP.0128.07

- Food and other important resources are concentrated in patches that can be monopolized and defended by the group rather than sparsely distributed.

- Neighbor groups with similar ecology are competing for the same resources.

- Efficient means for transportation of warriors to enemy territory are available.

- Efficient means of communication over distance are available.

- Efficient means of food storage and transportation are available.

- Efficient weapons are available.

- The climate is favorable.

In contrast, intergroup conflict is likely to be impeded or prevented when these factors are reversed, specifically when

- Food is sparsely distributed and the population density is low.

- Geographical barriers or logistic problems make travel difficult.

- A group has specialized in a unique niche that neighbor groups are unable to exploit.

- Extreme climate conditions make work and fighting more difficult.

It must be noted that these assumptions do not hold in modern industrial societies, where international economic interdependence or third party intervention can prevent war.

The next step in our prediction is that a culture will be regal when intergroup conflicts are frequent and severe, while we will expect to find a kungic culture where the level of intergroup conflict is low. We will predict an intermediate level on the regality scale when some factors point in one direction and other factors point in the opposite direction. Other environmental factors that threaten a sociocultural group, such as natural disasters and unpredictable food shortages, may also have an influence in the regal direction. The level of regality will be reflected in a number of social, political, and cultural indicators, as discussed in chapter 2.6.

The relevant predictions of regality theory can be summarized as follows:

- People will show a preference for a strong leader and a social structure with strict discipline in cases of a high level of intergroup conflict or other collective danger. They will show a preference for an egalitarian social structure in the absence of collective danger.

- These individual preferences will influence the social and cultural structure in the directions called *regal* and *kungic*, respectively, as reflected in the indicators listed in table 1 (chapter 2.6).

- A society with a high level of intergroup conflict will use both reward and punishment of its members to enhance its military strength (see chapter 2.2).

- Individual danger and *intra*group conflict will not have the same effect on social structure as collective danger and *inter*group conflict have. However, both kinds of danger will have the same effect on fertility.

- The level of intergroup conflict will depend on the geographic environment, the ecology, and the available technology, as described above.

We will now study a number of non-modern and non-industrial cultures to see how well they fit these predictions. Each of these cultures can be seen as an interesting case study illustrating the application of ecological theory and regality theory.

The methods used for theory testing in the social sciences can roughly be divided into two categories: (1) statistical testing on a large number of cases, and (2) comparative historical analysis where a small number of cases are compared and analyzed in more detail.

In the statistical method, you look at the input variables and the output variables, but rarely at what happens inside the 'black box' causing the input to affect the output. In the comparative historical method, you open the black box and look at the chain of events that is responsible for the causal relationship. The comparative historical analysis allows methods such as studying whether events are happening in the temporal order predicted by the theory, and looks for events that should or should not happen according to the *a priori* theory but not according to alternative theories.[1]

A combination of statistical testing and comparative historical analysis is the best method we have to test a causal theory of social phenomena. The political scientist Evan Lieberman recommends a method that starts with a correlational analysis on a large number of cases and then studies a few cases in more detail.[2] However, a correlational analysis can only confirm a relationship between variables that are already known to be relevant—it is unlikely to lead to the discovery of new interesting variables or relationships. Therefore, the present study uses case studies before statistical analysis. A number of ancient and non-industrial societies are analyzed below. A statistical summary and discussion of the results is provided in chapter 8.3. These cultures are selected according to a number of predefined criteria, explained in chapter 8.3, in order to avoid selection bias.

7.1. Andamanese

The Andamans is a group of islands in the Bay of Bengal, totaling 6,400 km². The islands are volcanic mountains with dense vegetation. The temperature is between 17°C and 36°C all year round. At the time of the first anthropological studies, the islands were populated by twelve tribes of Negritos speaking different languages. They lived from hunting, fishing, and gathering and had sufficient food at all times. Their only vehicles of transportation were canoes, which were used for fishing and traveling along the coasts of the islands. The canoes were unable to sail in open sea. The Andaman Islands had very little contact with the surrounding world until the mid-nineteenth century.[3] The Andamanese

---

1   Mahoney and Thelen (2015)
2   Lieberman (2015)
3   Man (1883), Radcliffe-Brown (1922)

are one of the oldest and most isolated human populations outside of Africa.[4]

Based on the geographic variables, we would expect the Andamanese culture to be moderately kungic. The mountains and dense jungle make traveling on land very difficult, and the available means of transportation by sea were insufficient for large-scale war. The natives had no knowledge of any people living more than 32 km away. The favorable climate and abundance of food can be expected to allow a moderately high population density, which would weigh in the regal direction. The possibility of natural disasters, such as cyclones, volcanic eruptions, and earthquakes, might also have a limited regal influence, but the mountains would offer some protection, at least against the cyclones.

The frequency of warfare is listed in the standard cross-cultural sample as continual or once every one to two years. However, this listing is very misleading due to a vague definition of war. There is no evidence that large scale fighting has ever taken place on the Andaman Islands. The literature agrees that the only evidence of fights is of 'brief and far from bloody skirmishes' where 'only a handful of warriors were engaged on each side and rarely more than one or two were killed'. These skirmishes were usually feuds between neighbor groups, belonging to the same or different tribes. The feuds may be ended after some time with a peace-making ceremony. No weapons of war have been found other than the bows and arrows used for hunting.[5]

While no evidence of large-scale war has been found, some inferences can be made. The many different tribes living in close proximity would quite possibly lead to conflicts over territory. One of the tribes, the Jarawa, who live on the South Andaman Island, are believed to be invaders from Little Andaman Island, because their language is similar to that spoken on Little Andaman. It is possible that the Jarawa took territory from the original inhabitants in a more serious conflict in a long forgotten past. The Jarawa are in constant conflict with their neighbors.

It is remarkable that so many tribes were able to coexist in a small area for as long as it took to develop different languages. This is indirect evidence that traveling was limited and that the Andaman tribes were

---

4   Thangaraj et al. (2003)
5   Man (1883, pp. 44, 135), Radcliffe-Brown (1922, p. 50ff)

willing to maintain peace and respect territorial boundaries, with the exception of the Jarawa.

The Andamanese were hostile to foreigners and have systematically killed all shipwrecked sailors who entered their land. In particular, the Jarawa were so hostile to foreigners that it has been impossible to study them. The Jarawa have survived for the very same reason, while most of the other tribes have perished after contact with modern settlers.[6]

Figure 13. Andamanese hunting turtles, ca. 1900. Photo by Bourne & Shepherd photographic studio. Smithsonian Institution, National Anthropological Archives.[7]

The Andamanese had no political organization above the local group. There were influential older men and women but no leaders with authority. There was no penal system, and neither children nor adults were punished for wrongdoing.[8]

Religious beliefs were fluid, flexible, and incongruent. The Andamanese believed in spirits and other supernatural beings who control diseases, weather, and other natural phenomena. E. H. Man found several similarities with Christian beliefs,[9] but these were

---

6   Pandya (2000)
7   Public domain, https://siris-archives.si.edu/ipac20/ipac.jsp?uri=full=3100001~!57153~!1
8   Man (1883, p. 25ff), Radcliffe-Brown (1922, pp. 22–77)
9   Man (1883, p. 89)

convincingly refuted by Alfred Radcliffe-Brown as projections of Man's own faith.[10] All other observers agree that no supernatural beings ruled the Andamanese or punished their misdeeds in life or in the afterlife.[11]

The relationships between neighbor tribes were friendly, except in times of feuds and except with respect to the Jarawa. Intermarriages occurred, as well as adoption of children across tribes.

Fertility was low, possibly because children were breastfed for three to four years. There was no birth control or infanticide. Mortality was high, and nobody lived longer than fifty years. Suicide was unknown. Children started to help their parents with various work from about the age of ten. Children as well as adults played many games.[12]

The marrying age was approximately 16 to 20 for women and 18 to 22 for men. Betrothals in childhood occurred, but in most cases young people were free to choose their spouses. There was no polygamy. Divorce was very rare. Sexual morals were lax. Premarital sex was almost universal, and adultery was frequent. Syphilis has spread fast.[13]

Men and women had almost equal status and influence. Children were named before birth, and the same names were used for boys and girls. The most important of their gods was described as female more often than as male.

Their art was simple. They adorned themselves with necklaces, body paint, and scars. The body painting consisted of simple patterns such as parallel lines or zigzag lines, with variations due to individual taste or changing fashions. The same adornments were used for men and women. Tools and canoes were decorated with similar painting or cuttings. Singing and dancing occurred on many occasions, often for ceremonial reasons. The songs were about everyday activities, not religion. Each singer had his own songs, while dancing was communal.[14]

The conclusion for the Andamanese is that most observations fit the expectations for a kungic culture. The observations regarding political organization, discipline, religion, fertility, sexual morals, marrying age, length of childhood, art, and music are in agreement with a kungic

---

10   Radcliffe-Brown (1922, pp. 141–197, 370)
11   Radcliffe-Brown (1922, p. 48), Cipriani (1966, p. 43), Ganguly (1961)
12   Man (1883, pp. 31, 43, 109), Radcliffe-Brown (1922, p. 78)
13   Man (1883, pp. 13, 67), Radcliffe-Brown (1922, pp. 50, 70), Cipriani (1966, p. 22), Nippold (1936), Pandya (1993, p. 15)
14   Man (1883), Radcliffe-Brown (1922), Cipriani (1966, p. 54), Portman (1888a)

culture. The absence of suicide, the absence of divorce, and especially the hostility towards strangers point more in the regal direction.

One explanation for the xenophobia of the Andamanese is their belief that dangerous spirits have a lighter skin color than themselves. They regard all light-skinned strangers as dangerous spirits.[15] However, this is just a proximate explanation with little theoretical value. The belief in dangerous light-skinned spirits is likely to have a historical origin in encounters with foreigners with firearms. Indeed, history gives the Andamanese ample reason to fear foreigners. Malay and Arab slave traders, shipwrecked sailors, and European and Indian settlers have all treated the Andamanese with deadly violence;[16] and the latest reports indicate that violent confrontations with the Andamanese have continued until the present time.[17] They were friendly to foreigners in previous times, according to some reports.[18]

The frequent feuds between neighbor groups, whether they belonged to the same or to different tribes, are easily explained by the absence of any political system for resolving conflicts. The fact that the Jarawa appear to be more hostile than the other Andaman tribes has no immediate explanation, since almost nothing is known about their history.

7.2. Arrernte

The Arrernte (Aranda, Arunta) tribe of Australian aborigines were a semi-nomadic people living in the desert-like areas near Alice Springs in Central Australia. They had no means of transportation other than walking naked and barefoot. The few navigable rivers sometimes dried out, and there was no timber suitable for making boats. The relatively flat land with sparse vegetation did not constitute any serious barrier to traveling, but the lack of efficient means for carrying water made it impossible to travel too far away from water sources except in occasional wet periods. The climate is dry with large variations in temperature. Day temperature can exceed 40 °C, and night temperature can go below

---

15   Man (1883)
16   Cipriani (1966, p. 3), Portman (1888b)
17   Pandya (2000)
18   Cipriani (1966, p. 4)

freezing point. They lived as hunters and gatherers and had no metals, no pottery, and no efficient means for food storage.[19]

Based on this information, we can expect the culture to be fairly kungic, because the population density must be low and because mass traveling was limited by lack of drinking water for the Arrernte as well as for the neighboring tribes. However, the droughts that appeared at unpredictable intervals constituted a significant collective danger that occasionally killed significant parts of the population. This may have pushed the culture somewhat in the regal direction.

Figure 14. Arrernte rain ceremony, ca. 1900.
Photo by W. B. Spencer and F. J. Gillen.[20]

Large-scale war was absent at the time of the first studies, but war was reportedly more common in earlier days before the spread of European diseases decimated the population. Deadly clashes could occur when droughts forced people to migrate into enemy territory. The political structure was very simple. Decisions were taken by deliberation among

19 Spencer and Gillen (1927, p. 1), Tindale (1974, p. 50)
20 Public domain, http://spencerandgillen.net/objects/50ce72e9023fd7358c8a8534

the old men. Some men had leadership status, but none had absolute power. There was no leader at the level of the whole tribe.[21]

Men often beat their wives. Physical punishment was common, and the death penalty was prescribed for several offences, including some rather trivial ones. The literature is not clear on how often the death penalty was actually executed in practice. Vendettas were very common.

Beliefs in life after death were weak or absent. Women became pregnant and gave birth as a result of totemic ancestors entering their body. The father played little or no role in this process, according to Arrernte beliefs. There was no supreme being ruling over the humans. Sickness, death, and misfortune were blamed on taboo violations, evil spirits, and, most commonly, on magic performed by enemies. Witch doctors had some political influence but no strong power. There were many painful rituals. Many religious traditions and objects were monopolized by men and kept hidden from women and children.[22]

People's identity was defined in terms of totem group and a complicated system of marriage classes. Tribal unity was not a strong factor. Foreigners were not allowed into their territory unless they had a peaceful purpose such as negotiation or barter. However, there were often friendly relations between neighbor tribes, and intertribal marriages were common.[23]

Fertility was sufficiently low to keep the population size stable. Abortion was known and infanticide was common. Twins were always killed at birth. Suicide was unknown. Children were breastfed for several years.[24]

The Arrernte had little or no idea of biological fatherhood and consequently no notion of illegitimate children. A man might therefore lend his wife to other men for a number of reasons, with or without her consent. There were plenty of rituals and traditions that allowed, or even prescribed, extramarital sexual contacts in certain situations. Premarital sex, however, was generally not tolerated. Homosexuality has been reported for both men and women. Girls married around

---

21 Spencer and Gillen (1927, p. 9), Tindale (1974, p. 33), Basedow (1925, p. 183)
22 Spencer and Gillen (1927, pp. 36, 443), Tindale (1974), Strehlow, T. (1947, pp. 5, 52, 93)
23 Spencer and Gillen (1927), Tindale (1974, pp. 35, 75), Strehlow, T. (1947, p. 41)
24 Spencer and Gillen (1927, p. 221), Strehlow, C. (1907–20, IV:I–II)

puberty, boys much later. Promises of marriage were often made for the expected future daughters of a woman, even before she got pregnant. Polygamy was common. Divorce occurred.[25]

Children were treated kindly, and much time was devoted to them. Punishment of children is not mentioned in the literature.[26]

Artistic expression was simple. Religious objects were embellished with lines, dots, and circles. Body adornments included necklaces, nose piercing, scarification, etc. Objects adorned with feathers and paintings were used in ceremonies. There were songs and dances for many occasions, but only few and very simple music instruments. Marriage ceremonies were rather simple, while initiation ceremonies were more elaborate, especially for young men.[27]

The conclusion for the Arrernte is that most observations fit the expectations for a kungic culture. The political organization, religion, group identity, fertility, sexual behavior, treatment of children, and art are all as expected for a kungic culture. The possible occurrence of severe punishment, intolerance of strangers, absence of suicide, and the strong male dominance point more in the regal direction. The marrying age is low for girls but high for boys. The amount of war before the spread of European diseases is difficult to estimate.

7.3. Babylonians

The alluvial plains along the rivers Tigris and Euphrates in what is now Iraq were the seat of many city-states and empires that waxed and waned over many thousands of years. We are considering only the period before the advent of monotheism, with the main focus on the era of King Hammurabi. The climate was dry and hot in the summer months, and the temperature went below freezing point in the winter. The fertile river plains were surrounded by desert. Boats of various sizes were available for transportation of goods and people along the rivers and the many canals that were dug. For transportation on land, the Babylonians had donkeys, camels, horses, and wheeled vehicles. Cuneiform writing was used for communication over long distances.

---

25 Spencer and Gillen (1927, p. 473), Strehlow, C. (1907–20, IV:I)
26 Basedow (1925, p. 69)
27 Spencer and Gillen (1927), Strehlow, C. (1907–20)

The Babylonians practiced intensive agriculture with a well-organized irrigation system. They had metals for tools and weapons.[28]

There are strong regal factors here. Nothing hindered the mobility of the people. They had means for storing and transporting food and water, and they could travel far by land as well as on the rivers. The intensive agriculture could sustain a high population density, and the area that could be traveled was large enough for including multiple states that could wage war against each other with efficient weapons. The only kungic factors we can find are the hot climate in the summer and the low population density in the surrounding desert. Since the regal factors are much stronger than the kungic factors, we can expect to find a rather regal culture here.

The political system was a monarchy where the king was the supreme ruler, though the priests also had a considerable influence. The religion was polytheistic, with a high number of gods and demons. Humans were created by gods for the sole purpose of doing service to these gods. Humans were thus servants of the gods, represented by the king. Humans were punished by the gods for their sins.[29]

Figure 15. Babylonian cylinder seal showing the king making an animal offering to the Sun God Shamash. Hjaltland Collection.[30]

---

28   Contenau (1954, p. 62), Nemet-Nejat (1998, p. 12), Saggs (1962, pp. 4, 158)
29   Saggs (1962, pp. 162, 300; 1965, p. 181)
30   CC BY-SA 3.0, https://commons.wikimedia.org/w/index.php?curid=31696169

War was frequent and the army was one of the most important institutions in Babylon. Armies numbered thousands of men, probably more than a hundred thousand, and war casualties were high.[31]

Slavery and slave trade were widespread. A man could be reduced to slavery if unable to pay a debt. Political and economic matters, as well as marriage and divorce, were regulated by a detailed set of laws written by the king. Discipline was harsh. The law prescribed the death penalty or mutilation for many common crimes. It is unknown whether legal practice followed the laws or whether law-breakers could get away with paying a fine. It is certain, however, that slaves and prisoners of war could be treated harshly, and there is ample evidence of torture.[32]

The cuneiform scripts and other archaeological sources tell us little about the life of children. Children were set to work and subjected to strict discipline, but the presence of toys indicates that there was also time for play. Children could be lent out as slaves for three years, perhaps longer, in order to pay their father's debt.[33]

Little is known about the fertility of the Babylonians. On average, between two and four children per family survived into adulthood. The marrying age has been estimated at 14 to 20 for women and 26 to 32 for men, though these estimates are uncertain. The sexual morals gave men more freedom than women. Girls were supposed to marry as virgins. Adultery between a married woman and a man was punished by drowning both, according to the law, while a man was free to have concubines and to visit prostitutes. Prostitution was widespread and regulated by law and by religion. Homosexuality was tolerated.[34]

Babylon had a large production of art, mainly under the patronage of the state and the priesthood. The numerous temples and palaces were highly embellished with bas-relief and monuments glorifying gods and kings. Favorite motifs were royal hunting scenes and battle scenes, supplemented with stylized religious symbols. Geometric patterns

---

31   Saggs (1965, p. 112), Delaporte (1970, p. 69)
32   Nemet-Nejat (1998), Saggs (1962), Pallis (1956, p. 556)
33   Contenau (1954, p. 18), Nemet-Nejat (1998, p. 130), Saggs (1962), Delaporte (1970, p. 80), Stol (1995, p. 490)
34   Nemet-Nejat (1998, p. 127ff), Saggs (1962, p. 185), Stol (1995, p. 488), Roth (1987)

reflecting a *horror vacui* tendency were common embellishments. Likewise, song, poetry, and literature often glorified gods and kings.[35]

The conclusion for the Babylonian culture is that the political organization, the discipline, the religion, the treatment of children, and the art all show typical regal characteristics. The sexual morals are ambiguous. We have insufficient information about the fertility.

7.4. Chiricahua Apache

The Chiricahua Apache Indians originally lived in Southern Arizona and New Mexico. They belong to the Na-Dene language group, having migrated from the Beringia later than the Amerind speakers.[36] The Apache had been displaced to reservations at the time of the anthropological studies. What is known about their original lifestyle is therefore based mainly on the recollections of elderly informants.[37]

The environment included semiarid plains and mountains. The Indians lived a seminomadic lifestyle and they were able to travel far, especially since the acquisition of horses in the seventeenth century.

We would expect a plains culture to be more regal than a mountain culture. The Apache culture is a mixture of these two.[38] The access to easy traveling on the plains and the possibility of simple communication by smoke signals makes war possible and likely. However, the relative scarcity of food sets a limit to the population density and to the scale of war. The maximal possible size of a war party was limited by the availability of food rather than by political factors.[39] We can therefore expect the culture to be moderately regal, though the scale of war is limited by the sparseness of food.

The tribe was divided into four autonomous bands at peace with each other. The bands were divided into camps, and each camp might include several families. There were leaders at the camp level and the band level, but there was no leader of the whole tribe. Leadership was based on common recognition. While the position of leader was

---

35   Contenau (1954), Nemet-Nejat (1998, pp. 66, 167), Delaporte (1970, p. 182), Pallis (1956), Galpin (1955)
36   Schurr and Sherry (2004)
37   Gifford (1940), Opler (1937, 1941)
38   Gifford (1940, p. 187)
39   Opler (1937, p. 177)

often inherited from father to son, it was not necessarily so. The leader was generally obeyed, especially in times of war, but his power was not absolute. An unpopular leader could always be replaced without violence.[40]

Discipline was strict, especially for children. Young children were strapped to a cradleboard where they could hardly move. Older children were taught obedience, and they might be subjected to physical punishment, including whipping, if their wrongdoing was severe. Adults were occasionally punished harshly, especially for such crimes as marital infidelity or murder. Deviants were often accused of witchcraft. Suspected witches might be tortured until they confessed, and then killed.[41]

Religious practices and shamanism were based on individual revelations with no orthodox teaching. Everybody could have supernatural powers and perform magic rituals. People could specialize in different rituals for different purposes. There were many supernatural beings, but none of these had a supreme position. The supernatural beings were much involved in human affairs. They gave power and were prayed to, and they could punish people with sickness and misfortune, but there was no punishment in the afterlife.

The most important ceremony was the girls' puberty rite. This was a very elaborate and expensive ceremony lasting for four days, where neighbors were invited for dancing and feasting. The initiation rite for boys was very different. The boy had to go as a novice on four raids or war expeditions before he was considered a man. He was trained to endure all kinds of hardship.[42]

Raiding and war was an integrated part of Apache life. They went on raiding expeditions to enemy tribes several times a year for the purpose of stealing horses and cattle or for avenging the deaths of lost warriors. They might take prisoners of war, who were tortured and killed. Captured boys might be adopted and raised to become warriors. Women were not captured.[43]

---

40   Opler (1941, pp. 73, 174), Schroeder (1974, IV)
41   Gifford (1940), Opler (1937, p. 191; 1941, p. 14)
42   Gifford (1940, p. 76), Opler (1937, p. 226; 1941)
43   Gifford (1940), Opler (1937, p. 230; 1941, p. 334), Schroeder (1974:VD)

Girls married at age 18 or 19, boys a little older. Marriages were arranged, but in most cases in agreement with the wishes of the young people. Sororal polygyny was practiced. Children were breastfed for three years. No information is available on the fertility or birth rate. Birth control was practiced by supernatural means and occasionally by abortion or infanticide. Suicide occurred.

Sexual morals were strict. Social contact between the sexes was limited. Nudity was not allowed, even when no members of the opposite sex were present. Masturbation was unknown. Girls were guarded for chastity. There was more sexual freedom for divorced women and widows than for married women and unmarried girls.

Art was elaborate. Clothing and utensils were decorated with colorful patterns or figures. Music and dance was used for ceremony as well as for social gatherings. It was the girls who chose their partners at social dances.[44]

The conclusion for the Chiricahua Apache is that this culture shows a mixture of regal and kungic signs. The cultural importance of war, the strict discipline, and the strict sexual morals are definitely regal signs. However, the political system was less hierarchic and dictatorial than one would expect for a regal culture. The low population density made it impossible to incorporate a large number of people under a single leader. Political leaders did not have absolute power, and there were no religious authorities either. The realm of supernatural beings also lacked a supreme ruler. It seems, nevertheless, that religion had a disciplining function in the sense that people feared supernatural punishment and witchcraft accusations.

There is no reliable information about fertility or family size, but the fact that birth control was practiced or attempted seems to indicate that the population size was controlled by the availability of food as much as by war. On the other hand, the practice of capturing young boys from enemy tribes indicates a clear desire to raise as many warriors as possible.

Art, body adornment, and ceremonial paraphernalia were elaborate, embellished, and so expensive that this may be interpreted as a moderately regal sign.[45]

---

44   Gifford (1940, p. 63), Opler (1937, p. 200; 1941, p. 131)
45   Opler (1941, p. 379)

Conflicts with white settlers and the availability of horses since the seventeenth century has no doubt increased the possibilities for war and raids, but the earliest available historical sources seem to indicate that the Apache had frequent conflicts with other tribes before the arrival of Europeans.[46]

7.5. Copper Inuit (Eskimo)

The Inuit, previously called Eskimo, are a nomadic people living in the Arctic regions. The focus in the Standard Cross-Cultural Sample is on the group that anthropologists have called Central Eskimo or Copper Eskimo, who live north of the tree line in Canada in the area around Coronation Gulf and Victoria Island. The environment is arctic tundra covered with snow most of the year and with very little vegetation. There are polar nights in the winter. The natives eat mainly fish, seals, and caribou, but hardly any vegetable food. Fuel is scarce and meat is often eaten raw. The native copper is hammered with stones to make tools such as spearheads, fishing hooks, and knives. Iron is sometimes found on shipwrecks or obtained by trade. People live in tents made of caribou skins in the summer. In the winter, they live on the ice in snow houses. People can travel far, but quite slowly, using dog sleighs.[47]

We can expect the culture to be kungic because food is sparse with no defendable resources (scramble competition) and because of the extreme climate. It is possible to travel very far, but people have to stop often and hunt for food during the travel, and this would make war difficult.

The concept of organized warfare is completely unknown to the Copper Inuit. There is no organized system of justice. Most crimes go unpunished. Murder may be revenged by the relatives of the victim, and this often leads to feuds unless the offender flees to another district. Theft is rare. There is no chief and no political system beyond the family unit.[48]

---

46 Schroeder (1974:IV)
47 Stefánsson (1914, p. 7), Boas (1964, p. 11), Jenness (1923, p. 13)
48 Stefánsson (1914, p. 131), Boas (1964, p. 173), Jenness (1917; 1923, pp. 93, 235; 1928, p. 85), Condon and Ogina (1996, p. 79), Rasmussen (1932, p. 68)

Figure 16. An Inuit family, 1917. Photo by George R. King.[49]

The fertility is low. Children are breastfed for at least two years and sometimes even up to six or seven years.[50] Infanticide is common. The total population in the area has probably remained at a stable level of below a thousand individuals for centuries. There is high mortality due to occasional famines and harsh living conditions. Infectious diseases were rare prior to European contact. Suicide is allowed and accepted, including assisted suicide.[51]

The population had no name for their own tribe until foreigners gave them the name Eskimo (of uncertain origin), or Inuit, which simply means 'human' in their language. Subgroups are named according to their home area, but an individual can belong to more than one home area. People have little interest in their own ancestry.[52]

The religion is animist, with many taboos that vary from place to place. Shamanistic séances are used for divination and for influencing the weather and hunting luck. Animals have souls just like human beings, and a shaman may just as well be possessed by the spirit of an animal as by the 'shadow' of a dead person. People are uncertain about what happens in the afterlife.[53]

49　Public domain, https://commons.wikimedia.org/wiki/File:Eskimo_Family_NGM-v31-p564.jpg

50　Stefánsson (1914, p. 255), Boas (1964, p. 158), Jenness (1923, p. 163), Dodwell (1936)

51　Boas (1964, pp. 172, 207), Jenness (1917; 1923, pp. 91, 166), Condon and Ogina (1996), Rasmussen (1932, pp. 17, 46), McGhee (1972, p. 116)

52　Stefánsson (1914), Jenness (1923), Condon and Ogina (1996)

53　Stefánsson (1914, p. 126), Jenness (1923, p. 177), Condon and Ogina (1996, p. 86), Rasmussen (1932, p. 72)

The sexual morals are lax. Polygyny and polyandry occur. Temporary wife swapping is a common way of building alliances, and it is not uncommon for a visiting man to be offered a woman for the night. Marriage appears to be based more on practical considerations than on love, with no elaborate ceremony to consolidate it. There are no love songs or other expressions of romanticism. Divorce is easy. Girls marry at the age of 14 to 16, boys somewhat later.[54]

The sex roles are not rigid. While hunting is the job of men and sewing is the job of women, it is not shameful to do a task that belongs to the opposite sex. There is no distinction between boys' names and girls' names.[55]

Children are treated with affection: they have toys, they are never punished severely, and they do not have to work hard.[56]

The most important form of art is singing and dancing. The only music instrument is a drum. People take turns singing and dancing, while the remaining people sit or stand in a circle around the dancer. Often, each person has his or her own songs and dancing style. The song texts are mostly about everyday occurrences and hunting events.[57]

People have tattoos identifying their home area. Tools and clothes are sometimes embellished with simple patterns.[58]

We can now conclude that all the cultural markers of the Copper Inuit fit the predictions for a kungic culture, except for the low marrying age.

7.6. E De (Rhadé)

The E De are one of several tribes that live in the highlands of southern Vietnam. The present study will focus mainly on the time before colonization. The mountain environment consisted of forests, dense bush and grasslands. They grew rice and other crops using slash and burn and shifting cultivation. Irrigation of rice fields was used only where the environment was favorable. They were herding buffaloes

---

54   Boas (1964, p. 171), Jenness (1923, 1925, 1928), Condon and Ogina (1996, p. 85), Rasmussen (1932, p. 50), Dodwell (1936)
55   Stefánsson (1914, p. 102), Boas (1964, p. 204), Jenness (1923, p. 167), Condon and Ogina (1996, p. 85), Rasmussen (1932, p. 42)
56   Stefánsson (1914, p. 154), Boas (1964, p. 163), Jenness (1923, p. 169)
57   Stefánsson (1914), Jenness (1923, 1925, 1928), Rasmussen (1932, pp. 16, 130), Dodwell (1936)
58   Boas (1964, p. 153), Jenness (1923, p. 117; 1946)

and several other animals, including a limited number of horses and elephants. They had tools of iron and other metals.[59]

The environment presented many obstacles to travel but, given the availability of horses and elephants, we would expect the E De people to travel far. An efficient means of travel combined with efficient food production makes us predict a regal culture.

The E De people lived in villages consisting of several longhouses. Each longhouse was the home of a matrilineal group of up to 300 people. The level of political integration varied considerably. The villages were often partially or fully autonomous under the leadership of a village chief. In some cases, a group of several villages formed a mini-state. In one period, the E De paid tribute to the closely related Jarai tribe. The E De were part of the Champa kingdom and paid tribute to the Champa king for a period of possibly several hundred years, although the Champa influence appears to have been very limited. Intervillage raids and wars were common.[60]

The E De society was highly stratified. The village chief and other rich people were very powerful, while up to two-thirds of the population lived in slavery or debt dependence.[61]

The justice system was based on a large set of orally transmitted laws. Prescribed punishments ranged from payment of fines and religious offerings to enslavement and (rarely) death. The animist religion supported the social stratification and the socially unequal justice system by the beliefs that wealth was sanctioned by the spirits and that misdeeds must be expiated by expensive offerings to the spirits. Unable to pay for these offerings, many people ended up in the possibly lifelong slavery-like conditions of debt dependence.[62]

Families typically had five to six children, and the population appears to have been growing. Infanticide was not allowed. Boys were married at age 16 or older, girls perhaps younger. Child marriage had been practiced in the past. Premarital sex was tolerated if the couple entered a secret marriage. The wedding ceremony was very simple.

---

59   Donoghue, Whitney and Ishino (1962, p. 3), LeBar, Hickey and Musgrave (1964, p. 251), Schaeffer (1979, p. 9), Special Operations Research Office (1965, p. 2)

60   Donoghue, Whitney and Ishino (1962, p. 38), LeBar, Hickey and Musgrave (1964, p. 254), Schaeffer (1979, p. 15)

61   Schaeffer (1979, p. 3)

62   Schaeffer (1979, p. 94), Sabatier (1940)

Extramarital sex might have been punished but was often tolerated as long as it did not have economic consequences. Polygyny was rare. Divorce was disapproved of. Children started helping their parents from age 7 to 8 and working in the fields from age 10 to 13.[63]

Little has been published about old E De art and music. Buildings and woven clothes sometimes had rich ornamentation. Rich people collected jewelry, decorated gongs, and other prestige items.[64]

Though the available data for precolonial E De are sparse, we can conclude that the cultural indicators are in accordance with a moderately regal culture, though not quite as regal as expected.

7.7. Ganda

The Ganda or Baganda are a Bantu people still living in the kingdom of Buganda, which comprises 50,000 km² of land to the north and west of lake Victoria in East Africa. Buganda is now part of the state of Uganda. The environment consists of flat hills, savanna, swamps, rivers, lakes, and forests at an elevation of 1,000 to 1,400 meters above sea level. The position on the equator provides a climate with only small seasonal variations. The average day temperature is between 25 °C and 28 °C all year round. There are two rainy seasons a year and no completely dry season.

The area of Buganda is very fertile and more suited for agriculture than the neighboring areas. The staple food is bananas or plantains, which are harvested all year round. Other crops are harvested twice a year. The diet is supplemented with protein from domestic animals, hunted game, fish, and insects.[65]

The traditional Baganda had good roads that were traveled by foot. They had no wheeled vehicles, and they did not use animals for work or transport. They had canoes for sailing the lakes and those of the rivers that were navigable. They used iron for making spearheads, hoes, and other tools.[66]

---

63   Donoghue, Whitney and Ishino (1962, p. 26), LeBar, Hickey and Musgrave (1964, p. 253), Schaeffer (1979, p. 136), Sabatier (1940)

64   Schaeffer (1979), Special Operations Research Office (1965, p. 9)

65   Kottak (1972), Roscoe, J. (1911, p. 4)

66   Roscoe, J. (1911, p. 239), Rusch (1975, p. 37)

Based on these facts, we can expect the culture to be regal. The population density was high, due to the fact that they had plenty of food all year round. They were able to travel far through the savannah and grassy hills as well as on rivers and lakes. Their range of movement, however, was limited by their inability to build bridges across large swamps and rivers.[67]

Buganda was the strongest military power in the region. Their power is generally attributed to the efficient production of plantains and to the availability of iron. They attacked and plundered the neighboring peoples on regular raids where they captured cattle, women, and slaves and killed as many men as they could.[68]

The political system was a highly hierarchical system of chiefs headed by the all-powerful king. The king ordered large numbers of men to be killed. Some were killed for minor offenses, some were sacrificed to the gods, and some were killed to honor the ancestor kings; but most of all, the king killed men just to confirm his power to kill.

All crimes and signs of disloyalty were severely punished. Theft, disobedience, adultery, and other crimes were punished by death, by cutting off a limb, or by heavy fines. The fines were to be paid in bark cloth, goats, cows, and women. The culprit might enslave himself or his relatives to pay the penalty. The question of guilt could be tried by a system of courts that might hear testimony and use divination, ordeals, or torture to reach a verdict.[69]

The Baganda worshipped various gods, most of whom were spirits of men who had served previous kings well while they were alive. Dead kings were worshipped in almost the same way as gods, but the living king did not have god status, and he could be killed and overthrown by rival princes. Anthropologists have discussed who was most powerful, the king or the gods. The king, and nobody else, could punish the gods by killing their priests and plundering their temples. The gods, in turn, could punish the king with disease and misfortune. There appears to be no clear winner in this contest of power. Most of the time, however, the king made rich sacrifices to the gods, and the gods in turn supported the king and gave him advice through mediums.

---

67   Roscoe, J. (1911, p. 319)
68   Kottak (1972), Roscoe, J. (1911, p. 4)
69   Roscoe, J. (1911), Kagwa (1934, p. 80)

The gods did not punish undetected sins, but some sins led to automatic punishment. For example, a child would die if its father committed adultery during the breastfeeding period. There was no punishment in the afterlife, except for the belief that a person who had had a limb cut off as punishment would continue to be maimed in his afterlife or next incarnation.[70]

The members of the kingdom were generally loyal and willingly sacrificed their lives in the frequent wars against neighboring peoples. In some cases, hostility towards strangers was an obstacle to trade, but for the most part they welcomed Arab traders.[71]

Polygamy was widespread. The later kings had hundreds, perhaps thousands, of wives and concubines. High-ranking chiefs had many wives as well. Such a degree of polygamy was possible because of a sex ratio of three women to one man. The surplus of women was due to the loss of men in war, the king's mass killings of men, and the capture of women in war. No fertility data are available for the period prior to the introduction of Christianity and the abandonment of polygamy in the late nineteenth century. By the mid-twentieth century, the birth rate was three to four children per mother. The common opinion of the natives is that fertility was higher in the old times, and this is in accordance with the few individual genealogies that have been reconstructed.[72] Suicide was probably rare. The only reports of suicide refer to the motive of shame.[73]

Girls married at age 13 or later, boys at 15 to 16. The marrying age increased after the introduction of Christianity. A man had the right to beat his wife and could even kill her with impunity. Divorce was relatively common, especially in connection with polygamy. Women were guarded carefully to preserve their chastity. It brought great shame on a girl to become pregnant before marriage. The punishment for adultery was severe, but even the risk of the death penalty was apparently not sufficient to completely deter this crime.[74]

---

70 Roscoe, J. (1911, p. 273), Kagwa (1934, p. 124), Ray, B. (1991, p. 42), Rusch (1975, p. 361), Southwold (1969, p. 90)
71 Roscoe, J. (1911, p. 346), Mair (1934, p. 191)
72 Roscoe, J. (1911, p. 83), Kagwa (1934, p. 69), Klein (1969, p. 135), Richards and Reining (1954)
73 Roscoe, J. (1911, p. 20)
74 Roscoe, J. (1911), Southwold (1969, p. 108), Mair (1934, p. 78)

Babies were sometimes nursed by maids so that their mothers were free to work. Children were often placed with relatives or chiefs so that they could receive a stricter upbringing. Children were taught politeness, etiquette, and cleanliness. Boys had time to play while they were herding animals, whereas girls spent their time making mats and baskets.[75]

The Ganda had very little material culture. All that people needed was a grass hut to live in, bark cloth for clothing, and a few tools for harvesting and cooking. The early kings did not even have chairs to sit on. Palaces, temples, and shrines were built of reed.

Figure 17. Kasubi Tomb, Uganda, nineteenth century. Royal buildings were large, but not as embellished as might be expected. Photo by Agner Fog, 2007.

Ganda art was more decorative than representational. High-ranking persons had some decoration and pomp to signify their status, and religious artifacts were embellished. These decorations were typically made of plant material, cowry shells, ivory, and bird feathers. Such embellishments may have impressed the natives, but they were quite simple compared to other cultures. The Ganda had many different

---

75   Roscoe, J. (1911), Southwold (1969, p. 107), Mair (1934, p. 31)

musical instruments, but we do not have sufficient information to evaluate the degree of embellishment in Ganda music and dance.[76]

The conclusion for the Ganda is that the observations regarding their political system, discipline, religion, fertility, marriage, sexual morals, and length of childhood are in agreement with a highly regal culture.

The prediction fails, however, on the question of art. Regal cultures typically make impressive pieces of art and architecture of durable materials. The Ganda culture has none of this. No royal or religious buildings, monuments, or sculptures were made of stone, brick, or metal. The fact that the art was non-naturalistic and the architecture perfectionist is in accordance with the theory, but it did not have the degree of embellishment that other regal cultures have. The low sophistication of their material art may be connected with the low sophistication of their material culture in general. There appears to have been very little impetus to technological innovation, because food was available everywhere for a minimum of work.

The lack of highly embellished art makes one suspect that the extremely regal political system was not fully internalized in the psychology of the people. While most of the cultural indicators are relatively regal, they are not fully on the same level as the political system. There were songs about military events, but no systematic glorification of kings, heroes, or gods in Ganda music and art. The religion supported the king by sanctioning the mass killings that his power depended on, but the disciplining function of the religion was less effective than we have seen in other regal cultures.[77] It is worth noting that sub-Saharan Africa is dominated by kungic cultures because of the dense vegetation that impedes traveling and a climate that is too hot for hard work. As a regal enclave, the Ganda culture may not have been able to evolve a cultural ethos that was too different from the surrounding cultures. We can thus observe a mixture of typically African traits with typically regal traits. We may speculate that the king had to compensate for this slight discrepancy in his culture by consolidating his power with mass killings of his own men.

---

76   Roscoe, J. (1911, p. 285), Kagwa (1934, pp. 90, 170), Rusch (1975), Mair (1934, p. 193), Lugira (1970)
77   Roscoe, J. (1911), Ray, B. (1991, p. 167), Mair (1934, p. 25), Kagwa (1934, p. 141)

7.8. Gilyak

The Gilyak lived in east Siberia along the river Amur and on the island of Sakhalin. Their main means of subsistence was fishing. The climate was very cold in winter. The vegetation was primarily taiga (primeval forest), which was difficult to penetrate. The means of transportation were limited to dog sleighs and snowshoes in the winter and rowing boats in the summer. The boats were unable to sail on the open sea.[78]

We can expect to find a kungic culture based on the facts that the means of transportation were too inefficient to make war likely and that the environment could not sustain a high population density.

The Gilyak culture had a moderately high level of internal conflict but no external wars. The Gilyak did not keep records of their own history and had no memories of any war. According to one source, there had probably been a conflict between the Gilyak that lived along Amur and those that lived on Sakhalin, as well as a conflict with the Russians. However, no solid information about these conflicts is available. Internal conflicts were relatively common. The main reason for such conflicts was rivalry over women.[79]

The Gilyak had no political system beyond the family or clan. There was no chief or ruler. Common decisions and resolution of disputes, if necessary, were made through deliberations where the most respected men had the most influence. Sources of respect were age and experience, as well as wealth and bravery. Crime was rare, and there was no formal system of policing. Punishments for wrongdoing were generally mild. Corporal punishment and capital punishment occurred only rarely.[80]

The religion was mainly based on beliefs in spirits and animals. There was no orthodox teaching, and not even the shamans were able to explain their beliefs to the level of detail that the explorers expected. The prophecies and teachings of the shamans were readily disputed or doubted. The Gilyak had a weak concept of a good god that they may have learned from Buddhists or Christians, but they did not pray, and this belief was completely void of influence on their daily life. There was no opposition to other religions. Some Gilyak even worshiped a few

---

78   Schrenck (1881, p. 655), Seeland (1882, p. 245)
79   Schrenck (1881), Seeland (1882)
80   Schrenck (1881, p. 663), Seeland (1882, p. 227), Shternberg (1933, p. 15)

church bells that they had found as well as the ruin of an old Buddhist temple.[81]

The Gilyak had no word for their own tribe other than the word for humans. The name Gilyak is used only when talking with foreigners. They had regular barter trade with the Chinese, Japanese, and Russians. The Gilyak were highly tolerant of foreigners and readily mixed with the neighbor tribes.

The fertility rate was consistently described as low, but there are no accurate accounts beyond the observation that there were no child-rich families. Children were breastfed for two to five years. Suicide is often mentioned in the literature on the Gilyak. No specific rates can be inferred, but at least it seems safe to conclude that the suicide rate was not low.[82]

The literature contains discrepant accounts of the sexual morals of the Gilyak. Leopold von Schrenck characterizes the sexual morals as strict, while Nicolas Seeland describes them as fairly lax, and L. Sternberg reports that the Gilyak were quite promiscuous.[83] We must recognize that sexual behavior is particularly difficult to study. There may be considerable differences between what people say and what they do. Sternberg's accounts are quite detailed and appear to be the most reliable. Schrenck mentions sexual morals only briefly and is more concerned with describing physical objects. Apparently, Schrenck was unable to penetrate through the initial facade of modesty that Sternberg described. It seems safe, therefore, to conclude that the sexual morals of the Gilyak were quite lax. The marrying age was variable. Bachelors aged 23 to 25 were commonly seen, but child marriages were also reported. Divorce was easy.

The Gilyak produced very little art. The most common products of art were small religious figures carved in wood. These wooden figures were quite simple and could be made by any member of the tribe. Women's clothes were often embellished with small brass plates and other adornments. Knives and other tools were sometimes embellished with inlaid pieces of brass, copper, or silver. These adornments were

---

81   Schrenck (1881, pp. 678, 739), Seeland (1882, p. 239)
82   Schrenck (1881, pp. 99, 641), Seeland (1882, p. 129), Shternberg (1933, p. 126)
83   Schrenck (1881, p. 638), Seeland (1882, p. 226), Chard (1961), Shternberg (1933, p. 124)

used as signs of wealth. Musical instruments were very primitive. Songs were simple and often improvised. It was quite common for each singer to have his own song. Dance was not observed, except for shamanic acts.[84]

The conclusion for the Gilyak is that the culture was quite kungic, in accordance with the geography, climate, and means of transportation. The predictions for a kungic culture are in excellent agreement with the observations of level of external conflict, political system, discipline, religion, tribal identification, tolerance of foreigners, fertility rate, suicide rate, sexual morals, art, and music. The occurrence of child marriages is in disagreement with the expectation of a high marrying age. The amount of adornment on clothes and tools is somewhat higher than expected. This adornment served the need to display wealth, which was an important source of political influence.

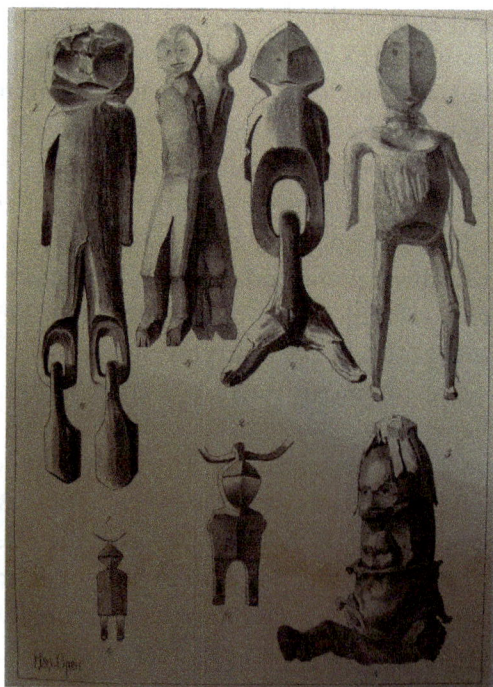

Figure 18. Gilyak wooden figures. From Schrenck (1881).[85]

---

84   Schrenck (1881, pp. 387, 681), Seeland (1882, pp. 109, 126, 234)
85   Public domain

Figure 19. Gilyak decorated tools. From Schrenck (1881).[86]

7.9. Hausa

The Hausa are a sedentary people living south of the Sahara in northern Nigeria and south-eastern Niger. The climate is hot and dry half of the year with a rainy season from May to September. The environment is mostly savannah, with patches of scrub and wooded watercourses infested by tsetse flies. Occasional droughts, famines, and epidemics have limited the population.[87]

The Hausa grow sorghum, millet, and several other grains and vegetables in the rainy season and store them for the dry season. They have various domestic animals, including horses and donkeys. The dry season is used for practicing many different crafts, including ironwork and long distance trade.[88]

86　Public domain
87　Smith, M. G. (1960, p. 1; 1965, p. 121)
88　Smith, M. G. (1965)

Based on this information, we can expect the Hausa to be quite regal. Intensive agriculture in defendable fields, food storage, no geographical obstacles to travel, horses and donkeys for transportation, and iron for making weapons are all factors that point in the regal direction. The only kungic influence is that shortage of water and intense heat may impede traveling armies.

For more than 1,500 years, the Hausa have lived in seven states, united by a common language. In early times, these states were ruled by kings or queens. Islam reached Hausaland in the mid-fourteenth century, and the kingdoms gradually became emirates. The Fulani, a pastoral people, waged jihad (holy war) on the Hausa from 1804 to 1810 for the purpose of purifying Islam in the population. Since then, most of the ruling class have been Fulani, though the majority of the population were Hausa. The population includes several other ethnic minorities. Slavery was common until Nigeria was colonized by England in the early twentieth century.[89]

There has been war more often than peace in the region from the earliest times and until the English colonization. Each kingdom or emirate was independent in some periods, at other times paying tribute to another state or empire. The majority of the population lived in cities surrounded by walls for protection.[90]

The political system was hierarchical and bureaucratic. A well-organized legal system suppressed internal conflicts. After Islamization, punishments were determined by Islamic law. Serious crimes were punished by death, theft and robbery by cutting off a hand. Suspected criminals might be whipped until they confessed. It is difficult to determine to what degree these severe punishments were actually carried out, because the old literature rarely mentions specific cases of punishment.[91] Many folktales tell of tricksters who got away with immoral behavior or who received lenient treatment.[92]

The literature is full of references to prostitution and other extramarital sexual affairs. Punishment for sexual crimes is rarely mentioned in the literature. Girls were supposed to be virgins at their

---

89   Smith, M. G. (1955, p. 91; 1978, p. 33), Palmer (1928)
90   Smith, M. G. (1955, p. 102; 1960, p. 2; 1978), Hogben and Kirk-Greene (1966)
91   Smith, M. G. (1960, p. 42; 1978, p. 39), Smith, M. F. (1954, p. 183), Palmer (1928)
92   Edgar and Skinner (1969), Tremearne (1913, p. 48)

first marriage, but premarital petting was often tolerated as long as pregnancy was avoided. Girls were married at the age of 13 to 14 (some say 18), boys at around 20 to 21. The first marriage was preferably between cousins, but these marriages rarely lasted. There were certain customs of avoidance and formalized speech between husband and wife. These customs also contributed to making marriages unstable and encouraging husbands to seek intimacy with prostitutes. It was normal for women to be divorced and remarried several times through their lives. Polygamy was widespread and desired. Rich men preferred to hold their wives in seclusion in their homes (purdah). Eunuchs served the ruler. No statistics for fertility before colonization are available, but many families mentioned in the literature were child-rich. The firstborn child was normally given away for adoption after weaning at the age of two, and there was formalized avoidance between parents and the firstborn child. This sometimes applied to subsequent children as well. Boys were circumcised and girls clitoridectomized.[93]

Children were expected to work, but they also had plenty of time for play. Some children earned money for themselves by selling things at the market.[94]

Suicide is not mentioned in the literature, except for a few cases where the reason was shame rather than anomie.[95]

The urban population was mostly Muslim, while some of the rural population preserved their indigenous religion and engaged in rituals of spirit possession. However, both groups were characterized by syncretism. The spirits in the 'pagan' beliefs were regarded as identical to the *jinn* (genies) mentioned in the *Quran*, and Allah was their king. The supernatural world of the urban Muslims was hierarchical, while the pagan spirits were organized around clans, where each clan worshipped their own local spirits. The spirits were responsible for disease and misfortune, but they were not perceived as agents of moral punishment. The Muslims believed in supernatural punishment—at least in theory—but apparently they did not fear hellfire much.[96]

---

93   Smith, M. G. (1955, p. 21; 1965, p. 139; 1978, p. 43), Smith, M. F. (1954, pp. 15, 60), Tremearne (1913, p. 77; 1914), Greenberg (1946, p. 16), Nicolas (1975, p. 182)

94   Smith, M. F. (1954, p. 55), Dry (1949)

95   Hogben and Kirk-Greene (1966, p. 286), Tremearne (1914, p. 143)

96   Tremearne (1913, p. 109; 1914), Greenberg (1946, p. 27; 1947), Nicolas (1975)

Individual feelings of identity were a mixture of clan membership, ethnicity, membership of a kingdom or emirate, and the general Hausa identity. Each of the ethnic groups was mostly endogamous.[97]

The most important form of art was praise singing. Formalized speeches and songs of praise were required at all major religious and political events. There were several different kinds of praise singers extolling artisans as well as the nobility. The oral tradition with pre-Islamic heritage is bombastic, boastful, and highly formalized, and it usually follows regular schemes for meter and rhyme. Written poetry follows Islamic tradition with mandatory doxologies.[98] The art of embroidery was also developed to perfection. Various objects were decorated with fine geometric patterns. Architecture was less developed. Royal palaces and mosques were beautiful and impressive, but their decoration lacked precision.[99]

Figure 20. Old Hausa building in traditional Tubali style.
Photo by Dan Lundberg, 1997.[100]

---

97   Smith, M. G. (1960, p. 2; 1978, p. 31)
98   Hiskett (1975)
99   Smith, M. G. (1978, p. 99), Tremearne (1913)
100   CC BY-SA 2.0, https://en.wikipedia.org/wiki/File:1997_277-10A_Agadez.jpg

We can conclude that the Hausa were regal, but perhaps not quite as regal as expected. In particular, the sexual morals were more permissive than expected. This may be due to the custom of avoidance between husband and wife that made the man seek warmth and intimacy elsewhere. The seclusion of women, clitoridectomy, and the use of eunuchs, however, point in the direction of strict sexual morality.

7.10. Inca

The Inca Empire had its center in the town of Cuzco in the Andes. During the hundred years prior to the Spanish conquest in 1533, the empire grew fast to reach the size of several million km², stretching from present Ecuador to Chile and part of Argentina. This vast area covered a wide variety of environments, including coastal areas, mountain plateaus, deserts, and rain forest. The day temperature in the Cuzco area was 21–22 °C all year round, while night temperature occasionally fell below freezing point. There are plenty of rivers running down the mountains, but these are not navigable.[101] Transportation by boat was possible only along the coast and in the lakes. The territory was connected by an elaborate system of roads totaling at least 23,000 km and probably more. Relay runners were placed along major roads for fast communication. The Incas had llamas for transportation of goods, but no animals capable of carrying humans were available. Wheeled vehicles were not available.[102] Agriculture and animal husbandry have been practiced in the mountain plains and basins for 3,000 to 4,000 years. Terraces and large irrigation systems were built to intensify agricultural production.[103]

Based on the geography alone, we would not expect to find a very regal culture in the Andes because of the many barriers to travel. However, knowing that water for irrigation, efficient food production technology, and communication technology were available, the prediction is a different one. The main crops — quinoa, potatoes, maize, and beans — provided highly nutritious and storable food in high yields. The environment and available technology therefore provided all the

---

101 Baudin (1961, p. 1), Mason (1957, p. 1), Means (1931, p. 3)
102 Mason (1957, pp. 140, 167), Means (1931, p. 329), Hyslop (1984)
103 Mason (1957, p. 137), Bauer (2004, p. 25)

factors necessary for developing a highly regal culture: efficient food production for sustaining a high population density, a concentration of food production in defendable irrigated areas, efficient transportation and communication, a climate that was suitable for hard work, and highly different environments leading to trade and wars between different cultures and different lifestyles.

The political system of the Inca Empire was highly centralized. Everything was planned and controlled by a large bureaucratic system. People had hardly any choice about where to live and what kind of work to do. At the same time, the administration took great care to make sure everybody had what they needed. Discipline was strict and punishment was severe. The worst crimes were punished by the death not only of the culprit but also of his entire village.[104] The religion justified the dictatorial rule by making the emperor a direct descendant of the sun. Rich sacrifices were made to the gods, occasionally including the sacrifice of humans.[105] Sinners were punished in the afterlife in a hell. Suspected witches might be tortured to confession and killed. Religious ceremonies were plentiful and more pompous than puberty rites, marriages, or burials.[106]

The Inca rule did not enforce total cultural uniformity, but they moved conquered people around to prevent rebellion. The different groups were allowed to maintain some of their cultural characteristics.[107]

Large families were desired, but infant mortality was high. Abortion was known but illegal; other means of birth control were unknown. Infants were strapped to a cradle day and night and were breastfed only three times a day for one or two years. They were never held in the arms or on the laps of their mothers but were breastfed by their mothers bending over them. Larger children were confined to a hole dug in the ground. Children were required to contribute to household and agricultural work as soon as they were able to do so.[108]

---

104 Baudin (1961, p. 34), Mason (1957, p. 170), Means (1931, p. 347), Brundage (1963, pp. 121, 151)

105 Mason (1957, p. 202), Malpass (1996, p. 101), Wilson et al. (2007)

106 Means (1931, p. 367), Brundage (1963, p. 57), Malpass (1996, p. 106)

107 Mason (1957, pp. 119, 197), Malpass (1996, p. xxvi)

108 Baudin (1961, p. 100), Mason (1957, p. 147), Means (1931, pp. 313, 363), Malpass (1996, p. 75)

Girls married at age 16 to 20 and boys in their mid-twenties. Some girls were selected for a religious life in celibacy or were given as rewards to high-ranking men. Most boys were probably able to choose their wife, but staying unmarried was not an option.[109] Polygamy was common among the nobility. Divorce from the first or principal wife was impossible, while divorce from secondary wives or concubines was easy. Trial marriages were possible in some areas, and the marriage ceremony made a distinction based on whether the bride was a virgin. Except for the institution of trial marriages, fornication and all other sexual sins were severely punished.[110]

Elaborate art was common. The Inca made large and impressive stone buildings, such as the famous Machu Picchu. Festive and ceremonial clothes were finely woven and richly decorated. The decorations on cloth and pottery were typically highly repetitive, with geometric patterns or stylized figures. Immense riches of gold and silver added to the glory of the empire.[111]

Figure 21. Inca carved gold leaf wall coating.
Photo by Manuel González Olaechea, 2007.[112]

---

109 Baudin (1961, p. 24), Means (1931, p. 358), Brundage (1963, p. 149), Malpass (1996, p. 37)
110 Mason (1957, p. 150), Brundage (1963, p. 225), D'Altroy (2002, p. 191)
111 Baudin (1961, p. 187), Mason (1957, pp. 144, 231), Means (1931, p. 105), Malpass (1996, p. 87)
112 CC BY 3.0, https://commons.wikimedia.org/wiki/File:Muro_de_oro.JPG

The conclusion for the Inca Empire is that the observations regarding political system, discipline, religion, fertility, sexual morals, length of childhood, and art are in agreement with a highly regal culture. There is no mention of suicide in the literature except in connection with war. The marrying age did not differ markedly from that in kungic cultures.

The observation that the Inca Empire could grow only because of the elaborate road system and the intensive agriculture raises a question: was the empire built because they had good roads and efficient food production, or did they build roads and irrigation systems because the power of the empire enabled them to do so? The answer is, of course, that the causality went both ways in a self-amplifying process. The area is more suited for a sedentary life than for nomadism, and the population had sufficient time to develop an advanced agriculture and a relatively high population density. The availability of water for irrigation and efficient crops such as potatoes and maize were also important factors. The first regal developments may have taken place in the coastal areas where the ocean provided ample food and means of transportation, or it may have taken place in fertile areas near rivers. There was trade as well as frequent wars between the coastal people and the mountain people for centuries. Numerous chiefdoms and kingdoms had waxed and waned at the coast as well as in the mountains for more than a thousand years before the Inca Empire flourished.[113] Our theory predicts that empires will collapse when they have reached the maximum manageable size, which is indeed what appears to have happened several times in this area. Each new empire may have grown bigger than the previous ones due to improved technology or political skills. The Inca Empire, too, was probably on the verge of collapsing at the time of the Spanish conquest.[114]

7.11. !Kung

The !Kung bushmen, also called San, have lived in the Kalahari Desert of southern Africa for many thousands of years, representing one of the oldest living human races.[115] They speak an old language containing click sounds; the '!' in their name is a click sound.

---

113 Means (1931, p. 47), Brundage (1963, p. 121), Malpass (1996, p. xxiii)
114 Means (1931, p. 47)
115 Chen et al. (2000)

The Kalahari Desert is dry for long periods of the year. The !Kung depended during the dry season on a few permanent water holes and on water-containing roots and melons. They traveled long distances on foot and had no more possessions than they could carry. No other means of transportation were part of !Kung tradition.

Our theory predicts that the !Kung culture must be kungic because the low population density, lack of water, and lack of efficient means of transportation make war impossible. The neighboring tribes had no desire to conquer their land because it was unsuitable for herding and agriculture. The !Kung adapted to the dry climate and were able survive the frequent droughts without starving. At the time of the first studies in the 1950s and 60s, they were happily unaware that their entire culture was threatened and that bushmen in more fertile areas had been exterminated. The !Kung were not refugees expelled from other territories.[116] Had they been refugees, they might have been more regal.

Figure 22. A !Kung bushman sucks up underground water through a straw and stores it in an ostrich eggshell. Photo by Jens Bjerre, 1958.[117]

The !Kung never engaged in war. They would usually run away and hide if foreigners entered their territory. The traditional means of subsistence

---

116 Lee (1979, p. 33)
117 All rights reserved, reproduced with permission

was hunting and gathering. At the time of the studies, some !Kung had started to work for neighboring black farmers who treated them well. Others worked for white farmers under slave-like conditions.[118]

The political organization of the !Kung is consistently described as fiercely egalitarian. There was no political unit beyond the individual settlements with fluid membership. Various observers disagree on whether settlements had chiefs. There was obviously some degree of leadership based on kinship, age, experience, and personal qualities.[119]

Sharing was very important in !Kung life. Anybody who had food was required to share with others. Hunting tools, clothes, and other possessions circulated widely through systematic gift exchange. Any attempt at boasting, self-promotion, or accumulation of wealth was despised and effectively prevented through social pressure.

There was no formal system of justice and punishment. Even the worst cases of wrongdoing could go unpunished, for children as well as for adults. Theft did not occur. Murder was occasionally revenged—in most cases as an outburst of rage, but in a few cases in a planned way. Severe conflicts were more likely to result in the social group splitting up than in any kind of disciplining. The most severe punishment described by the observers was that somebody sang a song about the misdeeds of the culprit. The need for social acceptance was sufficient to keep people in line. When the first schools were set up in the area, the !Kung did not accept the fact that corporal punishment was practiced in the schools.[120]

The !Kung beliefs were fluid, ambiguous, and incoherent. They had no clear distinction between natural and supernatural, between human and divine, or between human and animal. Their legends included traces of Bantu religion, Christianity, and even European fairy tales. Several observers have equated the most powerful of their mythological characters, the creator of everything, with the Christian God. But this god had few similarities with this monotheistic God. He was, in a mythological past, a vulgar trickster displaying all the immoral behaviors that humans abhor. The gods and spirits interacted with humans in good and bad ways but were not concerned with upholding human morality. Rather, they manipulated humans for their

---

118 Lee (1979, p. 77), Thomas (1959)

119 Lee (1979, p. 244), Thomas (1959), Marshall (1976, p. 181)

120 Lee (1979, p. 24), Thomas (1959), Marshall (1976, p. 289), Draper (1975)

own obscure purposes. The religious beliefs were used in storytelling, in the use of oracle discs, and in trance dances and healing. The exercise of these rituals was not monopolized by any religious authorities, although some people were obviously better healers than others.[121]

Anthropologists disagree on what to call the !Kung because they themselves had no clear name for their tribe or race. They did not recognize as tribe fellows other people that spoke the same language. They readily mingled with and worked for neighboring people of different races. Mixed marriages were common.[122]

The fertility of the hunting and gathering !Kung was so low that overpopulation was avoided and the natural resources were not overexploited.[123] Children were breastfed for three to four years, and birth spacing was three to five years. Infanticide occurred at a reported rate of 1–2%. No other means of birth control were used. Children did not work but were free to play all day. The average marrying age was 14 to 17 for women and 22 to 30 for men, according to some observers, but child marriages are also reported to have been common. Polygamy occurred, and a single case of polyandry was reported in a neighbor tribe. Divorce was easy. Suicide was rare.[124]

The frequency of premarital and extramarital sex is difficult to estimate. Lorna Marshall reports that illegitimate sex was rare because it was impossible to hide, but she also notes that spouse swapping was allowed. M. G. Guenther notes that their marriages were 'loose'. I. Eibl-Eibesfeldt writes that they were fairly liberal in sexual matters, but open promiscuity was not tolerated. G. Silberbauer reports for a neighbor tribe that people often could get away with adultery. R. Lee reports that gonorrhea was spreading fast. Rape was rare.[125]

The !Kung rarely produced pictorial art, because it didn't fit into their nomadic lifestyle. The main art forms were song, dance, and storytelling. Women and girls decorated themselves lavishly with strings of beads. Men were tattooed as an element of hunting magic. They used

---

121 Thomas (1959), Guenther (1999), Biesele (1976), Marshall (1962, 1999)
122 Thomas (1959)
123 Lee (1979, p. 44), Silberbauer (1981, p. 286), Marshall (1976, p. 146)
124 Lee (1979, pp. 265, 310), Thomas (1959), Marshall (1976, p. 166; 1999, p. 320), Silberbauer (1981, p. 149)
125 Marshall (1976, p. 279), Guenther (1999, p. 237), Eibl-Eibesfeldt (1978, p. 136), Silberbauer (1981, p. 156), Lee (1979, p. 417)

bows as musical instruments. Other simple musical instruments were introduced during the twentieth century. They appreciated individual inventiveness in art and readily borrowed elements of art from neighbor cultures. There were no big ceremonies, not even at marriage or burial. The most important socially unifying ceremony was the frequent trance dance.[126]

Figure 23. !Kung bushmen telling stories around the fire.
Photo by Jens Bjerre, 1958.[127]

The conclusions for the !Kung are that the culture was very kungic, in accordance with the geography, lack of efficient transportation means, and niche specialization that prevented territorial conflicts with neighbor tribes during the times of early observations. The predictions for a kungic culture are in excellent agreement with the observations of a low level of conflict, an egalitarian political system, lax discipline, non-authoritarian religion, low degree of group identification, tolerance of foreigners, low fertility rate, long childhood, and flexible art and music.

---

126 Marshall (1976, p. 34, 363; 1999, p. 63), Guenther (1999, p. 165)
127 All rights reserved, reproduced with permission

The expected high suicide rate has not been found. The sexual morals appear to be fairly liberal for a people without birth control.

Fertility has recently been increasing and egalitarian ideals decreasing for those !Kung who have changed to a settled lifestyle based on herding or farming, as can be expected from our theory.[128]

Previous literature uses the word 'kalyptic' for the opposite of regal.[129] As this word sounds somewhat awkward and may be misinterpreted as an abbreviation of 'apocalyptic', it was decided to replace it with the new term kungic in commemoration of the !Kung culture. There is no circular reasoning in the observation that !Kung culture is kungic, as the decision to use this term was made *ex post facto*.

7.12. Maasai

The Maasai are a seminomadic pastoral people living on the semiarid plains and plateaus of the Great Rift Valley in Kenya and Tanzania. They have migrated from the north since the fifteenth century and displaced other tribes as they expanded. Their territory reached its maximum size in the mid-nineteenth century. The population was decimated in the late nineteenth century due to rinderpest and other epidemics among their livestock, as well as smallpox and drought.[130]

They have large herds of cattle and goats and a smaller number of donkeys used as pack animals. Their diet consists almost entirely of milk, meat, blood, and honey. They avoid meat from wild animals. They have metal-headed spears which they use in warfare.[131]

To make a prediction for the Maasai, we first note that the ease of travel on the plains makes war possible. Their donkeys are useful for transport of material but not for war. The population density must be relatively low, due to the limited and periodic rainfall and frequent droughts. However, the use of cattle as a food source allows groups of several hundred people to live together. Scramble competition for grazing land points in the kungic direction, while contest competition for important water sources points in the regal direction. We will predict

---

128 Lee (1979, pp. 48, 324), Draper (1975), Katz and Biesele (1986)
129 Fog (1997)
130 Coast (2001, p. 24), Fosbrooke (1948), Huntingford (1953, p. 105)
131 Huntingford (1953, p. 107), Merker (1904, p. 157)

a moderately regal culture, since there are more regal than kungic factors in their environment and ecology.

The social and political organization of the Maasai is based on a peculiar age set system. When an age group of boys reach maturity, they are collectively circumcised and soon afterwards included in the class of warriors (moran). This event also initiates them into the age set system. A new age set is formed every fifteen years, while the members of the preceding age set mature and get married and a new group of warriors is formed. The cohort of men belonging to the same age set forms a lifelong political unit with its own leaders and spokesmen. The fifteen-year age span allows the warrior group to be big enough to form a strong military force.[132]

Figure 24. Maasai warriors. The patterns on their shields indicate their group and age set. Photo by Walther Dobbertin, 1906–1918. Bundesarchiv, Bild 105-DOA0556.[133]

The Maasai have no paramount chief. The political power above the division into age sets resides with leaders selected among the elders. A particularly powerful man is the medicine man and diviner, the laibon, who has many wives. Nobody can go to war or make other important decisions without first asking the laibon for ritual directions.[134]

The non-migratory ironsmiths form a separate caste, despised by the pastoralists.[135]

132　Fosbrooke (1948), Merker (1904, p. 70)
133　CC-BY-SA 3.0, https://commons.wikimedia.org/wiki/File:Bundesarchiv_Bild_105-DOA0556,_Deutsch-Ostafrika,_Massaikrieger.jpg
134　Fosbrooke (1948), Merker (1904, p. 18), Krapf (1860, p. 362), Whitehouse (1933)
135　Huntingford (1953, p. 109), Merker (1904, p. 110), Fokken (1917)

The entire life of the pastoral Maasai revolves around the accumulation of wealth in the form of livestock. The warriors make regular raids against neighboring tribes in order to steal livestock. Indeed, this is the main occupation and the *raison d'être* of the warrior group. Raids and wars against other Maasai sometimes occur. The Maasai have a strong sense of tribal identity and they despise all other tribes.[136]

There is a well-defined system for punishment of crimes against other Maasai. Most crimes, including murder, are punished by the payment of fines in the form of livestock. Revenge killing occurs, but is mostly prevented. Children work from an early age, and they can be punished with beating if they fail to do their work properly.[137]

The religion of the Maasai is monotheistic. Their god, Engai, is identified with the sun, rain, and other natural phenomena. Nobody knows or speculates about what Engai looks like. The Maasai believe that they are Engai's chosen people. Engai has given livestock to them and nobody else, and this gives them the right to attack other tribes and steal their livestock.[138]

The Maasai have little or no beliefs in spirits or an afterlife. Dead people are rarely buried—they are just left in the wild to be eaten by scavengers.[139] Ritual curses are used as a means of disciplining people. A person who breaks a curse by misbehaving or by failing to fulfill a duty will be punished by misfortune.[140]

The sexual morals of the Maasai are lax. Premarital and extramarital sex is allowed and considered normal as long as certain rules are obeyed.[141] This is unexpected for a moderately regal culture and requires an explanation. The social organization of the Maasai into age sets serves the function of generating a large group of young unmarried warriors with high prestige who can devote their time to warfare and raids without having to care about family life and children. It is unlikely that these powerful young men can be disciplined into spending the

---

136 Huntingford (1953, p. 107), Merker (1904), Krapf (1860, p. 354), Baumann (1894, p. 163)
137 Coast (2001, p. 32), Merker (1904, p. 203), Spencer, P. (1988, p. 51), Sankan (1971, p. 11), Saitoti (1986, p. 7), Hollis (1905, p. 203)
138 Merker (1904, p. 195), Krapf (1860, p. 360), Fokken (1917), Baumann (1894, p. 163), Spencer, P. (1988)
139 Fokken (1917), Baumann (1894, p. 164), Hollis (1905)
140 Spencer, P. (1988, p. 219), Saitoti (1986, p. 3)
141 Coast (2001, p. 115), Merker (1904, p. 44)

most virile years of their lives without access to women. This would have undermined any strict sexual morals that may have existed in a distant past.[142]

Men do not marry until they advance into the group of elders, while girls marry at a much younger age. Girls cannot marry before they have been clitoridectomized, which is a rite of passage analogous to the circumcision of boys.[143]

Most men will acquire additional wives as they grow older, if they can afford it. One may suspect that the economic aspects of marriage are more important than the romantic aspect. The society is patriarchal, and inheritance is patrilineal. The children that a woman bears belong to her husband, regardless of who the biological father is. Divorce is rare and difficult for a woman who has children. Due to the age difference between husband and wife, there are many widows, and widows rarely remarry although they may continue to have children. It is even possible to marry a dead childless man in order to give him a son that can inherit his property.[144]

Figure 25. Young married Maasai woman wearing iron spirals and other ornaments. Photographer unknown, ca. 1900. From Hollis (1905).[145]

---

142 Fosbrooke (1948)
143 Coast (2001, p. 80), Merker (1904, p. 44), Krapf (1860, p. 363)
144 Coast (2001, p. 86), Fosbrooke (1948), Huntingford (1953, p. 115), Sankan (1971, p. 47)
145 Public domain

The Maasai desire to have many children, and the fertility is high. They have a strong taboo against mentioning the name of a dead person. For this reason, it is very difficult to gather information about the occurrence of infanticide and suicide, though these are known to occur. The casualties of war are high.[146]

The conclusion for the Maasai is that most of the cultural markers fit the predictions for a moderately regal culture, with the exceptions of the political structure, which is less hierarchical than expected, the sexual morals, which are unusually permissive, and the late marrying age for men. The Maasai have developed a unique sociopolitical structure that has brought them military success, despite being different from the highly hierarchical and centralized political structures of other militarily strong societies.

7.13. Mbuti

Mbuti is a name for several groups of pygmies who lived in the Ituri forest, a tropical rain forest in the Democratic Republic of the Congo near the equator. The average day temperature was around 27 °C all year round. It was possible to pass through most parts of the forest on foot because of limited undergrowth. Rivers and swamps formed barriers to the movement of the Mbuti, who did not make boats. The Mbuti were nomads with a peculiar lifestyle alternating between two different cultures. For periods of several months, they lived in the forest, where hunting and gathering formed their subsistence. At certain times of the year, they lived near villages at the edge of the forest, where they traded meat for vegetables and tools with slash and burn agriculturalists of Bantu and Sudanese origins. The villagers had bound the Mbuti into a kind of serfdom, which was inherited on both sides — at least in the imagination of the villagers. However, the Mbuti could easily escape and change their allegiance, because the villagers were afraid to enter the forest. The relationship between Mbuti and agriculturalists has lasted for so long that the Mbuti have lost their original language but

---

146 Coast (2001, p. 135), Merker (1904, p. 333)

not their genetic uniqueness.[147] The only evidence that the pygmies ever lived independently in the forest is archaeological.[148]

Figure 26. Mbuti men hunting with net. Photo by Paul Schebesta, n. d. Bildarchiv Austria, Österreichische Nationalbibliothek, Nr. 10819171.[149]

To make a regality prediction for the Mbuti, we first note that the population density in the forest was low. Walking through the forest was slow, and there were rivers that the Mbuti could not cross. The environment was healthy and dangers were few. The forest provided good protection from the villagers, who would never pursue the Mbuti into the forest. These are factors that point in the kungic direction. Other factors point in the regal direction. The nomadic lifestyle could lead to territorial conflicts. The territory of the Mbuti was shrinking as forest was converted to agricultural land. Furthermore, the Mbuti were heavily influenced by the village culture, which we would expect to be more

---

147 Cavalli-Sforza (1986, p. 18), Harako (1976, p. 39), Schebesta (1938, p. 54), Turnbull (1961, p. 17; 1965, p. 23)
148 Bailey (1989), Mercader et al. (2000), Wilkie and Curran (1993)
149 All rights reserved, reproduced with permission

regal than the forest culture. Nevertheless, we will expect the kungic factors to dominate, because the Mbuti could escape any conflict with the villagers, who were unable to penetrate their niche. Consequently, the prediction is that the Mbuti culture would be moderately kungic.

The Mbuti had very little political organization. Some people, mostly older men, took leading roles in the camps, but egalitarian principles were held in high regard and everybody had a say in discussions. Conflicts were handled collectively. Everybody in a camp was responsible for avoiding or stopping conflicts between camp members. Young bachelors sometimes played a key role in handling conflicts between the older married people by means of mockery and ridicule. Conflicts were often diverted or dissolved into ridicule rather than settled with a formal decision. Retaliation for wrongdoing was common, but organized punishment was very rare. Serious wrongdoing could lead to the flight or ostracism of the perpetrator, but the banished person might return after a few days of hiding in the forest, where somebody had secretly brought him food. No case of permanent banishment is known.[150]

The Mbuti abhorred physical violence. They have been engaged by the villagers as scouts in wars with other tribes, but there is no evidence that the Mbuti have ever themselves taken any part in war beyond killing a few trespassers.[151]

The religious beliefs of the Mbuti varied considerably and lacked precision and dogmas. Different observers disagree on even the most fundamental aspects of Mbuti beliefs, such as the names and nature of their supernatural beings, whether they prayed and sacrificed, whether they performed witchcraft, and whether they believed in an afterlife. The villagers involved the Mbuti in their rites of passage in an attempt to exercise supernatural control over them. The Mbuti willingly participated in these rituals and pretended to share the beliefs of the villagers, which they could use to their own advantage. The villagers believed that the forest was full of malevolent spirits, and the Mbuti actively worked to reinforce such beliefs among the villagers in order to keep them out of the forest. When no villager was watching them, the Mbuti would ridicule the beliefs of the villagers and deliberately violate

---

150 Harako (1976, p. 89), Schebesta (1948, p. 332), Turnbull (1961, p. 102; 1965, pp. 26, 178; 1983, p. 44), Putnam (1950)

151 Schebesta (1948, p. 534), Turnbull (1965, p. 220), Putnam (1950)

the religious taboos of their villager hosts. When in the forest, the Mbuti had few religious rituals. It is uncertain whether the Mbuti honestly believed in any form of supernatural punishment, but we can say with reasonable confidence that religion was not a strong disciplining force in Mbuti life.[152]

Different Pygmy groups considered themselves as belonging to the same tribe, even if they spoke different languages (the languages of their respective village hosts). They did not allow trespassers. They were fearful of foreigners but also friendly toward them, and they were open to foreign cultural influences. They regarded the villages as hunting grounds and saw the villagers as 'animals' to exploit for their own advantage.[153]

Sexual morals were liberal among the Mbuti. Premarital intercourse was sanctioned in certain puberty rituals. Marital infidelity was not a rare occurrence, and divorce was easy. Polygamy occurred. They married after puberty with little or no ceremony. The average age of first marriage was 18 for boys, 16 for girls. A marriage was not considered fully established until the first child was born.[154]

Children were breastfed for at least one to three years. Birth spacing was three to four years. Birth control was obtained by a postpartum sex taboo and probably by other, unknown means. Abortion was known but rare. Infanticide was practiced after twin births. The fertility was low according to some accounts, higher according to others, but the total population size was stable. Suicide occurred.[155]

Children were treated with affection. They had considerable freedom, were rarely punished, and did not have to do hard work.[156]

The most important art forms were singing and dancing. The singing of the Mbuti and related Pygmy tribes is characterized as polyphonic. Everybody could join in or contribute in turns with possibly improvised harmonies in no predefined order and with no apparent hierarchy of voices. The result was a rich complexity of different voices to a common rhythm. The dances often formed mimicry of daily events. Erotic dances

152 Schebesta (1950, p. 14), Turnbull (1961, p. 75; 1965, p. 29)
153 Schebesta (1948, p. 512), Turnbull (1965, pp. 37, 82)
154 Cavalli-Sforza (1986, p. 37), Schebesta (1948, pp. 297, 359), Turnbull (1961; 1965, pp. 63, 122; 1983, p. 34)
155 Cavalli-Sforza (1986), Schebesta (1938, p. 112), Turnbull (1965, p. 128), Bailey (1989, p. 100)
156 Schebesta (1941, p. 118; 1948, p. 419), Turnbull (1961, p. 118)

were also common. There was no graphic art other than simple body painting, body adornment, and decoration of clothes.[157]

We can conclude that Mbuti culture was even more kungic than expected. All the cultural indicators point in the kungic direction. The further away they were from village influence, the more peaceful and egalitarian was their behavior.

7.14. Somali

The Somali live in the north-east of Africa at the Gulf of Aden. The climate is dry with temperatures ranging from below freezing point to above 40 °C. Droughts occur at irregular intervals. The environment is mostly semiarid plains and plateaus with little vegetation. Most of the population lived as nomads and herders, and milk was the staple food. There are only two permanent rivers in Somaliland. Agriculture with irrigation was practiced near the rivers. The coastal population lived off fishing, trade, and crafts. There was extensive trade with Arabia and with other parts of Africa.

The Somalis traveled around with camels and donkeys as pack animals. Lack of water was no big obstacle to traveling, because the camels could go long distances without water, and because the herders could drink camel milk on their travels. The sparse grazing resources for the animals set a limit to the size of troops that could travel together. The most important weapons were iron-headed spears. Horses were used in battle if available, but there were only few horses.[158]

Based on this information, we can expect the culture of the Somalis to be moderately regal. The relative ease of travel and the availability of metal weapons are factors that point in the regal direction. The occurrence of droughts at unpredictable intervals may also be a regal factor. However, the population density must necessarily be low, except near the coast and the rivers. This points in the kungic direction. The low-density nomadic population could not easily be incorporated into a larger political or military organization. The hot and dry climate also limited the fighting ability of the population.

---

157 Schebesta (1941, p. 243), Turnbull (1961, p. 76; 1983, p. 46), Arom (1983), Ichikawa (2004)
158 Lewis (1955, pp. 11, 75), Cassanelli (1982, p. 9)

Prior to European colonization, the political organization of the Somalis was fluid and ever changing. The political leader of a family, clan, or tribe was chosen partly by majority decisions of the adult men, partly by inheritance, and partly by power, influence, and age. Higher-level political structures were either absent or unstable. City-states and small sultanates were formed from time to time, but they often disintegrated again after some time. The political leader and the religious leader were seldom the same person. Arab immigrants had high prestige and considerable political and religious influence. Slavery was very common. However, slaves were reportedly treated well and many slaves were freed.[159]

War, feuds, raids, and plundering were frequent occurrences. The wars were of a relatively low scale with at most a few hundred deaths, but it is assumed that the territorial boundaries were ultimately defined by military strength.[160]

The religious belief of the Somalis was dominated by Islam due to a strong Arabic influence. The religion was a strong power, especially in moral and judicial matters. Legal matters were governed mainly by Islamic sharia law. Crimes were punished by the payment of a fine, for example 100 camels for homicide, which was paid partly by the offender and partly by his group. Other forms of punishment were rare.[161]

The literature gives no reliable information on fertility and population growth. Children were breastfed for two years, but were also given other food. Child mortality was high. Children were set to work from an early age. Children were well disciplined, but there are no reports about punishment of children. The marrying age varied from 12 to 20 for girls and 17 to 25 for boys. Child betrothal and forced marriages occurred but were not common. Men who could afford it practiced polygamy. The sexual morals were strict, except for the divorce rate, which was very high. Girls were infibulated to ensure chastity. Social contact between boys and girls was limited.[162]

Pictorial art was not very important because of the nomadic lifestyle of the majority of the population. Clothing and body adornment was

---

159 Cassanelli (1982), Lewis (1955, pp. 11, 99; 1961, p. 198), Luling (2002)
160 Lewis (1955), Haggenmacher (1876, pp. 11, 31)
161 Cassanelli (1982, p. 17), Lewis (1955, pp. 107, 140; 1961)
162 Cassanelli (1982, p. 13), Lewis (1955, p. 135; 1962), Luling (2002, p. 71), Haggenmacher (1876, p. 27)

simple, without excessive embellishments. Tools and objects were sometimes embellished with woodcarvings, showing simple figures and often repetitive geometric patterns. Dances were relatively simple, sometimes with a war theme. The most important art form was poetry, which was highly stylized and rule bound.[163]

The conclusion for the Somali is that the political organization is intermediate, whereas the justice system and the relatively mild punishments can be characterized as kungic. The religion, the sexual morals, and the widespread slavery are more on the regal side. The artistic expressions show mixed characteristics: the poetry in particular shows regal tendencies, while the clothing and body adornment look more kungic.

7.15. Warao

The Warao Indians lived mainly in the large delta of the Orinoco River in Venezuela, an area of approximately 18,000 km². The environment was swamps, mangroves, tropical rain forest, and a labyrinthine system of rivers. The staple food was the sago-like pith of the Moriche palm, supplemented by fish and small animals. Food was plentiful but seasonal. Food storing was practiced where necessary. In later times, the Warao built their huts on stilts for protection against floods. They went everywhere by canoe. Their canoes were capable of sailing along the coast of Venezuela and to the nearby island of Trinidad, where they went for trading. Traveling over land was virtually impossible due to the dense forests, lack of dry ground, and many rivers. They had no pottery and no metal tools, except in later times where modern tools were obtained by trade.[164]

Warao culture was adapted to a very specialized niche. No foreign tribe had the necessary skills to navigate the swampy delta, and much less to gather food and survive there. This gave the Warao good protection. Whenever the Warao were attacked by neighbor tribes, they found protection in the swamps and dense forests.

We would expect the Warao to be able to attack neighbor tribes because of the ability to travel far in their canoes, although they lacked

---

163 Cassanelli (1982, p. 267), Lewis (1955, p. 132), Luling (2002, p. 164), Puccioni (1936, p. 88)
164 Wilbert (1958, p. 272; 1993, p. 235), Wilbert and Layrisse (1980, p. 3)

efficient weapons and the population density was low. However, they would have little incentive to leave the area that fitted their specialized niche, and outsiders would have little reason to attack them. Hence, we can predict that the niche specialization gave the Warao a moderately kungic culture.

The Warao were peaceful people who detested physical aggression. They were rarely involved in war, but there is evidence that they were at war with neighbor Carib, who allegedly cannibalized the Warao or sold them as slaves. There was no war between Warao subtribes. Internal conflicts were dealt with by public hearings or ritualized competitions, and territorial transgressions were prevented by threats of black magic.[165]

The Warao were divided into subtribes. Each subtribe consisted of a number of bands, which were led by an elite of elders, patriarchs, or chiefs, possibly including women. Shamans belonged to the elite, and chiefs were often shamans. There was only a loose organization of the subtribe and no political organization above the level of the subtribe.[166]

The political leaders had little power to discipline or coerce people. Shamans did, however, have considerable power because of their contact with spirits who could cause fortune or misfortune. The Warao believed in a supernatural world populated by spirits or deified ancestors. People were rewarded in the afterlife if they had been skillful at their work. There was no dogmatic system of beliefs.[167]

The Warao traditionally avoided contact with other tribes, except for trade of necessary products. In modern times, however, they intermarried with foreigners, adopted Spanish names, migrated, and accepted foreign culture elements quite quickly.[168]

The fertility was very high. A woman who survived into old age would have almost nine children on average. It must be mentioned, however, that this fertility rate was measured at a time when modern medicine had reduced the—previously very high—mortality, and

---

165 Wilbert (1993, pp. 16, 244), Grohs-Paul (1979, p. 33), Heinen (1985, p. 14; 1994, p. 359), Kirchoff (1948, p. 878)
166 Wilbert (1993, p. 31), Wilbert and Layrisse (1980, p. 7)
167 Wilbert (1993, p. 74; 1996, p. 257), Wilbert and Layrisse (1980, p. 8), Heinen (1994, p. 358), Olsen (1996, p. 227)
168 Wilbert (1993, p. 93), Wilbert and Layrisse (1980, p. 180)

where a large fraction of the population had settled into an agricultural lifestyle. It is possible that infectious diseases stabilized the population size in earlier times. Children were breastfed for three to four years. Infanticide was practiced.[169]

Girls were usually married in their teens or early twenties; boys were a few years older when they married. A few girls were married as children. Polygyny occurred. Most marriages were within the same subtribe. Promiscuity was common in connection with certain rituals and dances. A man could have sex with the unmarried sister of his wife. Premarital trial unions were common. Divorce was rare for couples with several children.[170]

Children were treated well, and parents made toys for their children. Children spent most of their time playing, but they also helped with light work. Children were rarely scolded, and teenagers were not well disciplined.[171]

Some religious artifacts were decorated, while tools, canoes, and huts were rarely so. People wore necklaces and other adornment. The main forms of art were songs, dances, and narratives. Singers had much freedom for variation and improvisation. Typical narratives were fables, stories about magical transformations, and stories about the supernatural world. There were no narratives or songs exalting powerful leaders, gods, or battles, but there were tales about past invasions and a few fables told about jaguars that symbolized the enemy Carib. The main occasions for large social gatherings were religious ceremonies. Ceremonies for marriages and other rites of passage were not very elaborate.[172]

The conclusion for the Warao is that the level of war, political system, discipline, treatment of children, sexual behavior, and art are all in agreement with the predictions for a kungic culture. The religion was relatively kungic, though it did have a disciplining function. The fertility was atypically high.

---

169 Wilbert and Layrisse (1980, pp. 25; 61), Kirchoff (1948)
170 Wilbert (1958, p. 274; 1993, p. 33), Wilbert and Layrisse (1980), Heinen (1994, p. 358), Kirchoff (1948), Suarez (1968, p. 128)
171 Wilbert (1993, p. 40), Grohs-Paul (1979, p. 51), Kirchoff (1948)
172 Wilbert (1970, p. 28; 1993, p. 136; 1996, p. 25), Olsen (1996), Suarez (1968)

7.16. Yahgan

This section concerns the canoe Indians who lived in the southernmost part of South America, Tierra del Fuego, with the main focus on the Yahgan tribe. The climate was harsh and cold, with frequent storms. The territory of the canoe Indians was an archipelago of islands with high mountains and dense forests. The land was difficult to penetrate and provided only very little vegetable food. Therefore, the Indians lived as nomads traveling in bark canoes, eating mainly seafood. Their tools were made mainly of wood, bone, and shells. They had no metals and no pottery. As they had very few clothes, they always kept a fire for warmth.[173]

Figure 27. Yahgans 1896. Photo by Otto Nordenskjöld.
From Nordenskjöld (1898).[174]

Fishing in this environment could sustain only a low population density. The Indians could travel in their bark canoes from island to island but not in open sea. Large-scale war was probably impossible because of the low population density and because the bark canoes were too small and fragile. Small-scale clashes were more likely to have taken place. If we have to make a prediction based on the geography and technology alone, we will expect the Yahgan culture to be moderately kungic. The

---

173 Cooper (1917, p. 3; 1946, p. 81), Gusinde (1937, p. 403; 1974, p. 243), Lothrop (1928)
174 Public domain

niche culture, harsh climate, low population density, and absence of technology suitable for war are factors that point in the kungic direction. The ability to travel far and the risk of territorial conflict that come with a nomadic lifestyle are factors that point in the regal direction.

The Yahgan had no social organization beyond the family. Several families might live together, or meet for social and ceremonial occasions or to feast on a stranded whale, but there was no common leadership and no group organization. War was unknown, but small-scale conflicts and feuds were common. Revenge was not always deadly. There were no weapons other than the usual hunting tools. There was no organized policing. Punishment occurred in the form of revenge from the wronged party assisted by his or her family, for example beating for marital infidelity. In most cases, however, public condemnation was sufficient to deter undesired behavior.[175]

They had few religious activities other than rites of passage and a few ceremonies. Some sources say that they believed in a high god,[176] but other sources deny this.[177] Medicine men worked to cure diseases by means of self-induced trance or dreams, but they had little or no religious paraphernalia; they did not sacrifice, and they did not believe that supernatural beings could be influenced by prayer. They believed that diseases and death were punishments for wrongdoing. For this reason, we can say that the religion had at least some disciplining function.[178]

There were several tribes of canoe Indians with distinct tribal identities. They feared the neighboring Selk'nam, who lived on land rather than in canoes. They were surprisingly tolerant of Whites, considering that they had been treated very brutally by white seafarers, gold diggers, and settlers in the past.[179]

There are no reliable data on the fertility of the Yahgan, and the sources do not fully agree. Some sources tell that they had many children, but the mortality was also high. The population size may have been stable for hundreds or thousands of years. The population became

---

175　Cooper (1917, p. 174), Gusinde (1937, p. 887; 1974), Lothrop (1928, p. 160), Bridges (1949, p. 79), Hyades and Deniker (1891, p. 16)
176　Gusinde (1937, p. 1040), Lothrop (1928, p. 35)
177　Bridges and Lothrop (1950, p. 98)
178　Cooper (1946, p. 102), Gusinde (1937, p. 1393), Lothrop (1928, p. 160), Bridges and Lothrop (1950)
179　Gusinde (1937, p. 28; 1974, p. 578), Cooper (1917), Nordenskjöld (1898)

extinct after contact with Whites, mainly due to infectious diseases. Abortion and infanticide were known. Suicide was unknown.[180]

There are conflicting accounts of the sexual morals of the Yahgan. They tried to keep boys and girls apart from each other, but premarital sex was nevertheless common. Extramarital sex occurred. The marrying age was reportedly around 17 to 19 for boys and 15 to 16 for girls, but these figures may be inaccurate since they did not count their ages. Polygamy was rare in some areas, perhaps common in other areas. Authors disagree on whether divorce was frequent or rare.[181]

Children were treated with devotion. Weaning age was 3 years or older. Children were given miniature tools to play and practice with. Small children were never punished. Children were obedient and helped their parents with daily chores, but they were never forced to do hard work.[182]

Yahgan art was extremely simple. Body painting consisted of simple lines and dots without precision. There was little or no tattooing and body mutilation. Music consisted of just a few notes, sometimes only a single repeated note. There were no musical instruments other than percussion. Dances were imitations of animals. Legends were mostly fables, myths about culture heroes, and a few stories about revenge.[183]

The conclusion for the Yahgan is that the level of war, political system, discipline, treatment of children, and art are all in agreement with the predictions for a kungic culture. The religion was kungic, though it did have a disciplining function. The fertility was high.

7.17. Yanomamo

The Yanomamo lived in the Amazon rainforest in Brazil and Venezuela. They resided mainly in semi-permanent villages with 20 to 400 inhabitants. Slash and burn agriculture was their main source of food. They were capable of living a nomadic life as hunters and gatherers, and it is speculated that they might have done so in pre-Columbian times,

---

180 Gusinde (1937, p. 699; 1974, p. 437), Lothrop (1928, pp. 25, 163), Hyades and Deniker (1891, p. 189)
181 Cooper (1917, p. 164), Gusinde (1937, p. 633; 1974, p. 338), Lothrop (1928, p. 163), Bridges and Lothrop (1950, p. 106), Hyades and Deniker (1891, p. 377)
182 Gusinde (1937, p. 745), Bridges and Lothrop (1950, p. 107)
183 Gusinde (1937; 1966, p. 71; 1974), Lothrop (1928, p. 125)

since their main crop—plantains—was brought to the Americas by Europeans. They traveled almost exclusively on foot on narrow paths that were almost invisible to foreigners. Many trails were impassable in the wet season. They went barefoot and almost naked. They had little or no means for sailing on the rivers, and they could cross big rivers only with difficulty. They had no domestic animals besides dogs. Their main weapons were clubs and bows and arrows. Arrow tips were mostly made of wood or bone, poisoned with curare. Metal tools were obtained by barter.[184]

The cultivation of high-yield crops made room for large village populations, but the dependence on hunting for acquiring protein made it necessary that each village had sufficient hunting territory. This makes us predict that wars were likely, and the wars would be aggravated by the availability of strong arrow poison. On the other hand, the difficulty of traveling through the rain forest and the low overall population density are factors that reduced the opportunities for war. Considering these opposing factors, we will predict an intermediate level of regality.

Figure 28. Yanomami. Unknown photographer, 2007.[185]

---

184 Ferguson (1995, p. 61), Chagnon (1977, p. 18; 1992, p. 56), Peters, J. (1998), Smole (1976, p. 32), Valero and Biocca (1970)

185 CC-BY-SA3.0,https://commons.wikimedia.org/wiki/File:Yanomami_en_el_Estado_ Amazonas_(21).jpg

The Yanomamo have a reputation for fierceness in war, but there has been considerable controversy among anthropologists over how fierce the Yanomamo actually were, why they fought, and whether their raids can be classified as war. It is undisputed, however, that violence was a frequent cause of death in some areas. The level of fighting was much higher in the lowlands, where villages had more inhabitants, than in the highlands.[186]

In their own understanding, the Yanomamo conducted raids against other villages for purposes of retaliation and to capture women. They were well aware that the enemy might move away as a consequence of a conflict, but they rarely mentioned the desire to make the enemy move away as a motivation for attack.[187] An ethologist has suggested that we make a distinction between the proximate cause of Yanomamo warfare, which is the personal motivation, and the ultimate cause of the warfare, which is connected to the evolutionary function of securing a sufficiently large hunting territory.[188] Since Yanomamo conflicts had territorial consequences, we will classify them with regard to regality theory as wars.

The Yanomamo had no political organization above the level of the village, and the village headman had very little authority. There was no organized system of justice and punishment other than retaliation. Even rape and murder could go unpunished if there were no relatives to retaliate on behalf of the victim. Personal conflicts were handled by fighting duels. These duels were formalized and rule-governed. Afterwards, the issue was regarded as settled, regardless of the outcome of the duel. However, a duel might sometimes escalate into collective fighting or even war, especially if somebody died.[189]

The religious beliefs of the Yanomamo involved a close connection with animals and spirits. Many men were shamans. They took strong hallucinogenic drugs in order to get in contact with, or even become, spirits. They manipulated the spirits for purposes of curing diseases and

---

186 Ferguson (1995), Chagnon (1992, p. 105), Good (1987)
187 Ferguson (1995, p. 7), Chagnon (1979, 1992, p. 113), Peters, J. (1998, pp. 24, 214), Valero and Biocca (1970, p. 31)
188 Eibl-Eibesfeldt (1989, p. 108)
189 Chagnon (1977, p. 118), Cocco (1972, p. 366), Lizot (1976, p. 83)

harming their enemies. They had no supernatural rulers and only weak concepts of supernatural punishment after death.[190]

The fertility was controlled by breastfeeding, abortion, infanticide, and magical anticonception means. Birth spacing was two to five years. The overall fertility was four to eight children per woman. Extramarital sex occurred when detection could be avoided. It was common for a man to share his wife with his brothers. Rape was common.[191]

Most girls married soon after puberty. Girls might be married before puberty, but they did not live with their husbands until after first menstruation. Men married at age 18 to 30. They often married for economic and strategic reasons rather than romantic reasons. Girls had little self-determination about whom to marry. Polygyny and polyandry occurred. There was no marriage ceremony. Wife beating was normal. Divorce was common.[192]

Children were rarely disciplined, but boys were taught to be fierce and to retaliate. Children were given miniature tools and other toys. Girls were obliged to do work from a much younger age than boys.[193]

The Yanomamo produced little art. Body painting was made as much for magical reasons as for decoration. Different colors and patterns were used for war paint, hunting, and shamanism. Some patterns symbolized or imitated various animals. Piercings in ears, lips, and the nasal septum were decorated with long sticks or feathers. Arrow tips were often decorated, while clay pots were not. Dancing and singing were common. There were few or no music instruments. Myths and stories did not glorify war.[194]

The conclusion for the Yanomamo is that the level of war is intermediate, while the political organization, justice, discipline, religion, sexuality, and art all tend in the kungic direction. The strong male dominance may be interpreted as a regal indicator.

---

190 Chagnon (1977, p. 44), Smole (1976, p. 23), Zerries (1964, p. 237)
191 Chagnon (1977, p. 40), Peters, J. (1998, p. 119), Smole (1976), Cocco (1972), Lizot (1976, p. 22)
192 Chagnon (1977, p. 14; 1992, p. 145), Peters, J. (1998, p. 111), Smole (1976, p. 75), Lizot (1976, p. 27)
193 Chagnon (1992, p. 145), Peters, J. (1998, p. 131), Smole (1976, p. 74), Cocco (1972, p. 292), Lizot (1976, p. 102)
194 Chagnon (1977, p. 110), Valero and Biocca (1970, p. 179), Cocco (1972, p. 124), Zerries (1964, p. 103), Wilbert and Simoneau (1990, p. 110)

7.18. Yi (Lolo, Nuosu)

The Yi people live in the large mountainous region of southwestern China. The present study is focused mainly on the subgroup living in the Liangshan area and the period prior to the rise of communism in China.

The climate is temperate, but winters are hard. The subsistence was based mainly on slash and burn agriculture and animal husbandry. Transportation was difficult because of the steep mountains. There were no roads except narrow footpaths. Roads built by Han Chinese invaders were systematically destroyed to prevent invasion. The Yi used horses for transportation and in war, but many paths were difficult or impossible to travel on horseback. Most traveling was done by walking barefoot. There were many rivers, but few were navigable. The weapons included bows and arrows, iron-headed spears and tridents, and, in later times, primitive firearms.[195]

Attempting to predict a regality level based on this information, we get a somewhat mixed result. The rivers produce fertile valleys between the mountains and, although the agriculture was not intensive, it could sustain a considerable population. The efficient food production, the possibility of storing and transporting food, the large territory, and the availability of horses and metal weapons are all factors that make war likely. On the other hand, the harsh environment and the difficulty of traveling are factors that point in the kungic direction. The fertile valleys are separated by large mountains, steep cliffs, and torrential rivers, so that the average population density was low. The overall expectation is that the regality level will be intermediate.

The Yi in the Liangshan area had a persistent caste system. The ruling caste, known as the Black Yi, constituted about 10% of the population. The rest of the population belonged to the slave caste, known as the White Yi. The White Yi were of mainly Han Chinese descent, while the Black Yi were related to the Tibetans. These two castes were further subdivided into social classes. Newly captured slaves were lowest in rank. Over the generations, slaves could work their way up the hierarchy and become partly independent and even own slaves themselves. But

---

195 LeBar, Hickey and Musgrave (1964, p. 1ff), Dessaint (1980), D'Ollone (1912), Ko (1949), Pollard (1921), Yueh-Hua (1961)

the line between Black and White could never be crossed. Interbreeding between the two castes was prevented by all means. The caste system was dissolved centuries ago among other Yi groups living in areas with a stronger Han Chinese influence.[196]

The political organization of the Yi was not as regal as one might expect for a society based on slavery. Most decisions were taken by public deliberation in which even members of the slave caste had a say, and the leaders of villages and clans were chosen based partly on heritage and partly on merit. In some periods and in some areas, there were political leaders above the clan level, but these had limited influence in times of peace. For centuries or even millennia, the Chinese dynasties tried to control the Yi territory and used Yi chiefs as suzerains. But the Yi successfully rebelled time and again, so that, in effect, the Yi were mainly autonomous until the establishment of the People's Republic of China.[197]

Political organization above the clan level—if any—was not strong enough to prevent the almost incessant feuds, raids, and attacks between enemy clans or ethnic groups. The inter-clan conflicts were, however, of a relatively low scale: certain rules were obeyed, and the opposing parties generally agreed to end a battle episode after a few hours when a winner had been found. Battles with more distant enemies, and in particular with the Han Chinese, were much more violent and merciless. Strangers who entered Yi territory without protection were habitually killed or enslaved.[198]

The available literature contains very little evidence of policing or punishment. Most crimes were punished by requiring that the offender pay compensation to the victim. Newly captured slaves were brutally tortured, but there is no other evidence of corporal punishment. Serious crimes were dealt with by persuading the offender to commit suicide. If he refused to do so, he would be banished but rarely killed.[199]

Religious rituals, linked to animism and ancestor worship, were controlled by priests and shamans. The priests (bimo) always belonged

---

196 LeBar, Hickey and Musgrave (1964), Dessaint (1980), Ko (1949), Harrell (2001), Hui (2001)

197 LeBar, Hickey and Musgrave (1964), Dessaint (1980), D'Ollone (1912), Ko (1949), Harrell (2001), Ayi (2001), Hill (2001), Jingzhong (2001), Shimei and Erzi (2001)

198 LeBar, Hickey and Musgrave (1964), Dessaint (1980), Pollard (1921), Yueh-Hua (1961), Jingzhong (2001), Liétard (1913)

199 Hill and Diehl (2001), Shimei and Erzi (2001)

to the White slave caste. Their skills were transmitted mainly from father to son. Major political and judicial decisions required consultation of a priest and performance of certain rituals. The disciplining function of religion was probably limited. A few sources talk about supernatural punishments, but there is no agreement between the different sources regarding the kind of punishments or even whether they took place before or after death. Beliefs in supernatural punishments, if any, may be due to Chinese influence or Christian missionaries.[200]

The Yi had a written language, which was mastered by the priests and only rarely by anybody else. Writing was used only for preserving religious rituals and formulas, not for trade or political administration.[201] We can assume that the priests had considerable influence. Whether they used this influence mostly in the interest of their own caste or mostly in the interest of the Black leaders, we do not know.

There is no reliable information about the fertility or birth rate of the Yi. Children were breastfed for four to five years, but they were also given other food.[202] Suicide was customarily committed for different reasons relating to war, conflict, peacemaking, crime, infectious diseases, and love.[203]

The marrying age ranged usually from 9 to 21 for both girls and boys, though children as young as 4 or 5 could be married. Married girls stayed home with their parents until their first pregnancy. During this period, they were free to have sexual relations with anybody they liked. For this reason, the firstborn child often had a father who was not the woman's husband, but the latter had to accept the child as his own. Divorce was more common in the period before first pregnancy than after. Polygamy was rare.[204]

The sources tell us very little about the life of children and nothing about the disciplining of children. Children were treated well, and boys were taught to be brave. Parents did not make toys for their children, but various games are known. Girls helped spinning hemp from age 4 and boys helped tending animals from age 7 to 8.[205]

---

200 Ayi (2001), Shimei and Erzi (2001), Liétard (1913)
201 Ayi (2001)
202 Yueh-Hua (1961)
203 Dessaint (1980), Ko (1949), Pollard (1921)
204 LeBar, Hickey and Musgrave (1964), Dessaint (1980), Ko (1949), Yueh-Hua (1961), Hui (2001), Liétard (1913)
205 Liétard (1913), Yu (2001)

Clothing, tools, and houses were often embellished. The Yi were fond of colorful clothes as well as gold and silver. Before firearms made face-to-face battle superfluous, the inter-clan feuds were not only contests of strength, but also opportunities to compete on fancy clothing or flute playing. The sources show no evidence of the highly repetitive geometric patterns typical of regal art. Singing and dancing was common. Song texts were mostly about everyday topics, but there were also war songs.[206]

Evaluating the Yi culture, it is surprising how a slave system could persist for centuries when the political and religious means of disciplining the population were so weak. The caste system that allowed slavery may have been supported by ideology or religion, but the religion was managed by priests belonging to the slave caste. No slave rebellion has been reported. The agricultural production was also fully in the hands of the slave caste.[207]

The caste system and the pervasive slavery are the strongest indicators of regality in the Yi culture. The full-scale wars with the Han Chinese are also regal indicators, while the frequent inter-clan feuds and raids with limited casualties can hardly be considered regal indicators.

The political administration, the discipline, the religion, and the sexual morals appear to be mostly kungic. The use of body adornment in conflicts is a regal sign, but the adornment is hardly more elaborate than in other cultures. The almost complete absence of praise for gods or rulers in art and songs is an indication of a fairly kungic art.

Conclusion

We have now looked at eighteen very different societies of the past and compared their levels of internal and external conflict, their political systems, systems of justice and punishment, religions, group identities, fertility, occurrence of suicide, sexual morals, marrying ages, frequency of divorce, attitudes toward child labor, and art styles with predictions based on regality theory. The principles behind the predictions were explained in the beginning of this chapter. A statistical analysis of the results follows in chapter 8.3.

---

206 LeBar, Hickey and Musgrave (1964), Dessaint (1980), Pollard (1921), Liétard (1913)
207 Yueh-Hua (1961)

8. Statistical Testing of Regality Theory

8.1. Problems of cross-cultural statistics

We need to compare different cultures when we want to test the predictions of regality theory. However, the statistical comparison of different cultures is full of problems that we have to deal with. The first problem is the availability of data. Many data sources are available for modern cultures, but few for ancient cultures. The quality of the data is a matter of serious concern. Data collected from different cultures may not be fully comparable because of language problems, differences in definitions, or different frames of reference. There is a general problem with quantifying culture, because it is not always meaningful to put numbers on things like ideas, beliefs, emotions, or understandings. How can we know if concepts such as prestige or patriotism have the same meaning in different cultures?

The available data for contemporary modern western countries are generally of a satisfactory quality because the methods of data collection have been refined and standardized to a certain degree. But here we encounter another problem. Modern cultures are too similar. We need variation in the measured variables in order to get useful statistical results. Any similarity that we find between two countries may simply be due to cultural diffusion rather than to some causal factor that influences both countries in the same way.

The best way to overcome the problem of cultural similarity, known as Galton's problem, is to look at the more isolated cultures of the past.

© 2017 Agner Fog, CC BY 4.0 https://doi.org/10.11647/OBP.0128.08

No culture is completely isolated, but many of the non-modern and non-industrial cultures of the past were sufficiently autonomous to make cross-cultural comparisons useful. The Standard Cross-Cultural Sample, mentioned in chapter 8.2, is a database of cultures selected specifically with the purpose of minimizing Galton's problem. The data in this database have been obtained from the ethnographic literature.

Unfortunately, the quality of the data in the Standard Cross-Cultural Sample is often problematic. Some of the variables are poorly defined or interpreted differently for different cultures. Some variables refer to concepts that make sense only in particular cultural contexts. Data are coded in the database as missing when a variable is irrelevant for a particular culture or the information is not available. In the course of the present study, it was discovered that data were often missing for non-random reasons. Special care was needed to reduce the systematic bias caused by such missing data.

Many of the problems with data quality and unclear interpretation of data are best dealt with by qualitative methods. A qualitative study can look into the life situation of the informants to put the information into a meaningful context. It allows us to discover new phenomena or mechanisms that were not anticipated when the study was planned. A qualitative study can give us a better understanding not only of what people do but also of the motives behind. For example, it is important for regality theory to study not only whether people fight but also why they fight.

It has been suggested that a combination of qualitative and quantitative methods is a particularly fruitful way of doing cross-cultural research.[1] We shall do this here with a subsample of the Standard Cross-Cultural Sample. Relevant information is gathered for each culture in the subsample by studying the original ethnographic literature (chapter 7). The qualitative study assures that we can gather meaningful information that actually reflects what we want to measure. The advantage of this mixed method is that we get more reliable data and that we are able to better understand the context and mechanisms that lie behind these data. The disadvantage is that we have fewer cultures

---

1   Harding and Seefeldt (2013)

to compare, because it requires more work to gather the information about each culture.

A study with a large data set (large N) makes it possible to construct bigger statistical models with more independent variables and still have enough statistical power to get significant results; a smaller data set (small N) can support only a smaller model with fewer independent variables. Each method has its advantages and disadvantages. The present study uses several different statistical methods and data sources, and we must keep in mind the strengths and weaknesses of each method when interpreting the results. The following statistical studies are included here:

- A statistical study based on George Murdock and Douglas White's Standard Cross-Cultural Sample (SCCS), which contains data from 186 non-industrial societies around the world. This study includes an exploratory factor analysis to identify the most relevant variables, a structural equation model to analyze the causal network in the proposed model, and a multiple correlation analysis of various variables against the central variables in the structural equation model.

- A statistical test based on the societies that were analyzed in chapter 7. This is a subsample of the SCSS. The results of the detailed analyses of these societies are compared with the predictions of regality theory.

- A study of individuals in contemporary cultures using data from the Survey of World Views Project, based on a survey of data from 8,883 persons in 33 countries. This study includes a factor analysis and various correlation analyses based on relevant factors that emerge from the factor analysis.

- A comparison with previously published statistical studies of social, psychological, and cultural effects of collective danger.

- Various tests based on previous studies of tight versus loose cultures and other cultural variables presented in chapter 3.8.

We are using these methods to test the predictions of regality theory, explained in chapter 7. No other known theory generates the same set of predictions.

8.2. Ancient cultures, large sample

Regality theory predicts effects both at the individual level and at the level of whole societies. The predictions of social-level effects should preferably be tested on distinct sociocultural groups. We prefer to compare different societies that have as little connection with each other as possible in order to avoid spurious correlations due to cultural diffusion or common descent (Galton's problem). Since almost all contemporary cultures are heavily influenced by modern western culture, we have used ethnographic data from non-industrial cultures of the past that were less influenced by global trends. The available archaeological data are insufficient for statistical testing, so we have used data mainly from ethnographic studies of non-industrial cultures.

Description of the data

Murdock and White's Standard Cross-Cultural Sample (SCCS) is a database covering more than a thousand variables recorded for 186 non-industrial societies around the world.[2] These societies are intended to form a representative sample of world cultures studied at a time when cultural independence was higher than today. The data are based on ethnographic records, many of which were collected by early explorers in the nineteenth and early twentieth centuries.

The variables include information about subsistence ecology, intergroup relations, political organization, culture, beliefs, and child—rearing practices.[3] Relevant variables from the SCCS were used in this study. Statistical calculations were done in the R programming language.[4]

The quality of the data in the SCCS is far from perfect, but it is the best available. The data have been collected from the original ethnographic literature and coded into a database. This process involves many potential sources of error, despite elaborate precautions. The native informants that were interviewed, the anthropologists or explorers

---

2   "Standard Cross-Cultural Sample". *World Cultures*, 2008, http://www.worldcultures.org/

3   Murdock and White (1969), Murdock and White (2006), http://repositories.cdlib.org/imbs/socdyn/wp/Standard_Cross-Cultural_Sample

4   Dow (2003)

conducting the field studies, the coders who interpreted the original literature and coded it into predefined categories for the database, and the database designers who defined the variables and categories to include in the database—all of these people may have influenced the data by their own agendas, predilections, selectivity, taboos, and language problems. Most of the field studies were conducted many years ago, often by people without adequate scientific training, such as explorers and missionaries.

Re-reading some of the original ethnographic literature revealed several flaws in the database, mainly due to imprecise definitions of the categories in soft areas such as religion, morals, and psychology. The data in the SCCS appear to be severely influenced by the observers in the area of religion; some of the psychological variables depend on psychoanalytical interpretations; and the reliability of the data is questionable in taboo-ridden areas such as sexual behavior.[5]

Many data values in the SCCS are coded as missing, and it was suspected that data might be missing for non-random reasons. For some variables, more than half of the data values were missing. There is reason to suspect that the coders tended to err on the side of caution and entered the code for 'data not available' rather than 'trait absent' when a particular trait was not mentioned in the ethnographic literature for a particular society. This leads to a systematic bias, so that data are more likely to be coded as missing when a trait is absent than when it is present. This suspicion was tested statistically by correlating missingness for each of the variables against the factors that emerge from the factor analysis described below. This correlation was highly significant ($p < 0.001$) for several variables. It is quite understandable, for example, that the value for 'leadership during battle' is coded as 'data not available' if there is no battle, but the statistical models do not work correctly if data are missing for non-random reasons. Hence, it is better to replace 'data not available' with no 'leadership during battle' here. Similar replacements were made for the variables 'despotic bias in conflict resolution' and 'interpersonal violence'.

---

5   Broude and Greene (1976)

Exploratory factor analysis

Regality theory predicts that many different variables are related to regality. If these variables are sufficiently strongly correlated, then they might form a single factor in a factor analysis. In other words, we will expect cultural indicators of regality, such as hierarchy and discipline, to load on the same factor as war and other indicators of collective danger. Collective danger and individual danger have different effects on regality but similar effects on r/K life history strategy, as explained in chapter 3.3. It will be interesting to see whether these turn out as two different factors or as a single factor.

The database includes more than a thousand variables for each society, but not all variables were relevant for the present study. Factor analysis requires that the number of variables must be less than the number of societies in order to avoid the covariance matrix becoming singular. It was therefore necessary to exclude a large number of variables and retain only those considered most relevant to the theory. Variables that had little or no relevance to the purpose of the current analysis, such as language and geographic region, were excluded. Also excluded were: categorical variables that could not be made ordinal, variables where more than half of the values were missing, and variables with low variance or high uniqueness. Unfortunately, many of the variables that are relevant to regality theory had to be excluded from the factor analysis because they had too many missing values, especially psychological variables and variables relating to childrearing.

Where a group of variables all relate to the same subject area, either the redundant variables were excluded or closely related variables were combined into one. After this reduction, ninety-one variables remained for the factor analysis (see table 8). These variables were normalized to unit variance.

Missing data was a big problem because the factor analysis model has no standard way of dealing with missing data. Missing values were replaced with appropriate values for three of the variables in the manner explained above. The factor analysis was performed twice on the same data set, using two different methods for dealing with the remaining missing data: (1) replacement with the mean, and (2) multiple imputation by the hot-deck method with population density as auxiliary variable.[6] The second method gave higher variance and poorer correlation than the first method on those variables that had many missing values. The results of the factor analyses are shown in tables 6 and 7 respectively for these two methods.

Factor interpretation	% variance explained
Political complexity and population density	7.9
War	5.2
Other conflict and violence	4.9
Agriculture	4.1
Urbanization	2.7
Animal husbandry	2.6
Fishing	2.1
Gathering	1.9

Table 6. Factor analysis of sociocultural variables in 186 societies. Missing data values were replaced by the mean.

The number of factors was chosen to be eight based on a scree plot. Oblique rotation (Promax) was used in order to allow factors to be correlated. Detailed factor loadings are shown in table 8.

Factor interpretation	Occurrences in 25 runs	% variance explained	Standard deviation between runs
Political complexity and population density	25	9.8	0.9
Climate	25	5.1	0.3
War, conflict, and violence	25	4.9	0.3
Animal husbandry	25	3.3	0.7
Gathering	24	2.0	0.3
Fishing	23	2.8	0.9
Urbanization	21	2.8	0.1
Hunting	19	2.9	1.2
Agriculture	11	3.6	0.8
Uninterpreted	2	3.3	0.0

Table 7. Factor analysis of sociocultural variables in 186 societies. Same data as in table 6, but missing data values were replaced by multiple imputation in 25 runs.

---

6   Reilly (1993)

Results

The first factor analysis (Table 6) shows that political complexity is highly correlated with population density, and these variables combine to form the strongest factor. The second factor represents intergroup conflict as well as its correlates. This factor has high loadings on variables related to war, army, and bellicosity (Table 8). The third factor has high loadings on both internal and external conflicts as well as individual violence. In other words, we have two factors that relate to conflict. These two conflict factors are both positively related to polygamy. Other indicators of life history strategy show only weak and inconsistent correlations with the two conflict factors.

It is interesting that the second factor (the war factor) has its highest loadings on prestige and rewards for warriors and cultural valuation of war (Table 8). This confirms the prediction of regality theory that warriors are rewarded by their group. Other theories of war rely on punishment of defectors, while rewards are too costly to be explained by traditional evolutionary theories (see chapter 2.2).

The two conflict factors merge into one in the second factor analysis (Table 7). The difference between the results of the two factor analysis methods is a mathematical artifact. The method used in the second factor analysis has a higher tendency to form factors around variables with few missing values. This is seen in the formation of a factor of climate variables, which have no missing values. While the first method probably gives more accurate results, the second method is useful for estimating the inaccuracy due to missing data.

It is noteworthy that the war factor is always distinct from population density and political complexity. This means that the level of war is not simply determined by population factors. The prediction that internal and external conflicts have different psychological and cultural effects gets weak support from the first factor analysis and no support from the second. We must conclude that the factor analyses do not give a clear answer to the difference between internal and external conflicts, due to mathematical problems and poor data quality.

Comparison with other studies

A similar factor analysis made earlier by Elbert Russell found a war factor that correlates clearly with social hierarchy, slavery, polygyny, achievement, and crime.[7] Interestingly, the war factor in Russell's study is also significantly correlated with a number of variables relating to sexual restrictiveness, tough treatment of children, and early socialization. The correlation with religious variables is weak. An interesting finding in Russell's study is that people in warlike and hierarchical societies show a high level of boasting and display of wealth. The status competition and striving for wealth may lead to increased levels of theft and other crimes and conflicts.[8] This finding may contribute to our understanding of the connection between external and internal conflicts. Another possible explanation is that, in warlike societies, boys are socialized for aggression.[9] Other studies have found that the level of intergroup conflict is positively correlated with harsh and punitive treatment of children and socialization for aggression[10] and a moralizing religion.[11]

A comprehensive meta-analysis of wars throughout history has found a number of correlations of war with large civilizations and empires, weapon technology, grasslands, social stratification, slavery, authoritarianism, boastfulness, crime, punishment, discipline, sexual repression, games of chance and strategy, and artistic creativity.[12]

Some of the predicted correlations have also been observed in modern cultures. Pippa Norris and Ronald Inglehart find that existential insecurity is linked with religiosity and high fertility,[13] Nicholas Carnagey and Craig Anderson find increased pro-war attitudes,[14] and Richard Sipes finds a link with warlike sports.[15]

A caveat is in place here when interpreting factor analysis results. The factors and factor loadings are quite sensitive to the design of the factor analysis and, in particular, to the choice of variables included in the analysis. Other cross-cultural studies have found only weak correlations with a war factor, or no war factor at all.[16]

---

7   Russell (1972)
8   Russell (1972), Stewart (1971)
9   Ember and Ember (1994)
10   Ember and Ember (1994)
11   Roes and Raymond (2003)
12   Eckhardt (1992)
13   Norris and Inglehart (2011)
14   Carnagey and Anderson (2007)
15   Sipes (1975)
16   Rummel (1972), Sawyer and LeVine (1966)

Variable number	Variable name	F1	F2	F3	F4	F5	F6	F7	F8
V64, V156	Population density	0.30			0.50	0.20			0.15
V61	Fixity of settlement	0.23			0.67		-0.25	0.13	0.20
V62	Compactness of settlement	0.30	-0.15		-0.18				-0.11
V63	Community size					0.91			
V66	Large or impressive structures	0.39			0.25		-0.18	0.19	
V149	Writing and records	0.66		-0.20		0.11	0.11	-0.11	
V152	Urbanization			-0.13		0.93			
V203	Dependence on gathering		-0.11		-0.44		-0.14	-0.21	-0.92
V204	Dependence on hunting	-0.16			-0.74		-0.38	-0.27	
V205	Dependence on fishing	-0.11			-0.16		-0.21	0.94	0.22
V206	Dependence on animal husbandry				-0.17		0.98		0.1
V207	Dependence on agriculture	0.19			0.92		-0.29	-0.27	0.27
V677	Migration	-0.31		0.26				0.13	0.16
V732	Importance of trade in subsistence		0.13	-0.10					
V1265	Occurrence of famine		0.22					0.11	
V1267	Severity of famine	0.23			-0.12				
V1684	Threat of weather or pest disasters						0.14	0.17	
V1685	Chronic resource problems	-0.12			-0.11	0.15		0.15	
V1260	Total pathogen stress			0.21	0.42		-0.25		
V854	Niche temperature			0.21	0.42		-0.24		-0.24
V855	Niche rainfall	-0.17			0.41		-0.3	0.19	0.11
V921	Agricultural potential	0.19			0.30	0.13	-0.20	-0.13	-0.20
V1122	Total population	0.58			0.22		0.14	-0.13	
V157	Political integration	0.91							
V1132	Political integration, state	0.41	-0.12	0.15		0.12			
V158	Social stratification	0.79							
V72	Intercommunity marriage	0.10	-0.17	0.15		-0.39	0.12		
V861, V79	Polygamy	-0.26	0.37	0.38				-0.20	
V82	Trend in autonomy	0.51	-0.20						
V83	Levels of sovereignty	0.87			0.14	-0.11			
V237	Jurisdictional hierarchy	0.92							
V270	Class stratification	0.77							
V272	Caste stratification	0.21					0.22	-0.15	
V274	Type of slavery	0.11		0.22			0.22		0.12
V920	Proportion of slaves		0.14		0.12				
V79, V1133	Polygamy, maximum harem size	0.13	0.22	0.28	-0.31		-0.26		-0.12

V1134	Despotic bias in conflict resolution	0.41		0.32				-0.20	
V1743	Sanctions	0.18	-0.17	0.28		0.12			0.13
V1650, V774, V892, V893, V670	Frequency of external warfare	0.14	0.24	0.39	-0.12				
V892	Frequency of external war — Attacking	0.16	0.56	0.13	-0.19				
V893	Frequency of external war — Being attacked		0.46					-0.13	
V668	Wives taken from hostile groups			0.34	-0.12	0.10		0.14	
V894	Form of military mobilization	0.27	0.53						
V902	Leadership during battle	0.40	0.45					0.15	
V903	Prestige, soldiers or warriors	-0.13	0.82						
V905	Rewards for killing enemy	-0.22	0.73		0.17				
V907	Value of war		0.72						
V908	Military success		0.37					0.10	
V909	Subjugation of territory or people	0.61							
V1654	Pacification	-0.23	-0.23						
V1649, V1748, V773, V891	Frequency of internal warfare	0.12	0.17	0.34		-0.11	0.13		
V666	Interpersonal violence	-0.15	0.51						
V906	Expect violence to solve problems		0.54						
V1776	Intraethnic violence		-0.16	0.42			0.16		
V1665	Homicide	0.17		0.56					
V1666	Assault		0.13	0.63		-0.15			
V1667	Theft	0.16		0.47		-0.18			
V1676	Socially organized assault			0.36					
V1677	Socially organized theft			0.40			-0.14		
V1678	Socially organized trespass			0.23					0.20
V1721	Rich people		-0.13	0.31		0.13			0.20
V1726	Communality of land	-0.28	-0.16	0.40	-0.13				
V24, V25	Bodily restrictiveness in infancy	0.12	-0.16					-0.16	
V31	Infant crying response			0.24					-0.10
V33	Childhood pain infliction			0.26				-0.12	0.10
V41	Autonomy — encouragement in childhood			0.17				0.14	0.15
V43	Covering genitals — age	0.18		-0.10	-0.11				0.30

Variable number	Variable name	F1	F2	F3	F4	F5	F6	F7	F8
V44, V45	Weaning age	-0.11	-0.10	0.19				-0.11	0.13
V53, V54	Role of father, infancy/ early childhood		-0.25						
V242	Segregation of adolescent boys	-0.18		0.40					
V293	Duration of early childhood		-0.14	0.24					
V831, V832	Differentiation of adolescence from childhood			0.12				0.13	
V453, V454, V455, V456	Corporal punishment of boys/girls		0.35			0.18			
V667	Rape			0.30		0.12			
V34	Postpartum sex taboo			0.21	0.13			-0.14	
V671	Menstrual taboos			0.28		0.20			
V672	Male avoidance of female sexuality		-0.18	0.29			0.11		
V827, V828	Sexual expression in adolescent boys/girls	0.20		0.26	-0.17				-0.13
V829, V830	Lack of sexual restraint in adolescent boys/ girls								
V740	Marriage arrangements (female)	0.15		0.28			0.20		-0.11
V864	Rooming arrangement for wives		0.28	0.27	0.27	-0.14	0.11		0.11
V868	Multiple wives for leaders, headman, chiefs		0.14	0.37		-0.20			
V664	Ideology of male toughness	-0.11		0.32					
V657	Flexible marriage mores		-0.11	0.16	0.19		-0.17	0.12	
V661	Female political participation		-0.26	0.21			-0.17	0.16	-0.12
V665	Male segregation			0.41		0.15			
V238	High gods	0.13	-0.13				0.33	-0.13	0.10
V713	Classical religion	-0.22							
V529, V530	Adolescent initiation ceremonies	-0.13	0.12	0.33	-0.11			0.22	-0.25
V1188	Evil eye	0.14		0.12			0.44		0.11
V1694, V1695	Scarification	-0.11		0.19					
	Total factor loading, square sum[*]	7.20	4.72	4.49	3.71	2.42	2.40	1.91	1.74

Table 8. Factor loadings for factor analysis of table 6.
*Values below 0.1 are not shown in the table because they are
insignificant, but included in the square sum.

Structural equation model

The purpose of a structural equation model is to study latent variables that cannot be observed directly, but only through their causes and effects. The theoretical constructs 'regality' and 'life history strategy' are latent variables that we cannot observe or measure directly.

Figure 29 shows a structural equation model based on the same data as used above in the factor analysis. Several indicators related to intergroup war are combined into a war variable. The regality dimension is inserted as a latent variable based on the war variable and a number of cultural indicators according to the theory. The r versus K life history strategy dimension is also inserted as a latent variable based on internal and external war as well as other dangers and cultural indicators of life history strategy. 'Sexual morals' is included as another latent variable, because sexual behavior could rarely be observed directly. The influences of cultural regality and life history strategy on sexual morals are included in the model.

Missing values are replaced with appropriate values for those variables where missingness correlates significantly with the factors in the exploratory factor analysis and where the reason for data missing is clear, in the same way as for the factor analysis. Missing values are replaced by the mean for the remaining variables.

Relevant variables according to the theory are included, but variables that fail to reach significance at the 0.05 level are removed. Unfortunately, most of the variables used by Robert Quinlan as indicators of parental investment fail to reach significance in the present model.[17] A few variables that relate to parental investment remain significant, while only one variable related to art and three variables related to religion remain significant.

The removal of non-significant variables may have caused selection bias resulting in false significance. Ideally, the confirmatory test on the structural equation model should not use the same data set as the exploratory factor analysis. Unfortunately, no extra data set is available, since the number of relatively independent cultures is limited. This is a general problem in cross-cultural research.

The model is analyzed with the R package named 'lavaan'.[18]

---

17   Quinlan (2007)

18   "Lavaan: Latent Variable Analysis". The Comprehensive R Archive Network. http://cran.r-project.org/web/packages/lavaan/index.html

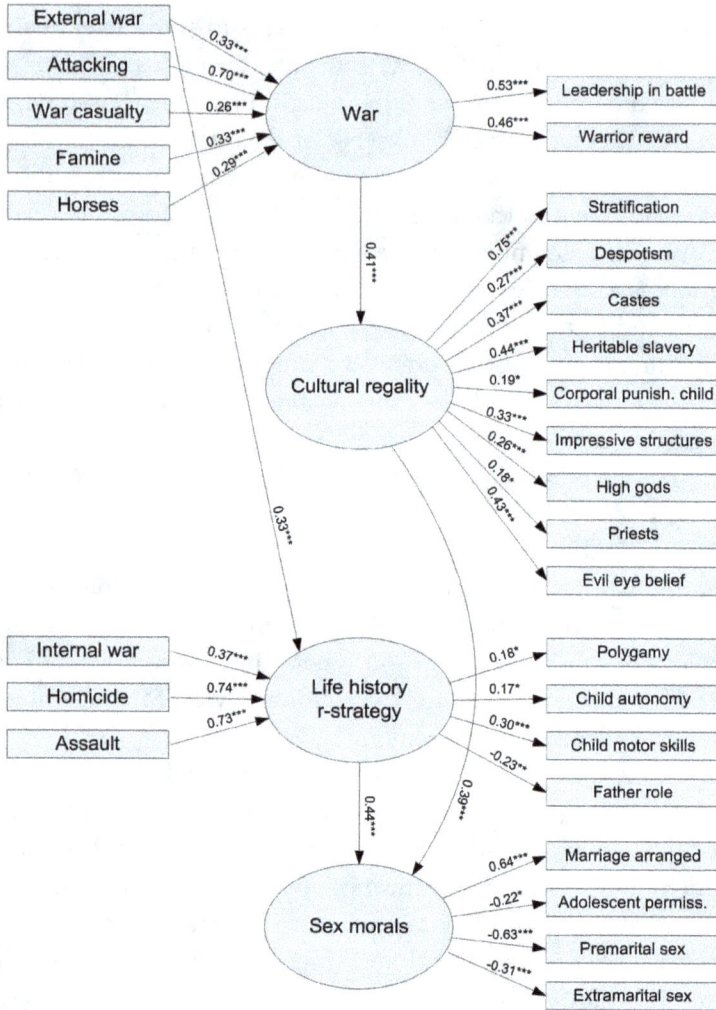

Figure 29. Structural equation model showing the influence of collective dangers and individual dangers on cultural regality and life history strategy, as well as the influence of both of these variables on sexual morals. Rectangles indicate measured variables while ovals indicate latent variables. The directions of the arrows indicate the assumed dominating direction of causality, but the statistical results are insensitive to the directions of these arrows. It is impossible to distinguish between cause and effect with this method. Numbers on the arrows are path coefficients. Levels of significance are indicated as: * $p \leq 0.05$, ** $p \leq 0.01$, *** $p \leq 0.001$. (Measures of fit: RMSEA = 0.076, 90% CI: 0.067–0.084, SRMR = 0.091, CFI = 0.61, AIC = 13782). By Agner Fog, 2017.

The structural equation model shown in figure 29 is based on the theory that war and other collective dangers influence the level of regality while both collective dangers and individual dangers influence the life history strategy. Both factors influence sexual behavior. Various cultural indicators of regality and life history strategy are included in the model.

The war factor emerges from the level of war and various indicators that we expect to be correlated with war. The results suggest that famine is a significant cause of war. The availability of efficient means of transportation (horses) appears to be a significant contributing factor as well.

The war factor is significantly correlated with rewards for successful warriors. There are few variables relating to cowardice and desertion, and these variables show no significant correlation. The database has no variable indicating punishment for defection in war. Therefore, we have more support for the theory that attaches importance to the rewarding of brave warriors than to the alternative theory that relies on the punishment of defectors only.

The cultural regality is modeled here as a latent variable that is influenced by war and other collective dangers. However, the regality is also influenced by unmeasured factors such as cultural traditions and subjective beliefs about dangers, including religious beliefs. A number of cultural indicators of regality are included in the model, and it is confirmed that they fit into the model with highly significant path coefficients.

The life history strategy is influenced by internal and external conflict as well as by other dangers. Only a few cultural indicators of life history strategy and parental investment are included in the model. Several other indicators of life history strategy were excluded because they failed to give significant coefficients.

Sexual morals are influenced by both the regality dimension and by the life history dimension, and reflected in various attitudes towards sexuality and marriage. No reliable measure of fertility was available, but we may assume that strict sexual morals lead to high fertility. The hypothesis is that strict sexual morals allow only reproductive sex within marriage, while sex for pleasure is suppressed. This puts pressure on young people to marry early and have many children because alternative outlets for their sex drive are blocked.[19]

---

19   Garcia and Kruger (2010)

The results from the structural equation model confirm that cultural regality is connected with political stratification, caste stratification, despotism, and slavery. These results are highly significant. The connection with disciplining of children is weaker, but still statistically significant ($p = 0.013$). Few relevant variables relating to art are available, but the presence of large or impressive structures (such as large buildings) has a highly significant connection to cultural regality. Three variables relating to religion are also significant. Finally, it is confirmed that sexual morals are influenced in the strict direction by both cultural regality and life history r-strategy. No direct path between regality and life history strategy was found.

Various correlation coefficients are given in table 9. The correlation between regality and life history strategy is moderate but highly significant ($p = 7 \cdot 10^{-4}$). Regality is also significantly correlated with population density and with various modes of subsistence.

Variable	Regality	Life history strategy	War
Regality		0.25***	0.49***
Population density	0.52***	0.11	0.20**
Political integration	0.73***	0.21**	0.43***
Urbanization	0.48***	0.04	0.31***
Agriculture	0.41***	0.13	0.15*
Animal husbandry	0.43***	0.09	0.17*
Hunting	-0.49***	-0.10	-0.09
Gathering	-0.50***	-0.13	-0.22**
Fishing	-0.25***	-0.11	-0.15*
Warrior reward	0.10	0.12	0.58***
Warrior prestige	0.18*	0.15*	0.61***

Table 9. Correlation coefficients of selected variables against cultural regality, life history strategy, and war. Level of significance: * $p \leq 0.05$, ** $p \leq 0.01$, *** $p \leq 0.001$.

While the structural equation model supports our theory, it is not a definite proof. It may be possible to construct other models that fit the data equally well.[20]

---

20   Breckler (1990)

Multiple correlation analysis

A simultaneous multiple correlation analysis was made for several variables according to the linear model

$$v = k0 + k1\,f1 + k2\,f2 + k3\,f3 + \varepsilon$$

where v is any of the variables in the standard cross-cultural sample. $f1$ represents cultural regality and $f2$ represents life history r-strategy, both obtained from the structural equation model. $f3$ is the population density, which was included in the model because it is a very likely confounding factor according to the results of the factor analysis. $k0$, $k1$, $k2$, and $k3$ are coefficients, and ε is the residue of unexplained variance. The levels of significance for $k1$, $k2$, $k3 \neq 0$ were calculated.

The results of the multiple correlation analysis for various variables, v, are shown in table 10. There is a correlation of regality with agriculture, which disappears when population density is controlled for, as we can see when comparing tables 9 and 10, while the correlations of regality with other means of subsistence remain significant. The correlation of sexual morals with both regality and life history strategy remains significant when population density is controlled for. Polygamy is significantly related to life history strategy but, contrary to expectation, not to cultural regality in this test. The expected correlation between regality and suicide was not found. The variable named 'classical religion' is a mixture of very different religions and therefore not as specific as we would wish, but it is included in table 10 because it has a significant correlation with regality.

Variable	Regality	Life history strategy	Population density
Political integration	0.59***	0.04	0.27***
Urbanization	0.26***	-0.07	0.46***
Agriculture	0.07	0.05	0.63***
Animal husbandry	0.47***	-0.01	-0.08
Hunting	-0.21**	0.01	-0.55***
Fishing	-0.20*	-0.06	-0.06
Gathering	-0.30***	-0.01	-0.37***
Sexual morals	0.42***	0.42***	-0.11
Trend in autonomy	0.24**	0.02	0.19*
Urbanization	0.26***	-0.07	0.46***

Class stratification	0.85***	-0.11**	0.04
Caste stratification	0.52***	-0.08	-0.08
Severity of famine	0.27**	-0.03	-0.04
Heritable slavery	0.63***	0.00	-0.17*
Polygamy	-0.04	0.24**	-0.13
Suicide	-0.01	0.38***	-0.09
Formal sanctions	0.23**	0.10	0.14
Horses	0.22*	-0.12	-0.18*
Classical religion	-0.24**	0.10	-0.09
High gods	0.41***	-0.05	-0.15
Evil eye beliefs	0.55***	0.03	-0.05

Table 10. Multiple regression of various variables against regality, life history strategy, and population density. Level of significance: * $p \leq 0.05$, ** $p \leq 0.01$, *** $p \leq 0.001$.

8.3. Subsample, 18 cultures

A subsample of eighteen of the cultures represented in the Standard Cross-Cultural Sample (SCCS) was selected for further study, and the desired data were extracted from the original ethnographic literature rather than from the SCCS database. This method was intended to reduce the problems with data quality and to obtain information that was not available in the SCCS.

The literature listed as sources for the SCCS[21] was supplemented by any additional literature published later. All coding of data was done by the author. Every conscious effort was made to avoid expectation bias. No funding was available for hiring extra raters or testing inter-rater reliability. A brief description of each society including an explanation of how it is evaluated and the literature used is provided in chapter 7.

The purpose of the subsample study is to test the prediction that certain environmental factors influence the level of intergroup conflict, which in turn influences the social and cultural indicators that we associate with regal and kungic cultures.

Many of the cultural traits that we would like to test—especially those that belong to soft areas such as religion, sexuality, discipline, treatment of children, and art—are poorly represented in the standard

---

21   White (1989)

cross-cultural sample. The relevant variables are either not included in the SCCS database, or they are unreliable, or they have too many missing values. The data in the database are coded into simple categories and values that may be poor representations of the complex behaviors, beliefs, and social structures in the cultures that we study. It is difficult to guess what the human realities are that lie behind a number in a database.

Firsthand observation and collection of the missing data is no longer a possibility, because the cultures in question have been heavily transformed by modernization if not completely annihilated or absorbed into modern cultures. We must resort to the original ethnographic literature and hope that we can find the missing data there. Using this literature has the advantage that the data are presented in a coherent and meaningful way. Therefore, it was decided to extract the relevant data from the original ethnographic literature for a subset of the cultures represented in the SCCS.

The subsample was selected on the basis of the following criteria:

1. The selected cultures must be geographically, culturally, and genetically distant from each other in order to minimize similarity due to cultural diffusion or common descent (Galton's problem).

2. The cultures should be different in terms of subsistence ecology. Where multiple cultures resemble each other too much, only one is included.

3. The population of each society must have lived in relative isolation long enough to develop a distinct culture and language.

4. The influence from modern cultures must be minimal.

5. The culture is not pacified by any colonial authority or other external power.

6. The culture must be well described by more than one ethnographer.

Many cultures had to be excluded for not meeting these criteria. In particular, criteria 4 and 5 were difficult to meet.

The subsample consists of eighteen cultures that were selected for further study. The ethnographic literature for the selected cultures was studied in order to extract sociocultural information relevant to the theory. A number of cultural variables were evaluated in order to test if the correlations predicted by our theory could be confirmed.

The different types of conflict, such as war, raiding, and feuding are often difficult to distinguish, and this is a serious problem for the testing of regality theory. Feuding is usually motivated by the desire to obtain justice by retaliating against the perpetrator of some wrongdoing or against his or her family. This takes place mostly where there is no organized system of justice, which is typical for a kungic culture. Territorial war, on the other hand, is expected mainly in regal cultures, according to our theory. Any confusion between feuding and territorial war will thus tend to blur the statistics. Unfortunately, it is difficult to determine whether the main motive behind a raid is retaliation, plundering, capturing women, or territorial expansion. For example, anthropologists cannot agree whether Yanomamo Indians fight over women, hunting territory, or steel tools.[22]

It was decided to use environmental, ecological, and technological parameters, rather than the level of war, as the independent variable in order to avoid the difficulties in determining the level of war and distinguishing between different motives for conflict. A statistical correlation cannot distinguish between cause and effect, but it seems reasonable to assume that the environment influences the culture more than the culture influences the environment, even when the available technology is included as an environmental factor. The semi-fixed status of ecological environment and available technology makes statistics based on these factors less ambiguous in terms of distinguishing between cause and effect.

The assumptions about environmental causation of conflict do not hold when the society is pacified by colonial rule or other foreign powers. This is the reason why pacified and modernized societies are excluded from our subsample.

To summarize, the prediction we want to test is that environmental and technological factors that enable intergroup conflict are positively

---

22   Ferguson (1995, p. 7), Chagnon (1979), Gat (2017, chapter 4)

correlated with those sociocultural variables that indicate a regal culture, as the theory claims. Other environmental factors that threaten a sociocultural group, such as unpredictable natural disasters, may also have an influence in the regal direction.

The statistical tests use rank correlation. This is a non-parametric test that makes no assumption about linear relationships or normal distribution, unlike the previous tests. Kendall's tau, τ, was calculated and the level of significance, p, was calculated for the one-tailed tests. There was no control for confounding factors. Most variables were evaluated on an arbitrary five-point scale as follows.

Expected conflict level. The *expected* conflict level is evaluated on a five-point scale, where 1 indicates that intergroup conflict is unlikely because of environmental factors or niche specialization, and 5 indicates that environmental factors make large scale war possible and likely.

War or intergroup conflict. The *observed* level of intergroup conflict is interpreted as the frequency, intensity, degree of violence, and casualty rate of conflicts against other social groups that do not have the same self-defined ethnic or tribal identity. Territorial conflicts are given higher weight than plundering and retaliation of injustice. (In the coding used, 1 = no intergroup conflict; 2 = only small raids, feuds, or vendettas; 3 = rare or low scale intergroup conflict; 4 = intermediate intergroup conflict; 5 = frequent large wars).

Internal conflicts and feuds. Internal or intragroup conflicts are conflicts between people with the same ethnic or tribal belonging. Examples are conflicts between families over properties or marriage, or conflicts between leaders and followers. Feuds are series of retaliations over real or alleged wrongdoing. The theory predicts that intergroup conflicts, but not necessarily internal conflicts, are positively correlated with regal indicators. (Coded as 1 = rare internal conflicts; 5 = frequent internal conflicts).

Political system. A hierarchical political system with strong centralized power is expected in a regal society. In a kungic society, we will expect an egalitarian political system with little or no hierarchy, or in extreme cases no formal system of leadership at all. (Coded as 1 = no leadership, or simple political system; 2 = formal leadership; 4 = complex or varying political system; 5 = strong and highly hierarchical political system).

Justice and punishment. We expect a strict justice system and harsh punishments in a regal society, but a mild or lenient degree of punishment in a kungic culture. (Coded as 1 = mild or lenient punishment; 2 = somewhat mild punishment; 4 = somewhat strict punishment; 5 = very strict punishment).

Religion. We expect the religion in a regal society to support the power structure by mirroring or being part of the hierarchical political structure and by legitimizing or exalting the ruler. The religion in a regal culture often has a disciplining function by enforcing strict rules, possibly with supernatural punishments. The religion in a kungic culture is typically non-discriminatory and not connected with political power, discipline, elitism, dogmatism, or strict rules, according to the theory. (Coded as 1 = mild, unorganized, or non-authoritarian religion; 2 = religion supports politics; 4 = some disciplining connected with religion; 5 = strict religion, legitimizes ruler).

Group identity. We expect people to be hostile to strangers and have a strong sense of ethnic or tribal identity in a regal society. We expect less distinction between 'them' and 'us' in a kungic culture. (Coded as 1 = low sense of group identity; 3 = hostility to strangers; 5 = high sense of group identity).

Fertility. We expect people to have more children in a regal than in a kungic society. Kungic cultures may keep the birth rate down by means of postpartum sexual taboos, long breastfeeding periods, contraceptive means, abortion, or infanticide. (Coded as 1 = low fertility; 3 = medium fertility; 5 = high fertility).

Suicide. We expect the incidence of suicide to be higher in kungic than in regal cultures.[23] This applies only to the types of suicide that Durkheim has labeled anomic and egoistic suicide,[24] not to culturally

---

23   Fog (1999)
24   Durkheim (1897)

prescribed suicide or self-sacrifice in battle. (Coded as 1 = rare or no mention of suicide; 3 = suicide occurs; 5 = suicide is common).

Sexual morals. We expect regal societies to have strict taboos and rules regulating sexual behavior, including bans on premarital and extramarital sex. Kungic societies are likely to be more tolerant of a variety of sexual behaviors. (Coded as 1 = lax, permissive sexual morals; 3 = mixed or intermediate sexual morals; 5 = strict sexual morals).

Marrying age. We expect the marrying age to be lower in regal than in kungic societies. The marrying age is recorded separately for men and women. Where the literature specifies a range for the marrying age, the median of this range is used in the statistics.

Divorce. We expect divorce to be easier and more tolerated in kungic than in regal societies. (Coded as 1 = divorce is rare, difficult; 3 =divorce occurs; 5 = divorce is easy, frequent).

Children work. We expect children to work harder, and from an earlier age, in regal than in kungic societies. (Coded as 1 = children do not work; 3 = children do some work; 5 = children work a lot, or from an early age).

Art. Different cultures produce different kinds of art. We are considering many different forms of art, including carving, painting, sculptures, decorated clothing, body adornment, architecture, and other material art, as well as poetry, tales, drama, music, singing, and dance. We expect the artistic production of regal cultures to be highly embellished and perfectionist, possibly glorifying representatives or symbols of power. We expect kungic cultures to produce a more individualistic art with less conformity. (Coded as 1 = simple, individualistic art; 3 = somewhat embellished art; 5 = embellished, repetitive, stylized art).

The evaluation of these variables for the eighteen societies is summarized in table 11. A rank correlation of these variables against the predicted conflict level is shown in the table. A review of the analysis of each culture is given in chapter 7.

Culture	Art	Children work	Divorce	Marrying age, women	Marrying age, men	Sexual morals	Suicide	Fertility	Group identity	Religion	Justice, punishment	Political system	Internal conflict, feuds	War or intergroup conflict	Expected conflict level based on environment
!Kung	1	1	5	15.5	26	1	1	1	1	1	1	1	1	1	1
Inuit	1	1	5	15	18	1	5	1	1	1	1	1	3	1	1
Gilyak	3	na	5	na	22.5	1	5	1	1	1	1	1	5	1	1
Mbuti	1	1	5	16	18	1	3	1	3	1	1	1	5	1	2
Yahgan	1	3	3	15	18	3	1	5	3	2	2	1	1	1	2
Warao	1	3	3	19	21	1	1	5	3	4	1	1	5	3	2
Andaman	1	3	1	18	20	1	1	1	3	1	1	1	5	3	2
Arrernte	1	1	3	na	na	1	1	1	1	1	4	1	5	3	2
Yi	3	3	3	13	13	1	5	na	3	1	2	2	5	5	3
Yanomamo	1	3	5	na	na	3	1	3	1	1	1	1	5	4	3
Apache	3	na	na	18.5	20	5	3	na	na	4	5	1	5	4	4
Maasai	3	5	5	17	28	1	3	5	5	2	2	2	3	5	4
Somali	3	5	5	16	21	5	1	na	na	5	2	4	5	5	4
E De	3	3	1	na	na	3	1	3	3	4	4	4	5	5	5
Hausa	3	3	5	15	20.5	3	1	5	3	4	4	5	1	5	5
Ganda	1	5	5	14	15.5	5	1	5	5	2	5	5	1	5	5
Inca	5	5	1	18	25	5	1	na	na	5	5	5	1	5	5
Babylon	5	5	na	17	29	3	1	na	na	5	5	5	1	5	5
τ	.57	.70	-.15	.04	.10	.62	-.34	.60	.61	.63	.72	.80	-.18	.85	
p	.004	.0009	.27	.45	.34	.002	.06	.01	.007	.001	.0002	.00006	.20	.00002	

Table 11. Rank correlation of various sociocultural parameters against the expected conflict level for eighteen non-industrial cultures. Variables are indicated on a scale of 1 to 5 as described above, except for marrying age, which is indicated by the median. τ = Kendall's tau; *p* = level of significance, one-tailed; na = data not available.

Table 11 is sorted by expected conflict level. At the peaceful beginning of the table, we find isolated cultures, such as the Inuit and the Gilyak, and niche cultures, such as the Mbuti pygmies and !Kung bushmen. At the high-conflict end of the table, we find empires such as the Babylonians and the Incas. This corresponds to the regality scale with kungic cultures at the beginning of the table and regal cultures at the end. Values of *p* below 0.05 are considered statistically significant. We can see that most of the correlations are in the predicted direction and highly significant.

First, we notice that there is a strong correlation between the expected conflict level and the observed level of war and conflict. This is a strong confirmation of the hypothesis that the level of war in non-modern societies is determined mainly by environmental factors. The cultures with a typical niche ecology, such as !Kung, Mbuti, and Warao, are all found to be peaceful and kungic. The same is the case for isolated cultures and cultures with scramble competition for food, such as Inuit and Yahgan. This indicates that competition over food sources is a major cause of conflict and that other reasons for violent intergroup conflict are less important in non-modern cultures.

The correlation for 'internal conflicts and feuds' shows a slightly negative, and non-significant, correlation with the expected conflict level. This confirms that the distinction between intergroup and intragroup conflicts is important, as predicted by our theory.

The correlations for 'political system', 'justice and punishment', 'religion', 'sexual morals', 'children work', and 'art' are all highly significant with *p* levels below 0.01. These results give strong support to regality theory. 'Group identity' and 'fertility' are significant at the 0.01 level.

Suicide seems to be more common in kungic than in regal societies, as predicted, but the correlation is only marginally significant ($p = 0.06$). It should be noted here that it was difficult to find reliable data for suicide and that it was difficult to distinguish between the different motives for suicide.

Divorce shows a weak correlation in the expected direction, but not a significant one. This correlation is weakened by the higher incidence of polygamy in regal societies.

There was no noticeable correlation between marrying age and expected conflict level. The data for marrying age were quite inaccurate, but this is probably not the only explanation for the absence of correlation.

It appears that marrying age is determined more by economic factors and level of education than by ecology.

The significant correlation of sexual morals with regality is interesting because cultural differences in sexual morals are difficult to explain by other theories. We observe that the sexual morals are permissive in kungic societies and strict in most regal societies. But not all regal societies have strict sexual morals. For example, premarital sex is common among the unmarried Maasai warriors, and extramarital sex is common among the Hausa. High-ranking and prestigious men in regal cultures have the power to give themselves more sexual opportunities than the morals would otherwise dictate.

8.4. Contemporary cultures, large sample

Data and methods

The Survey of World Views project is a psychometric study based on a survey of data from 8,883 persons in 33 countries carried out in 2012. The participants were mainly college students who were rewarded for filling out an online questionnaire with 293 questions. The methods are described in detail by Gerard Saucier and coworkers.[25] A copy of the raw data files was kindly provided by Saucier.

An exploratory factor analysis on the complete data set was performed first in order to identify relevant items with significant factor loadings. A scree plot, given in figure 30, shows that there are at least four important factors. These factors could be interpreted as follows:

1. Universalism ethics, benevolence, and self-direction

2. Cynicism, traditional family values, spirit beliefs, materialism, violence

3. Religiousness

4. Egalitarian society

---

25   Saucier et al. (2015)

The factor interpretations were less clear when more factors were included.

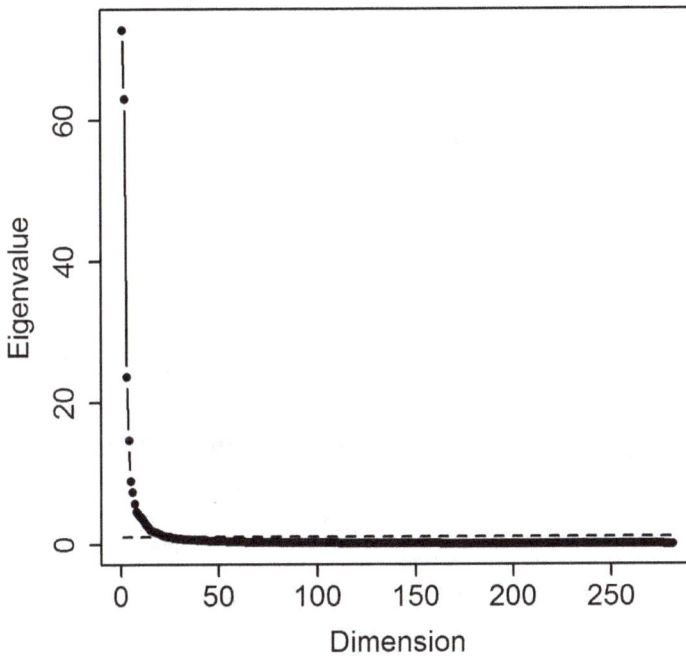

Figure 30. Scree plot for factor analysis of World Views Survey,
2012. By Agner Fog, 2017.

Construction of variables

Twelve composite variables were constructed for the purpose of the current analysis. Each variable was formed by a combination of questionnaire items. These items were selected based on their theoretical relevance and on their loadings in the exploratory factor analysis. Items with low factor loadings and items with ambiguous interpretations were not included. The composite variables are listed in table 12. The items were scored on various Likert scales with different numbers of points, some of them with a forced choice. All items and variables were scaled in order to obtain equal weighting.

Variable name	Items
Perceived danger ($\alpha = 0.52$)	IV2: The present-day world is vile and miserable. IV5: Our people are in danger: everyone is trying to divide us and hurt us. ISM45: Foreigners have stolen land from our people and they are now trying to steal more.
Want tough leader (no α)	ISM35: We need tough leaders who can silence the trouble-makers and restore our traditional values.
Bellicosity ($\alpha = 0.42$)	IV4: We should never use violence as a way to try to save the world. (reverse scored) IV10: If violence does not solve problems, it is because there was not enough of it. M1: I honor the glorious heroes among my people who sacrificed themselves for our destiny and our heritage.
Discipline ($\alpha = 0.71$)	P11: I would never take things that aren't mine. P17: I cannot imagine that I would engage in lying or cheating. P19: I shirk my duties. (reverse scored) P23: I steal things. (reverse scored) P25: I like order. IC7: It is my duty to take care of my family, even when I have to sacrifice what I want. FV08: Children should respect their grandparents. MF04: Respect for authority is something all children need to learn. MF05: People should not do things that are disgusting, even if no one is harmed. MF08: Justice is the most important requirement for a society.
Hierarchy ($\alpha = 0.71$)	NO9: In this society, rank and position in the hierarchy (in the social order) have special privileges. NO25: In this society, people in positions of power try to increase their social distance from less powerful people. NO28: In this society, power is concentrated at the top, rather than being shared throughout society. NO39: In this society, individuals generally attempt to be dominant over other individuals. NO42: Power is shared throughout this society, rather than being concentrated at the top. (reverse scored)
Egalitarianism ($\alpha = 0.63$)	ISM19: I support the rights and power of the people in their struggle against the privileged elite. ISM33: There should be increased social equality. ISM34: Wealthy people should have a higher tax rate than poor people. MF02: When the government makes laws, the number one principle should be ensuring that everyone is treated fairly.
Conformity ($\alpha = 0.65$)	P38: I stick to the rules. AM3: Breaking the rules and regulations is an unpleasant activity. ISM30: It is good to defy 'traditional family values' as feminists and homosexuals have done. (reverse scored) EN4: My first loyalty is to the heritage of my ancestors, their language and their religion. MF11: I would call some acts wrong on the grounds that they are unnatural. MF21: When you decide whether something is right or wrong, to what extent are the following considerations relevant to your thinking?: Whether or not someone conformed to the traditions of society.

	VAL1: Rate the following values as a guiding principle in your life: tradition, that is, respect for tradition, humbleness, accepting one's portion in life, devotion, modesty. VAL4: Conformity, that is, obedience, honoring parents and elders, self-discipline, politeness.
Xenophobia ($\alpha = 0.58$)	ISM11: I believe in the superiority of my own ethnic group. ISM21: My own race is not superior to any other race. (reverse scored) ISM29: People of different races and nationalities should live in different places apart from one another. MF14: When you decide whether something is right or wrong, to what extent are the following considerations relevant to your thinking?: Whether or not someone's action showed love for his or her country.
Religiosity ($\alpha = 0.92$)	DRI01: How often do you attend church, mosque, temple, or other religious meetings? DRI02: How often do you spend time in private religious activities, such as prayer, meditation, or study of religious scriptures? DRI03: In my life, I experience the presence of the Divine. DRI04: My religious beliefs are what really lie behind my whole approach to life. DRI05: I try hard to carry my religion over into all other dealings in life. Ax17: Belief in a religion helps one understand the meaning of life. Ax29: Religious faith contributes to good mental health. ISM8: There is no God or gods. (reverse scored) ISM12: I adhere to an organized religion. ISM17: I believe in predestination—that all things have been divinely determined beforehand. ISM22: Religion should play the most important role in civil affairs.
Cynicism ($\alpha = 0.72$)	AM10: I don't need to care about the problems of other people because nobody cares about me. AM2: The true purpose of stories about honesty and goodness is to make people confused and stupid. AM7: A person who helps others ahead of himself quickly comes to ruin. IV12: Modern governments have overstepped moral bounds and no longer have a right to rule. IV13: It is better to assume that people will deceive you than to always assume that they are telling the truth. IV15: It would be pointless to reduce one's possessions and live more simply.
Violence ($\alpha = 0.60$)	AM8: When someone insults me, I can go for days thinking of nothing else but revenge. IV1: Killing is justified when it is an act of revenge. IV6: A good person has a duty to avoid killing any living human being. (reverse scored) PA03: If a person insults me, it may happen that this person gets beat up (by me or someone else).
Patriarchy ($\alpha = 0.81$)	FV01: The father should be the head of the family. FV03: The mother's place is in the home. FV05: The father should handle the money in the house. FV07: The mother should accept the decisions of the father. MF10: Men and women each have different roles to play in society.

Table 12. Variables constructed from Survey of World Views items. α values are Cronbach's alpha, which is a measure of internal consistency in the variable.

The items included in each variable express the theoretical concepts less clearly than we could wish. This is a consequence of the current state of cross-cultural psychology, where it is not yet known which variables are most important.[26] Theories about causal relationships are not very developed. Many of the questionnaire items were formulated without any clear underlying theory and definitely without any knowledge of regality theory. A discussion of the quality of each variable is in place here:

- The items included in 'perceived danger' are somewhat abstract and probably do not capture the full diversity of collective dangers and the distinction between individual and collective dangers that regality theory presupposes.

- The variable 'want tough leader' depends on just a single item that expresses a rather extreme opinion.

- The variable 'bellicosity' does not include any items relating explicitly to war, as there were no such items in the questionnaire. The Cronbach's alpha is low, indicating some inconsistency.

- The variable 'discipline' includes items relating both to behavior and opinion. It covers the concept well.

- The variable 'hierarchy' reflects only a subjective evaluation of the hierarchy of the culture in question, with no standard or reference specified. People tend to evaluate things with their own culture as a frame of reference. Evaluating a culture with the same culture as reference makes little sense.[27] It is possible that participants have used some other culture as reference when answering the questions, but it is also possible that they have used their own idea of an ideal society as reference. Thus, a participant indicating that the level of hierarchy is high may actually mean 'higher than I would like'.

- It was expected that 'egalitarianism' would be the opposite of 'hierarchy' and that these two concepts would load on the same factor with opposite signs. However, the exploratory factor

---

26 Saucier et al. (2015)
27 Heine, Lehman, Peng and Greenholtz (2002)

analysis showed, quite surprisingly, that egalitarianism and hierarchy loaded on different factors and that they had in fact a significant positive correlation with each other, rather than a negative one. This paradox may be due to acquiescence bias: some people tend to answer 'yes' to anything. The items under 'egalitarianism' are general statements that are easy to agree with, and none of these items are reverse scored. The problems with 'egalitarianism' are further discussed on page 263.

- The variables 'conformity' and 'xenophobia' contain enough relevant items to express the concepts adequately.

- The variable 'religiosity' reflects the strength of religious beliefs and the involvement in religious activities, but perhaps not the strictness of religious observance or the disciplining function of the religion, which would be more relevant to regality theory.

- The variable 'cynicism' was included because it turned up strongly in the factor analysis with significant correlations to relevant variables. Cynicism is related to a pessimistic worldview, but the relationship of this variable with regality has not been fully investigated yet.

- The variable 'violence' does not adequately distinguish between organized or socially sanctioned violence and simple crime and rage, or between group-internal and external conflicts. However, it may be related to collective danger in its broadest sense.

- 'Patriarchy' includes values and opinions that have been thoroughly described in the literature. The questions are appropriate and consequently the alpha is high.

Correlations of the variables

There are significant differences between countries for all of the variables defined here. For example, many of the variables correlate strongly with the gross domestic product (GDP) and the human development index (HDI) of the countries, as shown in table 13. Therefore, we need to control for country-specific effects in the statistical investigations of relationships between these variables.

Variable	Human development index	log gross domestic product
perceived danger	-0.71***	-0.71***
want tough leader	-0.71***	-0.75***
bellicosity	-0.35*	-0.37*
discipline	-0.60***	-0.59***
hierarchy	-0.28	-0.32
egalitarianism	-0.46**	-0.45**
conformity	-0.73***	-0.75***
xenophobia	-0.60***	-0.66***
religiosity	-0.72***	-0.72***
cynicism	-0.71***	-0.74***
violence	-0.33	-0.39*
patriarchy	-0.63***	-0.66***

Table 13. Correlation coefficients (*r*) of the country average of each variable with HDI and GDP. Levels of significance: * $p < 0.05$, ** $p < 0.01$, *** $p < 0.001$.[28]

The country-specific effects are not limited to economic factors. Environmental factors and cultural differences are also very likely to influence these variables. We can get an indication of the influence of all country-specific factors on each variable by testing how much of the variance in each variable is accounted for by between-country differences. Table 14 lists the correlation of each variable with country differences. The correlation coefficients in table 13 are mostly higher than in table 14 because the former is looking at country averages while the latter is looking at differences between individuals.

Variable	Correlation with country differences
perceived danger	0.19***
want tough leader	0.53***
bellicosity	0.34***
discipline	0.39***
hierarchy	0.40***
egalitarianism	0.29***
conformity	0.48***
xenophobia	0.53***

---

28 Data sources: United Nations Development Programme (2013), New World Encyclopedia (2016), International Monetary Fund (2012)

religiosity	0.65***
cynicism	0.37***
violence	0.28***
patriarchy	0.55***

Table 14. Correlation coefficients (*r*) of each variable with country differences. Level of significance: *** $p < 0.001$.

The next step in our analysis is to study the correlations between the variables while controlling for country differences. Table 15 lists all pairwise correlations between two variables with country differences as control. We can see that most of the variables are correlated significantly with each other, even after controlling for confounding country differences. The averages of each variable within each nation are listed in table 16.

Variable	perceived danger	want tough leader	bellicosity	discipline	hierarchy	egalitarianism	conformity	xenophobia	religiosity	cynicism	violence	patriarchy
perceived danger	1	0.17***	0.11***	0.01	0.22***	0.22***	0.12***	0.13***	0.09***	0.37***	0.19***	0.16***
want tough leader	0.17***	1	0.08***	0.18***	0.02	0.07***	0.37***	0.24***	0.27***	0.20***	0.07***	0.30***
bellicosity	0.11***	0.08***	1	-0.35***	-0.07***	-0.27***	-0.10***	0.32***	-0.09***	0.51***	0.51***	0.19***
discipline	0.01	0.18***	-0.35***	1	0.11***	0.36***	0.47***	-0.10***	0.29***	-0.29***	-0.41***	0.08***
hierarchy	0.22***	0.02	-0.07***	0.11***	1	0.28***	0.02	-0.11***	-0.05***	-0.01	-0.08***	-0.05***
egalitarianism	0.22***	0.07***	-0.27***	0.36***	0.28***	1	0.14***	-0.22***	0.05***	-0.20***	-0.28***	-0.05***
conformity	0.12***	0.37***	-0.10***	0.47***	0.02	0.14***	1	0.23***	0.45***	0.05***	-0.08***	0.39***
xenophobia	0.13***	0.24***	0.32***	-0.10***	-0.11***	-0.22***	0.23***	1	0.16***	0.45***	0.37***	0.36***
religiosity	0.09***	0.27***	-0.09***	0.29***	-0.05***	0.05***	0.45***	0.16***	1	0.00	-0.07***	0.35***
cynicism	0.37***	0.20***	0.51***	-0.29***	-0.01	-0.20***	0.05***	0.45***	0.00	1	0.59***	0.31***
violence	0.19***	0.07***	0.51***	-0.41***	-0.08***	-0.28***	-0.08***	0.37***	-0.07***	0.59***	1	0.25***
patriarchy	0.16***	0.30***	0.19***	0.08***	-0.05***	-0.05***	0.39***	0.36***	0.35***	0.31***	0.25***	1

Table 15. Partial correlation coefficients between all pairs of two variables, controlling for country differences. Levels of significance: *** $p < 0.001$.

Nation	perceived danger	want tough leader	bellicosity	discipline	hierarchy	egalitarianism	conformity	xenophobia	religiosity	cynicism	violence	patriarchy
Argentina	0.52	0.56	0.19	0.81	0.77	0.76	0.60	0.33	0.40	0.25	0.17	0.37
Australia	0.40	0.37	0.31	0.72	0.60	0.74	0.54	0.28	0.33	0.19	0.20	0.33
Bangladesh	0.70	0.80	0.41	0.78	0.76	0.83	0.68	0.67	0.64	0.50	0.36	0.59
Brazil	0.54	0.29	0.16	0.74	0.81	0.85	0.52	0.22	0.49	0.21	0.15	0.23
Canada	0.43	0.38	0.29	0.71	0.61	0.72	0.54	0.29	0.37	0.21	0.21	0.35
China	0.41	0.58	0.37	0.72	0.70	0.77	0.60	0.56	0.31	0.39	0.31	0.46
Egypt	0.62	0.63	0.31	0.70	0.70	0.78	0.65	0.44	0.70	0.38	0.30	0.62
England	0.44	0.47	0.35	0.71	0.67	0.72	0.54	0.29	0.30	0.29	0.24	0.33
Ethiopia	0.52	0.63	0.42	0.77	0.61	0.76	0.67	0.43	0.68	0.37	0.29	0.48
Germany	0.38	0.14	0.24	0.70	0.54	0.69	0.45	0.21	0.27	0.18	0.12	0.20
Greece	0.63	0.35	0.20	0.73	0.76	0.82	0.58	0.34	0.38	0.30	0.19	0.31
India	0.59	0.71	0.34	0.76	0.72	0.82	0.66	0.49	0.50	0.37	0.29	0.53
Ireland	0.36	0.34	0.20	0.76	0.67	0.78	0.53	0.22	0.31	0.17	0.14	0.23
Japan	0.38	0.64	0.39	0.59	0.62	0.65	0.57	0.36	0.22	0.32	0.32	0.42
Kenya	0.58	0.70	0.24	0.83	0.75	0.84	0.74	0.43	0.67	0.30	0.21	0.60
Korea	0.46	0.58	0.43	0.69	0.70	0.72	0.63	0.53	0.46	0.36	0.42	0.50
Malaysia	0.63	0.77	0.29	0.81	0.58	0.76	0.76	0.52	0.82	0.39	0.23	0.72
Mexico	0.52	0.51	0.23	0.80	0.71	0.78	0.66	0.42	0.51	0.32	0.20	0.43
Morocco	0.56	0.63	0.32	0.71	0.62	0.76	0.67	0.48	0.73	0.39	0.30	0.63
Nepal	0.63	0.80	0.44	0.81	0.77	0.80	0.68	0.43	0.55	0.42	0.25	0.48
Netherlands	0.40	0.34	0.23	0.73	0.56	0.77	0.53	0.19	0.28	0.21	0.16	0.21
Peru	0.50	0.53	0.22	0.76	0.78	0.72	0.58	0.37	0.41	0.30	0.23	0.36
Philippines	0.51	0.67	0.39	0.82	0.65	0.74	0.72	0.54	0.68	0.36	0.27	0.58
Poland	0.41	0.31	0.22	0.68	0.68	0.71	0.57	0.21	0.41	0.23	0.16	0.37
Russia	0.46	0.56	0.34	0.73	0.82	0.67	0.58	0.38	0.35	0.30	0.23	0.50
Singapore	0.46	0.49	0.35	0.73	0.66	0.73	0.59	0.37	0.45	0.30	0.24	0.46
Spain	0.54	0.29	0.27	0.72	0.74	0.78	0.52	0.29	0.22	0.30	0.23	0.23
Taiwan	0.43	0.47	0.36	0.68	0.65	0.80	0.58	0.43	0.40	0.33	0.31	0.36
Tanzania	0.72	0.87	0.36	0.80	0.67	0.75	0.71	0.54	0.69	0.47	0.28	0.61
Thailand	0.53	0.69	0.31	0.79	0.72	0.80	0.68	0.44	0.57	0.36	0.25	0.61
Turkey	0.62	0.46	0.36	0.73	0.80	0.81	0.61	0.28	0.42	0.34	0.23	0.42
Ukraine	0.46	0.76	0.33	0.74	0.78	0.73	0.61	0.43	0.40	0.30	0.23	0.56
USA	0.43	0.45	0.34	0.70	0.66	0.68	0.57	0.36	0.44	0.26	0.25	0.41

Table 16. Nation averages for twelve variables.

Factor analysis on the variables

A factor analysis was carried out in order to search for a structure in the relationships between these variables. Mixed factor analysis was used in order to allow both quantitative variables and a qualitative variable ('nation') to be included.[29] Missing values were imputed.[30] The two strongest factors account for 9.5% and 6.5% respectively of the total variance. The contributions of the twelve variables to these two factors are shown in figure 31. Figure 32 shows how the nations are distributed along the dimensions of these two factors.

This analysis was repeated using multiple factor analysis. This is a two-step procedure which carries out first a principal component analysis on each group of items that make up the twelve variables in table 12, and then a factor analysis of the dimensions that result from the principal component analyses.[31] The results of the latter analysis were almost identical to figures 31 and 32 and are therefore not shown here.

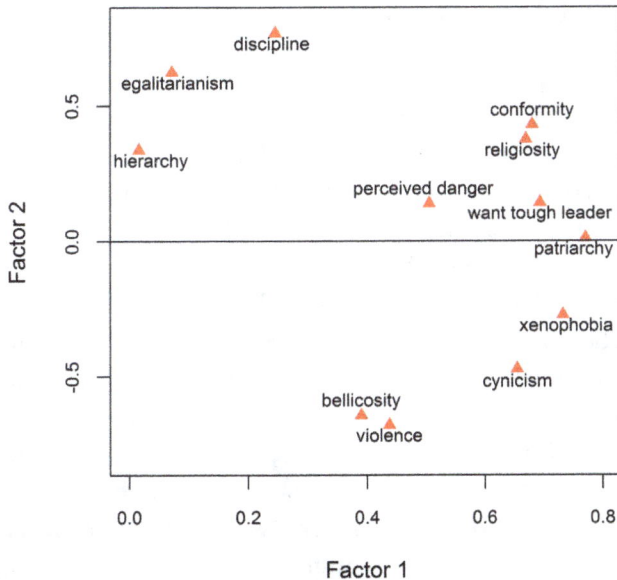

Figure 31. Mixed factor analysis. The figure shows the distribution of the variables along the two strongest factors. By Agner Fog, 2017.

---

29   Pagès (2004), Lê et al. (2008)
30   Audigier, Husson and Josse (2013)
31   Lê et al. (2008), Bécue-Bertaut and Pagès (2008)

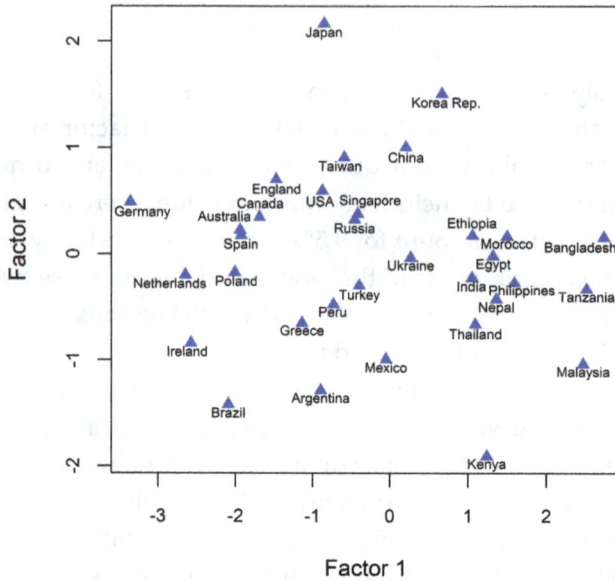

Figure 32. Mixed factor analysis. The figure shows the distribution of nations along the two factors. By Agner Fog, 2017.

Most of the variables have a significant loading on factor 1, as predicted. We can regard factor 1 as reflecting the basic aspects of regality, with a central focus on perceived danger and the desire for a tough leader.

Factor 2 is more difficult to interpret. It is dominated by discipline at the positive end and violence at the negative end, both of which are indicators of regality as evidenced by their positive loading on factor 1. The two factors are almost orthogonal, and it appears that factor 2 distinguishes between different variants of regal cultures. The interpretation of this factor is further discussed below.

Discipline is unrelated to perceived danger, according to table 15, but positively correlated with the desire for a tough leader. Discipline is strongly correlated with conformity, and conformity is strongly correlated with the desire for a tough leader and moderately correlated with perceived danger.

The variable 'hierarchy' has only an insignificant effect in the factor analysis (Figure 31), but it has a significant correlation with perceived danger (Table 15). The weak effect of this variable is very likely due to the reference group effect discussed on page 256. The respondents are probably evaluating the degree of hierarchy in their own culture with that same culture as reference.

It is puzzling that the variable 'egalitarianism' is positively correlated with hierarchy, while we would expect egalitarianism and hierarchy to be each other's opposites (see table 15). Egalitarianism is negatively correlated with bellicosity and xenophobia, as expected, but positively correlated with perceived danger, contrary to expectation. However, perceived danger is positively correlated with bellicosity and xenophobia. An explanation for this paradox can perhaps be found in the strong correlation of egalitarianism with discipline. People who score high on discipline may be more likely to express agreement than disagreement with any statement. People with such an acquiescent response style will score high on egalitarianism because this variable has no reverse scored items. An acquiescent response style may be linked to hierarchy, collectivism, and uncertainty avoidance, according to one study.[32]

Perhaps some people express egalitarian opinions as a reaction against a level of hierarchy that they perceive as too high. This may also contribute to the positive correlation between egalitarianism and hierarchy. It is also possible that people who agree with egalitarian values are conforming to an egalitarian ideology expressed by their political leaders. An egalitarian society requires some discipline to suppress the pursuit of egoistic goals. Note that conformity is positively correlated with egalitarianism and discipline but not correlated with hierarchy (Table 15).

The two variables that are called bellicosity and violence here do not actually make a proper distinction between internal and external conflicts, because the survey questions failed to make this distinction. Given the mainly peaceful environment, it is likely that these variables mostly express internal conflicts.

Correspondence with other theoretical constructs

It can be useful to compare the factors in figures 31 and 32 with similar measures found by other scientists. Table 17 lists the correlations of these two factors with the various cultural and psychological measures that have been listed for different countries by other authors, as discussed in chapters 3.7 and 3.8. It is clear that all these measures, under different names, have something in common. They are all correlated

---

32   Harzing (2006)

in the predicted direction with factor 1, and all these correlations are statistically significant, except for Lazar Stankov's harshness versus softness variable ($p = 0.053$, $n = 18$).

Other studies		Regality factors	
Author and theory	**Cultural measure**	**Factor 1: Perceived danger and desire for tough leader**	**Factor 2: Discipline versus violence**
Uz: Cultural tightness[i]	Tightness, domain specific	0.60**	0.06
	Tightness, domain general	0.54**	0.33
	Tightness, combined	0.78***	0.15
Gelfand et al.: Tight versus loose cultures[ii]	Tightness	0.60**	0.22
Inglehart and Welzel: Human development and values[iii]	Secular/rational versus traditional values	-0.44*	0.63***
	Self-expression versus survival values	-0.56***	-0.24
Welzel: Human empowerment[iv]	Secular values	-0.49**	0.32
	Emancipative values	-0.74***	0.18
Stankov, Lee and Vijver: Personality, attitudes, values, norms[v]	Conservatism versus liberalism	0.74***	-0.38
	Harshness versus softness	0.46	0.46
Stankov and Lee: Social attitudes[vi]	Nastiness	0.66***	0.14
	Religiosity	0.79***	-0.46**
	Morality	0.77***	-0.57***
United Nations	Life expectancy	-0.65***	-0.40*
	Human development index	-0.76***	0.35*

i   Uz (2015)

ii   Gelfand et al. (2011)

iii   Inglehart and Welzel (2005), Inglehart (2007)

iv   Welzel (2013)

v   Stankov, Lee and van de Vijver (2014)

vi   Stankov and Lee (2015)

Table 17. Correlation of factors from the present study with various other cultural measures at the country level. Levels of significance, two-tailed: * $p < 0.05$, ** $p < 0.01$, *** p < 0.001.[33]

---

33   Data sources: Inglehart and Welzel (2005), Gelfand et al. (2011), Uz (2015), Stankov, Lee and van de Vijver (2014), Welzel (2013), United Nations Development

If we accept the suggestion of Irem Uz and coworkers that life expectancy is a useful measure of threat,[34] then we can confirm that factors 1 and 2 are both significantly correlated with threat. The human development index (HDI) is a less accurate measure of threat and danger. While factor 1 is negatively correlated with HDI, factor 2 is positively correlated with HDI. This will be discussed later.

Discussion

The results of our analysis are useful for investigating relationships between important political and cultural phenomena, but it must be emphasized that it is not possible to infer causal relationships from correlations without time series data. The results are quite interesting despite the fact that many of the items in the Survey of World Views questionnaire are somewhat off the mark from the concepts that are relevant to regality theory. Future research will need to include questions, methods, and samples that are designed specifically to test the predictions of regality theory.

Most of the twelve variables that we defined in table 12 are significantly correlated with each other, as we can see in table 15.

The first and strongest factor in the factor analysis is centered around perceived danger and the desire for a tough leader, as shown in figure 31. This fits nicely with the central claim of regality theory that perceived danger leads to the desire for a strong leader, and that this affects many aspects of a culture. All the variables except hierarchy and egalitarianism have substantial loadings on this factor. Patriarchy has particularly strong loading because of the high internal consistency (α) of this variable. Factor 1 also has significant negative correlations with life expectancy and human development index, which can both be seen as rough measures of safety and absence of dangers.

Several scientists have tried to map cultural differences and have each developed their own methods and measures. It is interesting that all these different measures seem to be related. There must be some core phenomenon that is reflected in these different measures, and this is exactly what regality theory asserts.

Programme (2013), New World Encyclopedia 2016), Inglehart and Welzel (2013), Welzel (2016), United Nations Data Retrieval System (2012)

34   Uz (2015)

The phenomenon that we call regality is expected to have a strong influence on many different aspects of culture, such as politics, religion, morals, and art, as shown in table 1 (chapter 2.6). This makes it very likely that attempts to map cultural differences will reveal some of these connections, regardless of which methods are used. Factor 1 in the factor analysis is a reasonably good indicator of the core aspects of regality, and this factor is correlated significantly with almost all the other cultural measures, as we can see in table 17. Some of these correlations are due to confounding economic factors, of course, but we can see from table 15 that significant correlations persist when economic and other country-specific effects are controlled for.

Factor 2 is dominated by discipline versus violence, as we can see in figure 31. A possible interpretation of this factor relates to the way conflicts are solved. Internal conflicts are solved by official legal or political means in countries with strong and efficient moral or legal systems. However, people tend to use violence in the absence of other means of conflict solving. Efficient ways of disciplining people are seen not only in regal societies, but also in highly developed societies regardless of regality. Countries with low loadings on factor 2 have less efficient legal or moral means for disciplining people. We saw the same phenomenon in the study of non-industrial cultures (page 246 in chapter 8.3). There is a lot of internal violence in the form of feuds and vendettas in some moderately kungic cultures without strong legal systems. Regal societies, however, were able to suppress internal violence, but engaged in a lot of external violence against other societies. We can conclude that factor 2 is related to the disciplining aspect of regality, but also to other effects that have less to do with regality.

The distribution of nations across the two factors, shown in figure 32, has notable similarities with cultural maps published by other scientists using other constructs and methods. Figure 32 is very similar to a map of the distribution of countries along two dimensions named conservatism versus liberalism and harshness versus softness, published by Stankov and coworkers,[35] despite the non-significant correlation of their harshness/softness dimension with factor 2 of the present study ($p = 0.056$, see table 17).

The factor map of figure 32 also has striking similarities with Ronald Inglehart and Christian Welzel's cultural map in figure 6 (chapter 3.7).

---

35   Stankov, Lee and van de Vijver (2014)

It is interesting that factor 2 is strongly correlated with Inglehart and Welzel's secular/rational values, which do not measure conflict solving but industrialization and secularization of authority. Inglehart and Welzel's map can be seen roughly as a mirrored and somewhat rotated version of our factor map. Regal countries are found at the lower left corner of figure 6 and kungic countries at the upper right. Countries with a strong disciplining system are found at the top of figure 6 and countries with weak discipline at the bottom.

The most conflict-ridden countries were not included in the Survey of World Views study because of technical and political difficulties.[36] Therefore, we cannot see the full effect of war in these statistics. Fortunately, Inglehart and Welzel's study is more inclusive, and we can see that the most conflict-ridden countries tend towards the lower left corner of figure 6, which we have identified as the most regal.

The position of Germany at the kungic end of figure 32 and the upper right corner of figure 6 is evidence that the regality of a culture can change quite dramatically in just a few generations. In addition to the obvious explanation that Germans felt safe at the time the studies were made, we should also note that Germans are determined to distance themselves ideologically from the Nazism of their past.

Our evaluation of the present results must take into account that the survey has been made in modern cultures with student samples. Furthermore, the countries included here were dominated by relative peace and security at the time of the study. This is far from the environment under which the mechanism behind regality evolved. Nevertheless, we can see highly significant correlations between nearly all the variables we have studied. This confirms that the psychological mechanism of regality that evolved in a very different environment still operates in modern cultures today.

8.5. Evidence from existing studies

Many previous studies have found effects of collective danger that are relevant to our theory. Some of these studies are listed in table 18. All of these studies find correlates of collective danger that are in accordance with regality theory, though a few of them fail to find all the effects that we would predict.

---

36   Saucier et al. (2015)

Study	Level of effect	Type of culture	Type of danger	Sample	Source of variation	Observed effect	Theory
Russell[i]	culture	non-industrial	war	world cultures	natural	punitiveness	psychology
Sipes[ii]	culture	non-industrial	war	world cultures	natural	combative sports	cultural evolution
Roes and Raymond[iii]	culture	non-industrial	intergroup competition	world cultures	natural	religion	various
Inglehart and Welzel[iv]	individual and culture	modern	existential insecurity	world cultures	natural	nationalism, politics, religion, discipline, fertility	political psychology
Canetti et al.[v]	individual	modern	political violence	representative	natural	attitude to peace, xenophobia	political psychology
Jugert and Duckitt [vi]	individual	modern	economy, violence	students	imagined scenario	prejudice	authoritarianism
Feldman and Stenner[vii]	individual	modern	economy, war	representative	correlation study	hostility to minorities, etc.	authoritarianism
Wohl et al.[viii]	individual	modern	group extinction	self-selected	salience manipulation	strengthen group	social psychology
Riek, Mania, and Gaertner[ix]	individual	modern	various	meta-analysis	various	hostility to out-group	various
Ladd[x]	individual	modern	terrorism	representative	natural	support for president	political psychology
Carnagey and Anderson[xi]	individual	modern	terrorism	students	natural	war attitudes, punitiveness	psychology
Huddy and Feldman[xii]	individual	modern	terrorism	representative	natural	strict foreign policy	psychology
Bonanno and Jost[xiii]	individual	modern	terrorism	terrorism witnesses	natural	authoritarianism, militarism, patriotism, religiosity	social psychology
Echebarria-Echabe et al. [xiv]	individual	modern	terrorism	self-selected	natural	authoritarianism, prejudice	social psychology
Curşeu, Stoop, and Schalk[xv]	individual	modern	immigration	semi-random	correlation study	prejudice	integrated threat theory

i   Russell (1972)
ii   Sipes (1975)
iii   Roes and Raymond (2003)
iv   Inglehart and Welzel (2005)
v   Canetti et al. (2015), Hirsch-Hoefler, Canetti, Rapaport and Hobfoll (2014)
vi   Jugert and Duckitt (2009)
vii   Feldman and Stenner (1997)

viii   Wohl, Branscombe and Reysen (2010)
ix   Riek, Mania and Gaertner (2006)
x   Ladd (2007)
xi   Carnagey and Anderson (2007)
xii   Huddy, Feldman and Weber (2007), Huddy and Feldman (2011)
xiii   Bonanno and Jost (2006)
xiv   Echebarria-Echabe and Fernández-Guede (2006)
xv   Curşeu, Stoop and Schalk (2007)

Table 18. Previous studies that found correlates of collective danger.

Several of the studies listed in table 18 use non-representative samples and unrealistic manipulations of danger, and measure only short-term effects. The cited studies are based on other theories than the one discussed here, or on no theory at all. We need similar studies based on regality theory in order to make more detailed predictions that can be tested, including predictions that differ from alternative theories.

8.6. Conclusion of the statistical tests

In this chapter, we have tested the predictions of regality theory systematically in several different ways. The effects at the individual level have been tested on contemporary societies only, while the effects at the social level have been tested on both contemporary and ancient non-industrial societies, using a combination of comparative historical analysis and various statistical methods. Few social science theories have been tested as thoroughly as this. Most of the predictions of the theory were confirmed. We now have empirical support for all the links in the causal chain hypothesized by our theory:

- The geography, ecology, and available technology have a strong influence on the level of intergroup conflict in a non-modern setting.

- A high level of intergroup conflict influences the psychological disposition of the individual members of the society in the direction of a desire for a strong leader, bellicosity, xenophobia, conformity, discipline, patriarchy, and strict religiosity.

- The emergent effects of individual psychological preferences makes the political system more hierarchical, makes the justice system more strict and punitive, and shapes the culture in the direction of a strong group identity, strict sexual morals, strict treatment of children, and a preference for perfectionist and highly embellished art.

- The opposite situation, where the environment makes violent conflict difficult or unlikely, has the opposite effects, leading to peace and tolerance at both the individual and the social level.

Thus, the central claims of regality theory are strongly confirmed by the data. A few less important predictions were only weakly supported or not supported at all. The prediction that anomic suicide is more common in kungic societies than in regal societies was only weakly supported for the non-industrial cultures. It was not supported for modern cultures in a country-level statistic (not shown here). The predictions for divorce rate and marrying age were not supported.

While the regality level is strongly connected with the political complexity and population density, it cannot be reduced to these factors alone. The first factor analysis (Tables 6 and 7) and the multiple correlation analysis (Table 10) indicate that significant effects remain when population density is controlled for.

None of the tests in the present study includes a time sequence which could be used to verify the direction of causality, but several of the studies cited in table 18 demonstrate the direction of causality based on single events, such as terror attacks or laboratory experiments.

In the study of non-industrial cultures (chapter 8.3), we can infer the direction of causality from the influence of the environment. This relies on the assumption that these cultures had little influence on their own environment.

Radical changes in the environment are so rare and case-specific that it would be difficult to make a statistical time series analysis to study the influence of the environment on whole societies, but individual case studies can be quite indicative. One example is the Moriori people mentioned in chapter 3.1. These people became isolated on the Chatham Islands because the islands lacked materials for building boats. The Moriori culture changed from violent to peaceful in a few hundred years. When the population was finally attacked by invading Maoris, the Moriori were easily wiped out because they lacked an efficient culture of defense.[37]

The problems and weaknesses of the different statistical studies are discussed in chapter 8.1.

---

37   Endicott (2013)

9. Discussion and Conclusion

Figure 33. Kungic versus regal. Demonstration against the Vietnam War. Arlington, Virginia, USA, 1967. Albert R. Simpson, US Army Audiovisual Center, 1967.[1]

9.1. Summary of findings

Evolutionary theories of war and conflict are not new, but previous theories have several weaknesses. It is difficult to explain why warriors are willing to risk their lives fighting for their group. Previous

---

1   Public domain, https://commons.wikimedia.org/wiki/File:Vietnamdem.jpg

© 2017 Agner Fog, CC BY 4.0   https://doi.org/10.11647/OBP.0128.09

explanations have relied on evolutionary mechanisms such as kin selection, group selection, altruistic punishment, or benefitting from the spoils of war,[2] often attributing much more power to these mechanisms than the quantitative models bear out. Regality theory proposes a mechanism that better explains the parochial altruism of warriors without resort to such weak or controversial mechanisms as altruistic punishment or group selection.

Another problem with many of the previous theories is that they try to explain variation with a constant.[3] Some theorists claim that humans are peaceful by nature, while others claim that humans are inherently ethnocentric and xenophobic, but few have explained the extreme variability in human societies from the extremely peaceful and tolerant to the unfathomable cruelty of warlike societies. Regality theory explains this variability by showing that humans have a flexible psychology that allows us to adapt to our life conditions.

Humans will express an authoritarian reaction in the event of war or any other perceived collective danger that requires collective action. This includes the desire for a strong leader, strict discipline, a strong group identity emphasizing 'us' versus 'them', xenophobia, and intolerance. The social and political structure of the society will take shape according to these preferences. We will see a hierarchical political organization and harsh punishment of traitors and deviants. Such a society is called *regal*.

This mechanism provides an evolutionary solution to the collective action problem. The powerful leader can reward brave warriors and punish cowards and defectors. In the absence of such a leader, there would be more free riders than fighters. It is not worthwhile for anybody to fight for his group if everybody else is free riding on his actions. It is a better strategy to support a strong leader who can make sure that nobody is free riding (see chapter 2.2).

There is no reason to support a strong leader when there is no war or other collective danger. On the contrary, a strong leader is likely to be a tyrant who exploits everybody else. Therefore, people will show the

---

2   Thayer (2004, chapter 3), van der Dennen (1995, chapter 1.2)
3   Thayer (2004, p. 234)

opposite psychological tendencies in a situation of peace and security, preferring an egalitarian and tolerant society. This kind of society is called *kungic*.

Most societies will be somewhere in between these two extremes. We can imagine a continuous scale from the extremely kungic to the extremely regal, and each society or culture can be placed somewhere on this scale depending on the level of perceived collective danger. The regal-kungic scale does not only affect the political climate. We have observed that a lot of different aspects of a culture reflect its level of regality. The characteristics of regal and kungic societies are summarized in table 1 (chapter 2.6).

This model can explain a lot of different social phenomena. Most importantly, it can explain why some societies are warlike and authoritarian while others are peaceful and tolerant. The hierarchical and highly disciplined structure in a regal society is good not only for defense, but also for offensive purposes. A regal country will attack any neighbor country if it has a good chance of gaining more territory through war.

9.2. Three epochs in human history

Different factors in the environment and subsistence pattern can make war or violent conflict feasible or impossible, as we have seen. We can roughly divide human history into three epochs where different means of subsistence have led to different dynamics for war and regality.

In the first epoch, humans depended on hunting, gathering, and fishing. The amount of violent conflict was determined mainly by the ecology and geography. Where food was concentrated in rich patches, we would have contest competition and frequent intergroup conflicts over access to the most attractive food patches. The population density would be higher around these rich patches so that larger groups were able to fight together. Geographic barriers and niche specialization were factors that could prevent war. Thus, we would find the relatively regal cultures in areas with rich food patches and the more kungic cultures in remote areas where food was sparse, where travelling was difficult, or where a small group had adapted to a special niche that neighboring groups could not penetrate. When nomadic pastoralism was introduced, the dynamics were basically the same, but the groups that traveled together were larger, which increased their fighting ability.

In the second epoch, humans depended increasingly on agriculture to produce their own food. Where the lifestyle had previously been nomadic, it now became more and more sedentary. This development probably started with slash and burn agriculture, in which groups of humans traveled to a new place whenever the soil had been exhausted. The lifestyle became more and more sedentary with the Neolithic revolution, when fields were cultivated in fertile areas such as river valleys. The population grew in these areas. As villages grew into cities, it became possible to organize and build irrigation canals to increase the agricultural yield.

The more the production of food became concentrated in fertile areas near water sources, the more these areas became the subject of violent conflicts. A self-amplifying cycle of increasing regality, territorial expansion, increasing food production, and population growth was started. Villages and small cities grew and became city states, nations, kingdoms, and finally large empires. The political structure became more and more hierarchical and centralized while the culture focused increasingly on the importance of military virtues and new conquests.

When such an empire had finally reached the practical limits to expansion, the opposite development set in. The wars and conquests that had driven the regal spiral were no longer feasible, the state was economically bankrupt, and a kungic development set in. The empire started to disintegrate into smaller states, some of which were later absorbed by neighboring empires.

The level of regality in this second epoch was now determined not only by ecological and geographic factors but increasingly by the dynamical rise and fall of empires. This cyclical process, which is explained in chapters 4.1 and 4.2, led to an alternation between regal and kungic developments over periods of several hundred years.

The third epoch is one of industrialization and a globalized economy. Large scale territorial wars have become less common since World War II due to the influence of international organizations, alliances, deterrence, economic costs, and economic interdependence. Political leaders may still have imperial ambitions, but their strategies are now based less on brute-force military dominance and more on proxy wars, economic structures, resource extraction, debt dependence, and the domination of culture, ideology, science, and mass media.[4]

We still have violent conflicts, but on a lower scale. In most cases, one or both parties in a conflict depend on foreign support (often clandestine). These conflicts can be characterized as proxy wars or foreign-supported attempts at inducing regime change. Most violent conflicts today are asymmetric wars where the traditional theories of war do not apply. The weaker party in an asymmetric conflict may turn to terrorism tactics when the traditional means of warfare are not available to them. The actual damage done by acts of terrorism is much smaller than in traditional warfare, but the regalizing effect can be very high because of the intense media attention (see chapter 4.4 and 6.1).

In fact, the mass media play a major role in the regal-kungic dynamics of modern societies. Most people depend on the mass media for information about political violence and other dangers, because they do not witness these events themselves. The image presented by the mass media is often exaggerated, though, because the media are forced by fierce economic competition to focus on dramatic and highly emotional stories about crime, violence, and mayhem. The 'mean world syndrome' caused by media competition is in fact a major regalizing force in modern society (see chapter 5.1). Politicians sometimes contribute to this effect by fabricating threats and conflicts (see chapter 6.3).

The mass migration of refugees out of conflict zones can cause regal reactions in the countries that receive these refugees. In the future, climate change, natural disasters, overpopulation, and ecological collapse may have similar effects.

Economic factors also contribute to regal developments when an unstable economic system causes crises and growing global inequality. The regal effect of an economic crisis depends on who is blamed for the crisis and whether the politicians are perceived to be in control of the situation (see chapter 5.2).

---

4   Altheide (2014)

Despite these regal influences, we are now seeing strong kungic developments in many parts of the world due to the improved living conditions, peace, and stability that are connected with the demographic transition (see chapter 3.4).

The state of the world today can be characterized by the observation that the pattern of war has changed (see chapter 4.4), but some wars are still being fought based on obsolete principles. Today, we are seeing mostly asymmetric conflicts which tend to be intractable and long-lasting. Asymmetric wars are often fought with unconventional means, including terrorism. Most violent conflicts today can be characterized as proxy wars. The level of conflict is increased here by foreign support in the form of weapons, training, and other resources.

On the surface, these asymmetric wars and proxy wars appear to be conflicts over ideology or religion, but the political or religious extremism of the fighters is mainly a consequence of the conflict itself. The fighters in proxy wars become radicalized as a regal reaction to the violence and also as a result of manipulation by their leaders. Behind the scenes, the sponsors who are pulling the strings are more motivated by a need to secure their access to critical resources and by geopolitical strategy (see chapter 4.3, page 75). They use religion, ideology, and a rhetoric of good and evil mainly to rally support for their cause. The weaker party in asymmetric conflicts is in most cases fighting against the suppression and exclusion of a minority. The religious and ethnic identity of this minority is amplified by the conflict.

The third epoch that we are now in has lasted for less than a century, and the driving forces behind the social and political developments are changing faster than the theories about their workings. The globalized economy is still changing shape with new dynamic forms of international cooperation and competition, concentration of wealth and power, economic booms and busts, security and insecurity, economic regulation and deregulation, and new conflicts and alliances. Likewise, the political climate is changing, with new forms of dynamics formed by new forms of mass communication, competition for attention, economic influences on the mass media, and new forms of propaganda and deception. As scientists, we have a lot of work to do to catch up with these developments in order to understand in which direction these processes are taking our society.

9.3. The regal/kungic dynamics and human social development

While it is certainly more pleasurable for ordinary people to live in a kungic society than a regal one, we should not forget the role that regal periods have played in the history of human development. Regal regimes have certainly caused unfathomable amounts of cruelty and human suffering, but without the historical periods of regality we might still be living in the Stone Age.

The first condition for technological innovation and development is a political and economic structure that allows division of labor, where some people produce food while others can devote their time to being artisans, intellectuals, and other specialists. Scientific progress requires a wealthy government that can support education and pay people for doing research.

A rich and powerful government is necessary for the construction of the large infrastructure that industrial development and large scale international trade depends on. However, this is not the friendliest environment for individual innovation and free enterprise. Throughout most of our history, rich and powerful governments have also been regal. Regal governments are likely to tax or confiscate new enterprises that appear to be profitable. Such governments are extractive in the sense of Daron Acemoglu and James Robinson's theory, as explained in chapter 5.2. Furthermore, regal societies are typically intolerant of the deviance and experimentation that may be necessary for social and cultural innovation.

On the other hand, a kungic society with a lax system of justice may fail to protect private entrepreneurs against theft and blackmail, or fail to provide the accumulation of wealth necessary for investment in new businesses. The societies that have been most conducive to technological progress have been those that had a strong legal system to protect private property, but that also encouraged diversity and individual inventiveness.[5]

The present state of human technology and social organization is the result of a development that has depended on different levels of

---

5   Acemoglu and Robinson (2012, p. 43)

regality at different times and in different places. Most states were created through violence.[6] Regal periods have created larger political units, division of labor, and the accumulation of wealth that has been required for the investment in infrastructure, education, and science. Subsequent kungic periods have fostered individual creativity and inventiveness, and allowed the political, cultural, and religious deviance that was connected with inventiveness, while perhaps the previously accumulated capital made the investment in technological inventions possible. The most fruitful environment for technological and industrial innovation is one that combines democracy and individual freedom with a strong law enforcement system to protect private property. Different periods with different levels of regality have created windows of opportunity for different innovations and developments. The changes in regality level and the conflicts between regal and kungic forces have been the root of much artistic and cultural creativity and innovation.

Today, the possibility of establishing a well-functioning and reliable government and legal system no longer requires a strong regal government; but the decoupling between regality and efficient law enforcement is historically new, as it came with the third epoch described in the previous chapter.

The regal periods in human history have not only produced war, tyranny, imperialism, slavery, cruelty, and mayhem, but also formed the preconditions for the highly developed society that we live in today. Most of the basic principles of law that are necessary for a civilized society to function today were developed in regal periods. Classical music, as well as many of the magnificent pieces of art and architecture that we are impressed by today, was created under the regal regimes of the past. We admire old fairy tales without realizing that they were written to glorify sovereign kings and emperors and to make unambiguous distinctions between friend and foe, between good and evil. And we enjoy the fruits of past scientific, technological, and political progress without thinking of the incredible hardship that made it possible.

---

6   Tilly (1992)

9.4. New explanations of well-known phenomena

Regality theory can contribute to a better understanding of many well-known phenomena, for example the rise and fall of empires, as discussed in previous chapters.

Altruism is a behavior that has always been difficult to explain in evolutionary theory. Many different theories have been put forth, but none of the previously proposed mechanisms seems to be strong enough to explain why people are willing to sacrifice themselves in war. Regality theory offers a model that predicts a stronger effect than the previous theories can account for (see chapter 2.2).

Another common phenomenon that receives a new theoretical explanation is authoritarianism. The psychological concept of authoritarianism has often been criticized for being politically biased and poorly defined. For many years, there has been confusion and disagreement over whether authoritarianism was a fixed personality trait, a psychopathology, a reaction to fear and danger, a consequence of pessimistic world views, or just a divergent political opinion. Regality theory explains the phenomenon that psychologists have called authoritarianism as an evolved pattern of response to perceived collective danger. This new explanation removes the weaknesses of authoritarianism theory and gives a good explanation of the relationship between individual reactions and sociopolitical phenomena (see chapter 3.5).

The question of whether humans are violent or peaceful by nature has often been discussed. Regality theory has the answer: it depends on the environment. The endless debates over whether human differences are due to genes or culture have often totally missed a third possibility, that the genes define a flexibility or plasticity that allows the individual to show different behaviors depending on the environment.

In folk psychology, people often believe that war and violent conflicts are caused by hate, fanaticism, nationalism, or religious extremism. Regality theory turns this argument upside down. Conflict leads to regality, which is typically expressed as xenophobia, fanaticism, an increased sense of national identity, and religious strictness and zeal. While the causal mechanism often goes both ways in a self-amplifying spiral of violence, it is fair to say that xenophobia, fanatical patriotism, religious extremism, and so on are the consequences of conflict more than the causes of it. This insight has important implications for peacemaking efforts. Attempts to root out, for example, a particular expression of religious extremism by force will most likely be counterproductive and create more regality, while third-party interventions that guarantee peace and security will have a kungic effect and make the extremism slowly fade away.

The fundamental dogmas of a religion are quite resistant to change. However, people have a peculiar ability to interpret their religion in a way that matches their psychological propensities or political agenda so that they can justify almost any action—violent or peaceful—by reference to carefully selected parts of their religious principles.

The statistical results show a quite significant correlation between religion and various measures of war and regality in both modern and non-industrial societies. The religion of regal societies typically supports the hierarchical political power structure and serves a disciplining function by emphasizing supernatural rewards and punishments. This insight can improve our understanding of religious differences.

Regality theory improves our understanding of several other cultural phenomena as well, including some phenomena that are difficult to explain with other theories. One phenomenon that has hitherto been difficult to explain is sexual morals. The statistics presented above show that the level of regality explains remarkably well why different cultures have such strikingly different sexual morals. Cultural prohibitions against birth control can be seen as a means by which a regal society tries to increase the population growth. A ban against premarital sex is likely to force young people to marry early and have many children when all other outlets for their sexual drive are closed. Many regal cultures even have bans against such harmless, but non-procreative, sexual behaviors as masturbation and homosexuality. More research is needed to study the possible psychological mechanisms that connect regality with the desire to make rules for the sexual behavior of others.

The theory also improves our understanding of patriarchy. Regal cultures tend to be patriarchal and male-dominated, while kungic cultures have more equality between the sexes.

Regality theory also offers a completely new way of explaining differences in artistic style and taste. Regal cultures have a marked preference for a highly embellished, refined, and perfectionist style, while the members of kungic cultures appreciate the more individualistic, imaginative, and less rule-bound artistic expressions. Art is a form of communication that reflects the social structure as the artist sees it or the social structure that the artist would like to see. This tendency to reflect the level of regality appears to be the same in many different kinds of art, including music, dance, literary fiction, drama, painting, body adornment, and architecture.

Finally, regality theory contributes to a better understanding of democracy. In international relations, it is often believed that democracy is the road to peace, but the statistics show that the democratization of a country almost always comes *after* the borders have been settled and peace established.[7] This observation fits perfectly with regality theory. Unstable or contested borders are a collective threat that drives the political climate of a country in the regal and autocratic direction, while peace and secure borders will promote a kungic political culture, which leads to democracy. In other words: peace comes before democracy.

This also explains the paradoxical phenomenon that the voters in a democracy sometimes elect undemocratic candidates. A return from democracy to autocracy is possible in the face of national threats — real or fabricated. We can expect democracy to be adversely affected by border conflicts, proxy wars, fabricated dangers, highly emotional threats such as terrorism, and by a concentration of valuable resources in areas that can be monopolized and fought over.

9.5. Integration with other theories

A common way of promoting a new theory is to prove previous theories wrong. This naive form of falsificationism has not been very fruitful in the social sciences.[8] While regality theory explains our observations better than alternative theories do, this does not mean that previous theories

---

7   Owsiak (2013), Gibler and Owsiak (2017)
8   Lakatos (1974)

should be completely rejected. Previous theories may still contain some important insights that might be remodeled and combined with the new discoveries in a synthetic process that makes our knowledge evolve and expand its explanatory power. Regality theory may improve previous theories by contributing to a deeper theoretical understanding of both individual behavior and social structures.

There is a long-standing debate in sociology about the relative importance of individual agency and social structure. The present book argues for a theory where individual preferences shape the social structure, and where the social structure determines the power relations between people and determines who is able to get into powerful positions and what interests they serve. The social structure and the environment also have a strong influence on the psychological preferences of individual people. A conflict-filled and dangerous environment fosters authoritarian preferences in people, while a peaceful and safe environment makes people prioritize egalitarian values. These individual preferences then feed back into the social structure in such a way that people tend to build hierarchical political systems in dangerous environments and egalitarian political structures in safe and peaceful environments.

The actor-centered explanation of historical events can be criticized for seeing only proximate causes. For example, Ronald Inglehart and Christian Welzel criticize explanations of democracy that can be reduced to the observation that democracy has been implemented because pro-democratic actors were stronger than anti-democratic actors. It is always clear in retrospect that the winners were stronger than the losers, but this gives no insight into the social forces that explain why the winners were stronger. The actor-centered approach often tries to explain social processes by actions that are endogenous to the processes themselves without seeing the deeper structural mechanisms.[9]

The present theory improves the opportunities for a scientific understanding of history based on environmental and technological factors and intergroup relations. Rather than explaining a war as caused by the whims of a particular bellicose leader, we can start to study why this leader was bellicose and, more importantly, why he had enough

---

9   Inglehart and Welzel (2005, p. 224)

supporters to stay in power, and why the population did not overturn this despot and replace him with somebody more peaceful (see chapter 6.4).

Many social developments are unpredicted or unintended. We cannot explain unintended developments adequately by idiographic accounts of the personal decisions of specific influential people. Unplanned developments are often the emergent effects of the actions of a large number of people, perhaps without anybody understanding how their own decisions contribute to the big picture of macroscopic change. Regality theory is only one among many theories that can explain emergent social developments. Other theories that may contribute to the understanding of unplanned developments include systems theory with feedback models, cultural selection theory, economic market mechanisms, social cognition theory, and media effects theories combined with media economy (see chapters 4.1, 5.1, and 5.2).

9.6. Policy lessons

Regality theory can be useful for guiding political decisions and for predicting the consequences of political actions.

There may be many reasons why one wants to push a society and culture in the kungic direction. A kungic development is the road to peace, tolerance, and democracy. It can also be an important tool for saving the environment. A kungic culture is usually coupled with a *K*-like life history strategy, which implies better education and reduced population growth (chapter 3.3). The current level of consumption of natural resources exceeds the carrying capacity of the planet, and so does the level of pollution.[10] We have often observed that the protection of the environment has very low priority in the event of war but higher priority in times of peace. A kungic development combined with demographic transition is probably the most efficient way to avoid ecological collapse.

The following policies may be useful for pushing a society and culture in the kungic direction:

---

10   Steffen et al. (2015)

- Protect the borders of countries. Prevent violent territorial conflicts. Borders should be created, removed, or moved only by peaceful negotiations, or decided by public referendum in the affected areas. Strong, fair, and accountable international organizations should oversee this.

- Remove violence, particularly against civilians, and reduce threat perceptions.[11]

- Resolve asymmetric conflicts, which tend to be intractable and produce terror. International mediation is often necessary for solving such conflicts in a way that is acceptable to both parties.

- Prevent proxy wars and expose clandestine foreign support for conflicting factions. Do not buy oil, diamonds, drugs, or other valuable products from conflict areas. Prevent foreign support for violent regime changes. Prevent and expose the deliberate fabrication of conflicts by third parties.

- Create economic stability and safety. The money system should not generate unpayable debt or inescapable debt dependence. Restrain the Matthew effect. Regulate the kinds of international trade that would lead to a race to the bottom.

- Reduce the dependence on critical resources, such as oil, rare minerals, and water. Make sure that all people and all countries have fair access to vital resources.

- Establish social security systems and safety nets that guarantee the basic necessities of life and health for everybody and make people feel safe.

- Promote good education for all citizens. Support education for the poor. Make education attractive by providing jobs for the educated.

- Support free mass media that do not have to rely on fearmongering and exaggeration of dangers for reasons of economic competition. This requires alternative sources of funding for 'public service' news media.

---

11   Hirsch-Hoefler, Canetti, Rapaport and Hobfoll (2014)

- Prevent overpopulation. The above policies will in fact advance the demographic transition and help to prevent overpopulation.

This theoretical discussion would not be complete without also describing policies that can be used to push a society in the regal direction. A society that is too kungic will be vulnerable to attacks from more regal forces. Karl Popper, citing Plato, expressed this as the paradox of freedom. Freedom in the sense of absence of restraining control 'makes the bully free to enslave the meek'. Popper also explained the paradox of tolerance: 'If we extend unlimited tolerance even to those who are intolerant, if we are not prepared to defend a tolerant society against the onslaught of the intolerant, then the tolerant will be destroyed, and tolerance with them'.[12] The optimal position of a society on the regal-kungic dimension depends on the dangers it faces from its neighbors as well as on the need for solidarity and collective action.

Regalizing policies have often been used to strengthen a government against the interests of the majority of the population, as we have seen in chapters 6.2 and 6.3. Such policies can also be used to strengthen a country against a likely attack from a militant neighbor, to increase solidarity in times of crisis, to nurture the will to cultural self-preservation, or to strengthen a military organization. Actions and inactions that can make societies more regal include the following:

- War, including proxy war and fabricated conflicts; political influence of the weapons industry; weak and corrupt international organizations.

- An unregulated economy depending on debt, exploitation of other countries, increasing inequality, perpetual growth, and overconsumption of natural resources.

- Climate change, overexploitation, and destruction of the environment.

- Exaggerated stories about all kinds of dangers in the mass media; a highly competitive media market.

---

12   Popper (1945)

- Political control of artistic production, especially the suppression and destruction of kungic art.

Evidently, tyrants and despots have used regalizing techniques such as these over the millennia to increase their own power, even if they did not know why these techniques worked. The list of regalizing techniques provided here will hardly add anything to the repertoire of power-hungry autocrats. It can, however, be useful to a political opposition and to the general population because it makes it easier to see through the hidden motives of rulers who want to strengthen their own power or abuse their power to enrich themselves.

The insights provided by regality theory can also be useful for identifying ineffective and counterproductive policies. For example, both parts in an asymmetric conflict should avoid escalating the violence because this will lead to further regalization and make the conflict intractable.

This theory also explains why so many attempts to establish democracy by force have failed. International interventions in trouble-filled countries have too often had a focus on regime change. While the professed aim has been to establish democracy, the result has invariably been less liberating. No matter how tyrannical the leader of an undemocratic country is, he will have supporters who see any foreign-backed attempts to remove him by force as an attack on their country. This has a regalizing effect that is likely to make the country less democratic, not more. The analyses of conflicts need to focus on systemic causes rather than hunting down individual 'bad guys'. Removing an 'evil dictator' by force will only regalize the population of his country and make sure his successor will be even more militant and despotic. Peace and stability must come before democracy.[13]

Counterinsurgency and counterterrorism have too often been counterproductive. They ignore the grievances that provide the driving force behind all insurgency. They also ignore the fact that, in most cases, there are no military targets that the insurgents can meaningfully attack with any chance of success, so they can see no other option than to attack less legitimate targets. This makes it easy for a suppressive regime to call them terrorists. We can be pretty sure that they have already tried less radical forms of protest with disappointing results.

---

13   Owsiak (2013)

It should be no surprise that insurgent movements become more radical and militant the stronger their grievances and the more they are suppressed. Their radical religious and political ideologies are likely to lose adherents if their grievances are dealt with in a sensible way. A suppressive and kleptocratic regime may be the main cause of their grievances. International intervention is unlikely to bring lasting peace if it only helps a despotic regime suppress the insurgents while ignoring their grievances. Intelligence organizations with their traditional focus on secret operations and paying secret informants are only likely to exacerbate the corruption that lies at the core of the conflict. Intelligence gathering should instead focus on identifying the root causes of a conflict.[14]

International policies that are likely to fail include the following:

- Blaming insecurity and political problems on specific political leaders; hunting down 'evil dictators' while ignoring the basic reasons why certain countries have despotic leaders.

- Blaming conflict on religion. Fanatic religiosity is a consequence of conflict as much as, or rather more than, a cause of it. People will automatically become less fanatical and less bellicose when their environment is made more secure. Putting the blame on 'Islamists', 'Zionists', or 'Crusaders' will only make them feel that their religion is threatened and make them more regal.

- Trying to solve international security problems by forcing a regime change in other countries by violent means.

- Proxy war; covert support for coup attempts in other countries; opportunistic support for one faction or another in foreign countries based on one's own economic and geostrategic interests.

- Focus on violent conflict rather than state building; trying to impose democracy on other countries by force without focusing on the root causes of why the country has an autocratic rule; trying to establish democracy before stability.[15]

---

14 Chayes (2015, p. 154)
15 Inglehart and Welzel (2005, p. 297), Owsiak (2013), Ahram (2011, p. 139)

9.7. Supporting evidence

Let us review the most important support we now have for regality theory:

1. First, we have a solid theoretical basis. Evolutionary theory has always had a problem explaining why people are willing to fight for their group. Many models have been proposed, but the effects always seemed to be too weak to account for the high costs and dangers of fighting. Regality theory offers a model with a stronger potential effect because it involves a strong leader who has the power to coordinate, reward, and punish. Common support for a strong leader makes sure that everybody will be fighting rather than free riding. A hypothetical genetic code that says 'support a strong leader in times of collective danger, but not otherwise' would be evolutionarily stable (see chapters 2.2 and 2.3). The actual observations are in good agreement with this model.

2. The extensive research that has been carried out in connection with authoritarianism theory confirms that people become authoritarian when they perceive collective danger. This involves support for a strong leader and strict discipline (see chapter 3.5).

3. Many previous studies have found various effects of collective danger that are in accordance with regality theory (see table 18, chapter 8.5).

4. Regality theory provides an explanation for the rise and fall of empires that is in very good agreement with the theory and observations behind Peter Turchin's historical dynamics (see chapter 4.2).

5. Many scientists have tried to map cultural differences. Different scientists have assigned different names to the cultural factors that they discovered, such as cultural tightness, harshness, traditional values, survival values, emancipative values, conservatism, and nastiness. All these constructs—whatever they are called—seem to be closely related to each other (see table 17, chapter 8.4). The fact that different scientists, relying on different theories and methods, make similar discoveries

suggests that there is a common underlying reality behind all these findings. Most of these different constructs are significantly correlated with regality at the country level. The various studies of cultural factors have found correlations with many different aspects of culture. These correlations are in very good agreement with the predictions of regality theory (see chapter 3.7 and 3.8).

6. A large statistical analysis of 8,883 persons in 33 contemporary countries found highly significant correlations between perceived danger, desire for a tough leader, bellicosity, and several other variables in accordance with the predictions of regality theory (see table 15 and figure 31, chapter 8.4).

7. A statistical analysis on ancient, non-industrial societies has been made based on data from the 186 cultures in Murdoch and White's Standard Cross-Cultural Sample. A structural equation model supports the predicted relationships between war, leadership, rewards for warriors, various cultural indicators of regality, life history strategy, and sexual morals (see figure 29, chapter 8.2).

8. A subsample consisting of eighteen of these ancient, non-industrial societies was investigated in more detail in a comparative historical analysis. The level of intergroup conflict in each of these societies was in very good agreement with the conflict level that could be predicted based on geographic, ecological, and technological factors. The level of hierarchy, discipline, and several other cultural indicators agreed very well with the predictions of regality theory for these societies and confirmed the direction of causality. A statistical summary of these results showed that most of the predicted relationships were highly significant (see table 11, chapter 8.3).

Such a broad and solid range of support is quite unusual for a social science theory. This is an excellent basis for further research and exploration of the many aspects of regality theory and the many social and cultural variables that are influenced by the level of regality, according to this theory.

9.8. What regality theory can be used for

This book has explained how regality theory was developed with contributions from many different scientific disciplines. In this conclusion, it is time to suggest how regality theory can give something back to these disciplines. The following list shows examples of how regality theory can make useful contributions to other scientific disciplines:

- In history, to explain war and peace, despotism, the rise and fall of empires, and the development of democracy.

- In archaeology, to get an image of the social and political structure of a society based on physical remains. A regal culture is likely to leave large, impressive, perfectionist, and embellished artifacts made from durable materials, while the artifacts of kungic cultures are likely to be less rule-bound, to reflect individual fantasy, and to be made of perishable materials. This causes a sampling bias in the archaeological record (see chapter 2.6).

- In social psychology, to explain authoritarianism, racism, xenophobia, punitiveness, and tolerance; to suggest a revision of the politically biased authoritarianism theory (see chapter 3.5).

- In conflict and peace research, to identify the conditions that lead to conflict or peace and to understand the social and psychological consequences of different kinds of conflict, including international war, civil war, proxy war, insurgency, revolution, terrorism, and spirals of violence.

- In the history and sociology of religion, to relate changes in religious sentiments, beliefs, and rituals to social and political changes and ecological factors.

- In art history, to relate artistic innovations and genres of art to social and political developments.

- In cultural studies, to understand the connections between social conditions, political ideas, artistic style, music genres, religious movements, and so on.

- In sexology, to relate sexual behaviors, lifestyles, morals, and tolerance or intolerance to the social and political climate.

- In media studies, to understand how the economic mechanisms in a competitive media market shape the quality of the media and the number of titillating or scary stories, and how this in turn may create a 'mean world syndrome' that influences the general level of regality in the society (chapter 5.1).

- In political demography, to understand the connection between ecological environment, *r/K* life history strategies, demographic transition, and political changes along the regal-kungic dimension.

- In political science, to understand the social and psychological factors that make authoritarian or egalitarian ideas popular, and to predict the likely consequences of political decisions.

- In futurology, to predict future developments in the political climate.

9.9. Further discussion

A forum for the discussion of regality theory is provided online at http://www.regality.info

10. Bibliography

Aagaard, Charlotte. *I Nationens Tjeneste: Frank Grevil, Majoren, der fik nok*. Copenhagen: Informations Forlag, 2005.

Abbas, Hassan. *Pakistan's Drift into Extremism: Allah, the Army, and America's War on Terror*. New York: M. E. Sharpe, 2005.

— *The Taliban Revival: Violence and Extremism on the Pakistan-Afghanistan Frontier*. New Haven, CT: Yale University Press, 2014.

Abelson, Robert P. "Psychological Status of the Script Concept". *American Psychologist* 36, no. 7 (1981): 715, https://doi.org/10.1037/0003-066x.36.7.715

Abrahms, Max. "Why Terrorism Does Not Work". *International Security* 31, no. 2 (2006): 42–78, https://doi.org/10.1162/isec.2006.31.2.42

— "Does Terrorism Really Work? Evolution in the Conventional Wisdom since 9/11". *Defence and Peace Economics* 22, no. 6 (2011): 583–94, https://doi.org/10.1080/10242694.2011.635954

Abrahms, Max and Karolina Lula. "Why Terrorists Overestimate the Odds of Victory". *Perspectives on Terrorism* 6, nos. 4–5 (2012).

Abrahms, Max and Matthew S. Gottfried. "Does Terrorism Pay? An Empirical Analysis". *Terrorism and Political Violence* 28, no. 1 (2014): 72–89, http://dx.doi.org/10.1080/09546553.2013.879057

Acemoğlu, Daron and James A. Robinson. *Economic Origins of Dictatorship and Democracy*. Cambridge: Cambridge University Press, 2005, https://doi.org/10.1017/cbo9780511510809

— *Why Nations Fail: The Origins of Power, Prosperity, and Poverty*. London: Profile Books, 2012.

Adams, Henry E., Lester W. Wright, and Bethany A. Lohr. "Is Homophobia Associated with Homosexual Arousal?" *Journal of Abnormal Psychology* 105, no. 3 (1996): 440–45, https://doi.org/10.1037//0021-843x.105.3.440

Adams, James. *The Financing of Terror*. Sevenoaks: New English Library, 1986.

Adorno, Theodor W., Else Frenkel-Brunswik, Daniel J. Levinson, and R. Nevitt Sanford. *The Authoritarian Personality*. New York: Harper and Brothers, 1950.

Agee, Philip. *Inside the Company*: *CIA Diary*. London: Allen Lane, 1975.

Ahram, Ariel Ira. *Proxy Warriors: The Rise and Fall of State-Sponsored Militias*. Stanford, CA: Stanford University Press, 2011.

Allen, Mark W. and Terry L. Jones. *Violence and Warfare among Hunter-Gatherers*. Walnut Creek, CA: Left Coast Press, 2014.

Alonso, Rogelio. "The Modernization in Irish Republican Thinking toward the Utility of Violence". *Studies in Conflict and Terrorism* 24, no. 2 (2001): 131–44, https://doi.org/10.1080/10576100151101641

Altemeyer, Bob. "The Other 'Authoritarian Personality'." *Advances in Experimental Social Psychology* 30 (1998): 48–92, https://doi.org/10.1016/s0065-2601(08)60382-2

Altheide, David L. *An Ecology of Communication*: *Cultural Formats of Control*. New York: Aldine de Gruyter, 1995.

—*Creating Fear*: *News and the Construction of Crisis*. New York: Aldine de Gruyter, 2002.

—*Terrorism and the Politics of Fear*. Lanham, MD: AltaMira Press, 2006.

—"Moral Panic: From Sociological Concept to Public Discourse". *Crime Media Culture* 5, no. 1 (2009): 79–99, https://doi.org/10.1177/1741659008102063

—*Media Edge*: *Media Logic and Social Reality*. New York: Peter Lang, 2014.

Amara, Emmanuel. *Nous Avons Tué Aldo Moro*. Paris: Patrick Robin, 2006.

Amnesty International. "The Amnesty International Report". Oxford: Amnesty International Publications, 2004, https://www.amnesty.org/en/documents/pol10/0004/2004/en

Anderson, Cameron, and Robb Willer. "Do Status Hierarchies Benefit Groups? A Bounded Functionalist Account of Status". In *The Psychology of Social Status*, edited by J. T. Cheng and Robb Willer, 47–70. New York: Springer, 2014. https://doi.org/10.1007/978-1-4939-0867-7_3

Anderson, Tim. *The Dirty War on Syria*: *Washington, Regime Change and Resistance*. Montreal: Global Research, 2016.

Apt, Wenke. *Germany's New Security Demographics*: *Military Recruitment in the Era of Population Aging*. New York: Springer, 2014.

Aptekar, Pavel. "Casus Belli". Raboche-Krest'ianskaia Krasnaia Armiia, 2001, http://rkka.ru/analys/mainila/mainila.htm

Arom, Simha. "Musicologie et Ethnomusicologie". In *Encyclopédie des Pygmées Aka*: *Techniques, Langage et Société des Chasseurs-Cueilleurs de la Forêt Centrafricaine*, Vol. I-1, edited by Jacqueline M. C. Thomas and Serge Bahuchet. Paris: SELAF, 1983.

Arreguín-Toft, Ivan. "Contemporary Asymmetric Conflict Theory in Historical Perspective". *Terrorism and Political Violence* 24, no. 4 (2012): 635–57, https://doi.org/10.1080/09546553.2012.700624

Asbrock, Frank and Immo Fritsche. "Authoritarian Reactions to Terrorist Threat: Who Is Being Threatened, the Me or the We?" *International Journal of Psychology* 48, no. 1 (2013): 35–49, https://doi.org/10.1080/00207594.2012.695075

Audigier, Vincent, François Husson, and Julie Josse. "A Principal Components Method to Impute Missing Values for Mixed Data", arXiv.org, 2013, no. 1301.4797v2, http://arxiv.org/pdf/1301.4797v2.pdf

Auty, Richard. "The Oil Curse. Causes, Consequences, and Policy Implications". In *Handbook of Oil Politics*, edited by Robert E. Looney, 337–48. London: Routledge, 2012.

Ayi, Bamo. "On the Nature and Transmission of Bimo Knowledge in Liangshan". In *Perspectives on the Yi of Southwest China*, edited by Stevan Harrell, 118–31. Berkeley, CA: University of California Press, 2001, https://doi.org/10.1525/california/9780520219885.003.0009

Babb, Sarah. "The Social Consequences of Structural Adjustment: Recent Evidence and Current Debates". *Annual Review of Sociology*, 2005, 199–222, https://doi.org/10.1146/annurev.soc.31.041304.122258

Baek, Young Min and Magdalena E. Wojcieszak. "Don't Expect Too Much! Learning From Late-Night Comedy and Knowledge Item Difficulty". *Communication Research* 36, no. 6 (2009): 783–809, https://doi.org/10.1177/0093650209346805

Bailey, Robert C. and Irven DeVore. "Research on the Efe and Lese Populations of the Ituri Forest, Zaire". *American Journal of Physical Anthropology* 78, no. 4 (1989): 459–71, https://doi.org/10.1002/ajpa.1330780402

Baily, John. *War, Exile and the Music of Afghanistan. The Ethnographer's Tale.* Farnham: Ashgate, 2015, https://doi.org/10.4324/9781315466934

Balaresque, Patricia, et al. "Y-Chromosome Descent Clusters and Male Differential Reproductive Success: Young Lineage Expansions Dominate Asian Pastoral Nomadic Populations". *European Journal of Human Genetics* 23 (2015): 1413–22, https://doi.org/10.1038/ejhg.2014.285

Bale, Jeffrey M. "The May 1973 Terrorist Attack at Milan Police HQ: Anarchist 'propaganda of the Deed' or 'false-Flag' provocation?" *Terrorism and Political Violence* 8, no. 1 (1996): 132–66, https://doi.org/10.1080/09546559608427337

— "Terrorists as State Proxies: Separating Fact from Fiction". In *Making Sense of Proxy Wars*: *States, Surrogates & the Use of Force*, edited by Michael A. Innes, 1–29. Washington, D.C.: Potomac, 2012.

Ballentine, Karen and Jake Sherman. *The Political Economy of Armed Conflict*: *Beyond Greed and Grievance*. Boulder, CO: Lynne Rienner Publishers, 2003.

Banks, William E., Francesco d'Errico, Harold L. Dibble, Leonard Krishtalka, Dixie West, et al. "Eco-cultural Niche Modeling: New Tools for Reconstructing the Geography and Ecology of Past Human Populations". *PaleoAnthropology* 4 (2006): 68–83.

Barber, Nicola. *Afghanistan*. London: Arcturus Publishing, 2008.

Barfield, Thomas. *Afghanistan: A Cultural and Political History*. Princeton, NJ: Princeton University Press, 2010.

Basedow, Herbert. *The Australian Aboriginal*. Adelaide: F. W. Preece and Sons, 1925.

Basow, Susan A. and Kelly Johnson. "Predictors of Homophobia in Female College Students". *Sex Roles* 42, nos. 5–6 (2000): 391–404, https://doi.org/10.1023/a:1007098221316

Bateson, Patrick and Peter Gluckman. *Plasticity, Robustness, Development and Evolution*. Cambridge: Cambridge University Press, 2011.

Baudin, Louis. *A Socialist Empire: The Incas of Peru*. Princeton, NJ: D. Van Nostrand, 1961.

Baudrillard, Jean. "L'esprit du Terrorisme". *Le Monde*. November 2, 2001.

Bauer, Brian S. *Ancient Cuzco: Heartland of the Inca*. Austin, TX: University of Texas Press, 2004.

Bauman, Zygmunt and Carlo Bordoni. *State of Crisis*. Cambridge: Polity Press, 2014.

Baumann, Oskar. *Durch Massailand zur Nilquelle: Reisen und Forschungen der Massai-Expedition des deutschen Antisklaverei-Komite in den Jahren 1891–1893*. Berlin: Dietrich Reimer, 1894.

Becher, Tony and Paul R. Trowler. *Academic Tribes and Territories*. 2nd ed. McGraw-Hill, 2001.

Beck, Ulrich. *Risk Society: Towards a New Modernity*. London: Sage, 1992.

Beckerman, Stephen, Pamela I. Erickson, James Yost, Jhanira Regalado, Lilia Jaramillo, Corey Sparks, Moises Iromenga, and Kathryn Long. "Life Histories, Blood Revenge, and Reproductive Success among the Waorani of Ecuador". *Proceedings of the National Academy of Sciences* 106, no. 20 (2009): 8134–39, https://doi.org/10.1073/pnas.0901431106

Bécue-Bertaut, Mónica and Jérôme Pagès. "Multiple Factor Analysis and Clustering of a Mixture of Quantitative, Categorical and Frequency Data". *Computational Statistics and Data Analysis* 52, no. 6 (2008): 3255–68, https://doi.org/10.1016/j.csda.2007.09.023

Beed, Clive and Cara Beed. "Is the Case for Social Science Laws Strengthening?" *Journal for the Theory of Social Behaviour* 30, no. 2 (2000): 131–53, https://doi.org/10.1111/1468-5914.00123

Belsky, Jay, Gabriel L. Schlomer, and Bruce J. Ellis. "Beyond Cumulative Risk: Distinguishing Harshness and Unpredictability as Determinants of Parenting and Early Life History Strategy". *Developmental Psychology* 48, no. 3 (2012): 662–73, https://doi.org/10.1037/a0025837

Bendahan, Samuel, Christian Zehnder, François P. Pralong, and John Antonakis. "Leader Corruption Depends on Power and Testosterone". *The Leadership Quarterly* 26, no. 2 (2014): 101–22, https://doi.org/10.1016/j.leaqua.2014.07.010

Benes, Jaromir and Michael Kumhof. "The Chicago Plan Revisited". IMF Working Paper, no. 202 (2012), https://doi.org/10.5089/9781475505528.001

Bennett, W. Lance and Shanto Iyengar. "A New Era of Minimal Effects? The Changing Foundations of Political Communication". *Journal of Communication* 58, no. 4 (2008): 707–31, https://doi.org/10.1111/j.1460-2466.2008.00410.x

Ben-Yehuda, Nachman. "The European Witch Craze of the 14th to 17th Centuries: A Sociologist's Perspective". *American Journal of Sociology* 86, no. 1 (1980): 1–31, https://doi.org/10.1086/227200

— *The Politics and Morality of Deviance*. Albany, NY: State University of New York Press, 1990.

Berger, Jonah. *Contagious: Why Things Catch On*. New York: Simon and Schuster, 2013.

Berry, John W. "Independence and Conformity in Subsistence-Level Societies". *Journal of Personality and Social Psychology* 7, no. 4 (1967): 415–18, https://doi.org/10.1037/h0025231

Berry, Steven T. and Joel Waldfogel. "Free Entry and Social Inefficiency in Radio Broadcasting". *RAND Journal of Economics* 30, no. 3 (1999): 397–420.

— "Do Mergers Increase Product Variety? Evidence from Radio Broadcasting". *Quarterly Journal of Economics*, 2001, 1009–25, https://doi.org/10.1162/00335530152466296

Besançon, Marie L. "Relative Resources: Inequality in Ethnic Wars, Revolutions, and Genocides". *Journal of Peace Research* 42, no. 4 (2005): 393–415, https://doi.org/10.1177/0022343305054086

Best, Joel. *Threatened Children: Rhetoric and Concern about Child-Victims*. Chicago, IL: University of Chicago Press, 1990.

Betzig, Laura L. *Despotism and Differential Reproduction: A Darwinian View of History*. New York: Aldine, 2008.

Bezemer, Dirk and Michael Hudson. "Finance is Not the Economy: Reviving the Conceptual Distinction". *Journal of Economic Issues* 50, no. 3 (2016): 745–68, https://doi.org/10.1080/00213624.2016.1210384

Biesele, Megan. "Aspects of !Kung Folklore". In *Kalahari Hunter-Gatherers*, edited by Richard B. Lee and Irven DeVore, pp. 302–24. Cambridge, MA: Harvard University Press, 1976.

Bird, Rebecca. "Cooperation and Conflict: The Behavioral Ecology of the Sexual Division of Labor". *Evolutionary Anthropology* 8, no. 2 (1999): 65–75, https://doi.org/10.1002/(sici)1520-6505(1999)8:2%3C65::aid-evan5%3E3.0.co;2-3

Bird, Tim and Alex Marshall. *Afghanistan: How the West Lost Its Way.* New Haven: Yale University Press, 2011.

Blix, Hans. *Disarming Iraq: The Search for Weapons of Mass Destruction.* London: Bloomsbury, 2004.

Boas, Franz. *The Central Eskimo.* Lincoln, NE: University of Nebraska Press, 1964.

Boesch, Christophe, Grégoire Kohou, Honora Néné, and Linda Vigilant. "Male Competition and Paternity in Wild Chimpanzees of the Taï Forest". *American Journal of Physical Anthropology* 130, no. 1 (2006): 103–15, https://doi.org/10.1002/ajpa.20341

Boesch, Christophe. "Patterns of Chimpanzee's Intergroup Violence". In *Human Morality and Sociality: Evolutionary and Comparative Perspectives*, edited by Henrik Høgh-Olesen, 132–59. New York: Palgrave Macmillan, 2010.

Bonanno, George A. and John T. Jost. "Conservative Shift among High-Exposure Survivors of the September 11th Terrorist Attacks". *Basic and Applied Social Psychology* 28, no. 4 (2006): 311–23, https://doi.org/10.1207/s15324834basp2804_4

Boone, James L. "Competition, Conflict, and The Development of Social Hierarchies". in *Evolutionary Ecology and Human Behavior*, edited by Eric A. Smith and Bruce Winterhalder, 301–37. New York: Aldine de Gruyter, 1992.

Boyd, Robert, Herbert Gintis, Samuel Bowles, and Peter J. Richerson. "The Evolution of Altruistic Punishment". *Proceedings of the National Academy of Sciences* 100, no. 6 (2003): 3531–35, https://doi.org/10.1073/pnas.0630443100

Bowles, Samuel and Herbert Gintis. *A Cooperative Species: Human Reciprocity and Its Evolution.* Princeton, NJ: Princeton University Press, 2011, https://doi.org/10.1515/9781400838837

Bradley, Brenda J. "Levels of Selection, Altruism, and Primate Behavior". *Quarterly Review of Biology* 74 (1999): 171–94, https://doi.org/10.1086/393070

Brand, Laurie A. "Palestinians and Jordanians: A Crisis of Identity". *Journal of Palestine Studies* 24, no. 4 (1995): 46–61, https://doi.org/10.2307/2537757

Breckler, Steven J. "Applications of Covariance Structure Modeling in Psychology: Cause for Concern?" *Psychological Bulletin* 107, no. 2 (1990): 260–73, https://doi.org/10.1037//0033-2909.107.2.260

Bridges, E. Lucas and Samuel K. Lothrop. "The Canoe Indians of Tierra del Fuego". In *A Reader in General Anthropology*, edited by Carleton Stevens Coon, 84–116. London: Jonathan Cape,1950.

Bridges, E. Lucas. *Uttermost part of the Earth: Indians of Tierra del Fuego*. New York: Dutton, 1949. Reprint: Dover, 1988.

Brinson, Mary and Michael Stohl. "From 7/7 to 8/10: Media Framing of Terrorist Incidents in the United States and United Kingdom". In *The Faces of Terrorism: Multidisciplinary Perspectives*, edited by David Canter, 227–44, 2009, https://doi.org/10.1002/9780470744499.ch13

Brodie, Richard. *Virus of the Mind: The New Science of the Meme*. Seattle, WA: Integral Press, 1996.

Broude, Gwen J. and Sarah J. Greene. "Cross-Cultural Codes on Twenty Sexual Attitudes and Practices". *Ethnology* 15, no. 4 (1976): 409–29, https://doi.org/10.2307/3773308

Brundage, Burr Cartwright. *Empire of the Inca*. Norman, OK: University of Oklahoma Press, 1963.

Bryant, Jennings and Mary Beth Oliver, eds. *Media Effects: Advances in Theory and Research*. 3rd edition. New York: Routledge, 2009.

Buhaug, Halvard and Päivi Lujala. "Accounting for Scale: Measuring Geography in Quantitative Studies of Civil War". *Political Geography* 24, no. 4 (2005): 399–418, https://doi.org/10.1016/j.polgeo.2005.01.006

Buhaug, Halvard. "Relative Capability and Rebel Objective in Civil War". *Journal of Peace Research* 43, no. 6 (2006): 691–708, https://doi.org/10.1177/0022343306069255

Bull, Martin. "Villains of the Peace: Terrorism and the Secret Services in Italy". *Intelligence and National Security* 7, no. 4 (1992): 473–78, https://doi.org/10.1080/02684529208432181

Burt, Jo-Marie. "Playing Politics with Terror: The Case of Fujimori's Peru". In *Playing Politics with Terrorism*, edited by George Kassimeris, 62–100. New York: Columbia University Press, 2008.

Butt, Tahir Mehmood. "Social and Political Role of Madrassa: Perspectives of Religious Leaders in Pakistan". *South Asian Studies* 27, no. 2 (2012): 387–404.

Buzbee, William W. "Recognizing the Regulatory Commons: A Theory of Regulatory Gaps". *Iowa Law Review* 89 (2003): 1, https://doi.org/10.2139/ssrn.447700

Calabrese, Andrew. "Casus Belli. US Media and the Justification of the Iraq War". *Television & New Media* 6, no. 2 (2005): 153–75, https://doi.org/10.1177/1527476404273952

Canetti, Daphna, Carmit Rapaport, Carly Wayne, Brian J. Hall, and Stevan E. Hobfoll. "An Exposure Effect? Evidence From a Rigorous Study on the Psycho-Political Outcomes of Terrorism". In *The Political Psychology of Terrorism Fears*, edited by Samuel Justin Sinclair and Daniel Antonius, 193–212. New York: Oxford University Press, 2013, https://doi.org/10.1093/acprof:oso/9780199925926.003.0011

Canetti, Daphna, Julia Elad-Strenger, Iris Lavi, Dana Guy, and Daniel Bar-Tal. "Exposure to Violence, Ethos of Conflict, and Support for Compromise Surveys in Israel, East Jerusalem, West Bank, and Gaza". *Journal of Conflict Resolution* 61, no. 1, 2015, https://doi.org/10.1177/0022002715569771

Canetti-Nisim, Daphna, Eran Halperin, Keren Sharvit, and Stevan E. Hobfoll. "A New Stress-Based Model of Political Extremism: Personal Exposure to Terrorism, Psychological Distress, and Exclusionist Political Attitudes". *Journal of Conflict Resolution* 53, no. 3 (2009): 363–89, https://doi.org/10.1177/0022002709333296

Cappella, Joseph N. and Kathleen Hall Jamieson. *Spiral of Cynicism: The Press and the Public Good.* New York: Oxford University Press, 1997.

Carlile, Lord of Berriew and Caris Owen. "The Impact and Consequences of Terrorist Legislation in the United Kingdom Since 2001: A Review". In *Investigating Terrorism: Current Political, Legal and Psychological Issues*, edited by John Pearse, 11–30. Chichester: Wiley-Blackwell, 2015.

Carment, David and Yiagadeesen Samy. "The Future of War: Understanding Fragile States and What to Do about Them". In *Failed States and Fragile Societies: A New World Disorder?*, edited by Ingo Trauschweizer and Steven M. Miner, 3–27. Athens, OH: Ohio University Press, 2014.

Carmody, Pádraig. *The New Scramble for Africa.* Malden, MA: Polity Press, 2011.

Carnagey, Nicholas L. and Craig A. Anderson. "Changes in Attitudes towards War and Violence after September 11, 2001". *Aggressive Behavior* 33, no. 2 (2007): 118–29, https://doi.org/10.1002/ab.20173

Carter, James, M. "The Lessons of the last War are Clear: The Military-Industrial Complex, Private Contractors, and US Foreign Policy". In *Failed States and Fragile Societies: A New World Disorder?*, edited by Ingo Trauschweizer and Steven M. Miner, 62–88. Athens, OH: Ohio University Press, 2014.

Cashman, Greg. *What Causes War?: An Introduction to Theories of International Conflict.* 2nd ed. Lanham, MD: Rowman & Littlefield, 2013.

Cassanelli, Lee V. *The Shaping of Somali Society: Reconstructing the History of a Pastoral People, 1600–1900.* Philadelphia, PA: University of Pennsylvania Press, 1982.

Cavalli-Sforza, Luigi L. *African Pygmies.* Orlando, FL: Academic Press, 1986.

Cederman, Lars-Erik, Andreas Wimmer, and Brian Min. "Why Do Ethnic Groups Rebel? New Data and Analysis". *World Politics* 62, no. 1 (2010): 87–119, https://doi.org/10.1017/s0043887109990219

Cederman, Lars-Erik, Kristian Skrede Gleditsch, and Halvard Buhaug. *Inequality, Grievances, and Civil War.* Cambridge: Cambridge University Press, 2013.

Çelik, Selahattin. *Türkische Konterguerilla: Die Todesmaschine.* Cologne: Mesopotamien Verlag, 1999.

Cento Bull, Anna. *Italian Neofascism: The Strategy of Tension and the Politics of Nonreconciliation*. New York: Berghahn Books, 2007.

Chagnon, Napoleon A. *Yąnomamö: The Fierce People*, 2nd ed. New York: Holt, Rinehart and Winston, 1977.

— "Protein Deficiency and Tribal Warfare in Amazonia: New Data". *Science* 203, no. 4383 (1979): 910–13, https://doi.org/10.1126/science.570302

— "Reproductive and Somatic Conflicts of Interest in the Genesis of Violence and Warfare among Tribesmen". In *The Anthropology of War*, edited by Jonathan Hass, 77–104. Cambridge: Cambridge University Press, 1990.

— *Yanomamö: The last Days of Eden*. San Diego, CA: Harcourt Brace Jovanovich, 1992.

Chapman, Tracey. "Sexual Conflict and Evolutionary Psychology: Towards a Unified Framework". In *The Evolution of Sexuality*, edited by Todd K. Shackelford and Ranald D. Hansen, 1–28. Springer, 2015.

Chard, Chester S. "Sternberg's Materials on the Sexual Life of the Gilyak". *Anthropological Papers of the University of Alaska* 10 (1961): 13–23.

Chayes, Sarah. *Thieves of State: Why Corruption Threatens Global Security*. New York: W. W. Norton & Co., 2015.

Chen, Daniel L. "Club Goods and Group Identity: Evidence from Islamic Resurgence during the Indonesian Financial Crisis". *Journal of Political Economy* 118, no. 2 (2010): 300–54, https://doi.org/10.1086/652462

Chen, Yu-Sheng, Antonel Olckers, Theodore G. Schurr, Andreas M. Kogelnik, Kirsi Huoponen, et al. "mtDNA variation in the South African Kung and Khwe—and their Genetic Relationships to Other African Populations". *The American Journal of Human Genetics* 66, no. 4 (2000): 1362–83, https://doi.org/10.1086/302848

Choi, Jung-Kyoo and Samuel Bowles. "The Coevolution of Parochial Altruism and War". *Science* 318, no. 5850 (2007): 636–40, https://doi.org/10.1126/science.1144237

Chowanietz, Christophe. "Rallying around the Flag or Railing against the Government? Political Parties' Reactions to Terrorist Acts". *Party Politics* 17, no. 5 (2010): 673–98, https://doi.org/10.1177/1354068809346073

CIA. "The World Factbook". 2016, https://www.cia.gov/library/publications/the-world-factbook

Cikara, Mina and Jay J. van Bavel. "The Neuroscience of Intergroup Relations: An Integrative Review". *Perspectives on Psychological Science* 9, no. 3 (2014): 245–74, https://doi.org/10.1177/1745691614527464

Cincotta, Richard P., Robert Engelman, and Daniele Anastasion. "The Security Demographic: Population and Civil Conflict after the Cold War". Washington, D.C.: Population Action International, 2003.

Cincotta, Richard P. and Elisabeth Leahy. "Population Age Structure and Its Relation to Civil Conflict: A Graphic Metric". Washington, D.C.: Wilson Center, 2007.

Cincotta, Richard P. and John Doces. "The Age-Structural Maturity Thesis: The Impact of the Youth Bulge on the Advent and Stability of Liberal Democracy". In *Political Demography: How Population Changes Are Reshaping International Security and National Politics*, edited by Jack A. Goldstone, Eric P. Kaufmann, and Monica Duffy Toft, 98–116. New York: Oxford University Press, 2012.

Cipriani, Lidio. *The Andaman Islanders*. London: Weidenfeld and Nicolson, 1966.

Clark, William R. *Petrodollar Warfare: Oil, Iraq and the Future of the Dollar*. Gabriola Island, BC: New Society Publishers, 2005.

Clarke, Colin P. *Terrorism, Inc.: The Financing of Terrorism, Insurgency, and Irregular Warfare*. Santa Barbara, CA: Praeger, 2015.

Coast, Ernestina. "Maasai Demography". Ph.D. Thesis. University of London, 2001.

Cocco, P. Luis. *Iyëwei-teri: Quince años entre los Yanomamos*. Caracas: Escuela Técnica Popular Don Bosco, 1972.

Coco, Vittorio. "Conspiracy Theories in Republican Italy: The Pellegrino Report to the Parliamentary Commission on Terrorism". *Journal of Modern Italian Studies* 20, no. 3 (2015): 361–76, https://doi.org/10.1080/135457 1x.2015.1026148

Cohen, Stanley. *Folk Devils and Moral Panics*. 3rd ed. New York: Routledge, 2011.

Cohrs, J. Christopher. "Threat and Authoritarianism: Some Theoretical and Methodological Comments". *International Journal of Psychology* 48, no. 1 (2013): 50–54, https://doi.org/10.1080/00207594.2012.732699

Collier, Paul and Anke Hoeffler. "On Economic Causes of Civil War". *Oxford Economic Papers* 50, no. 4 (1998): 563–573, https://doi.org/10.1093/oep/50.4.563

— "Greed and Grievance in Civil War". *Oxford Economic Papers* 56, no. 4 (2004): 563–95, https://doi.org/10.1093/oep/gpf064

Collier, Paul and Nicholas Sambanis, eds. *Understanding Civil War: Evidence and Analysis*. Vols 1and 2. World Bank, 2005.

Collier, Paul, Anke Hoeffler, and Nicholas Sambanis. "The Collier-Hoeffler Model of Civil War Onset and the Case Study Project Research Design". In *Understanding Civil War: Evidence and Analysis*, Vol 1, edited by Paul Collier and Nicholas Sambanis, 1–33. World Bank, 2005.

Collier, Paul, Anke Hoeffler, and Dominic Rohner. "Beyond Greed and Grievance: Feasibility and Civil War". *Oxford Economic Papers* 61, no. 1 (2009): 1–27, https://doi.org/10.1093/oep/gpn029

Condon, Richard G. and Julia Ogina. *The Northern Copper Inuit: A History.* Norman, OK and London: University of Oklahoma Press, 1996.

Contenau, Georges. *Everyday Life in Babylon and Assyria.* London: Edward Arnold, 1954.

Cooke, T. "Paramilitaries and the Press in Northern Ireland". In *Framing Terrorism: The News Media, the Government, and the Public,* edited by Pippa Norris, Montague Kern, and Marion Just, 75–92. New York: Routledge, 2003.

Cooper, John M. *Analytical and Critical Bibliography of the Tribes of Tierra del Fuego and Adjacent Territory.* Washington, D.C.: Government Printing Office, 1917.

—"The Yahgan". In *Handbook of South American Indians,* Vol. 1, edited by Julian Haynes Steward, 81–106. Washington, D.C.: Smithsonian Institution, 1946.

Copeland, Dale C. *Economic Interdependence and War.* Princeton, NJ: Princeton University Press, 2015.

Corson, William R. *The Armies of Ignorance: The Rise of the American Intelligence Empire.* New York: Dial Press, 1977.

Côté, Stéphane, Paul K. Piff, and Robb Willer. "For Whom Do the Ends Justify the Means? Social Class and Utilitarian Moral Judgment". *Journal of Personality and Social Psychology* 104, no. 3 (2013): 490–503, https://doi.org/10.1037/a0030931

Cramer, Christopher. "*Homo Economicus* Goes to War: Methodological Individualism, Rational Choice and the Political Economy of War". *World Development* 30, no. 11 (2002): 1845–64, https://doi.org/10.1016/s0305-750x(02)00120-1

Crédit Suisse. "Global Wealth Report 2015". Zurich: Crédit Suisse, 2015.

Crofoot, Margaret C. and Richard W. Wrangham. "Intergroup Aggression in Primates and Humans: The Case for a Unified Theory". In *Mind the Gap: Tracing the Origins of Human Universals,* edited by Peter M. Kappeler and Joan B. Silk, 171–95. Berlin: Springer, 2010. https://doi.org/10.1007/978-3-642-02725-3_8

Cronin, Audrey Kurth. *How Terrorism Ends: Understanding the Decline and Demise of Terrorist Campaigns.* Princeton, NJ: Princeton University Press, 2009.

Cummins, Joseph. *Why Some Wars Never End: The Stories of the Longest Conflicts in History.* Beverly, MA: Fair Winds Press, 2010.

Curşeu, Petru Lucian, Ron Stoop and René Schalk. "Prejudice Toward Immigrant Workers among Dutch Employees: Integrated Threat Theory Revisited". *European Journal of Social Psychology* 37, no. 1 (2007): 125–40, https://doi.org/10.1002/ejsp.331

D'Altroy, Terence N. *The Incas.* Malden, MA: Blackwell, 2002.

Davies, Philip H. J. "Daniele Ganser, Nato's Secret Armies". *Journal of Strategic Studies* 28, no. 6 (2005).

De Dreu, Carsten K. W. "Oxytocin Modulates Cooperation within and Competition between Groups: An Integrative Review and Research Agenda". *Hormones and Behavior* 61, no. 3 (2012): 419–28, https://doi.org/10.1016/j.yhbeh.2011.12.009

De Dreu, Carsten K. W., Lindred L. Greer, Gerben A. Van Kleef, Shaul Shalvi, and Michel J. J. Handgraaf. "Oxytocin Promotes Human Ethnocentrism". *Proceedings of the National Academy of Sciences* 108, no. 4 (2011): 1262–66, https://doi.org/10.1073/pnas.1015316108

De Graaf, Beatrice. *Evaluating Counterterrorism Performance: A Comparative Study.* Abington and New York: Routledge, 2011.

De Jesus, José Duarte. *A Guerra Secreta de Salazar em África: Aginter Press: Uma Rede Internacional de Contra-Subversão e Espionagem Sediada em Lisboa.* Vol. 12. Alfragide, Portugal: Publicações Dom Quixote, 2012.

De Lutiis, Giuseppe. *I Servizi Segreti in Italia.* Roma: Editori Riuniti, 1998.

De Soysa, Indra, and Eric Neumayer. "Resource Wealth and the Risk of Civil War Onset: Results from a New Dataset of Natural Resource Rents, 1970–1999". *Conflict Management and Peace Science* 24, no. 3 (2007): 201–18, https://doi.org/10.1080/07388940701468468

Delaporte, Louis. *Mesopotamia: The Babylonian and Assyrian Civilization.* London: Routledge and Kegan Paul, 1970, orig. 1925.

Delsing, Marc J. M. H., Tom ter Bogt, Rutger Engels, and Wim Meeus. "Adolescents' Music Preferences and Personality Characteristics". *European Journal of Personality* 22, no. 2 (2008): 109–30, https://doi.org/10.1002/per.665

DeRouen, Karl, Jr. *An Introduction to Civil Wars.* Washington, D.C.: CQ Press, 2014.

Dershowitz, Alan M. *Why Terrorism Works: Understanding the Threat, Responding to the Challenge.* New Haven, CT: Yale University Press, 2002.

Desjardins, Jeff. "All of the World's Money and Markets in One Visualization". *The Money Project*, 2015, http://money.visualcapitalist.com/all-of-the-worlds-money-and-markets-in-one-visualization/

Dessaint, Alain Y. *Minorities of Southwest China: An introduction to the Yi (Lolo) and Related Peoples and an Annotated Bibliography.* New Haven, CT: HRAF Press, 1980.

Dimaggio, Paul. "Culture and Cognition". *Annual Review of Sociology* 23 (1997): 263–87, https://doi.org/10.1146/annurev.soc.23.1.263

D'Ollone, Vicomte. *In Forbidden China.* London: T. Fisher Unwin, 1912.

Dobbs, Richard, Susan Lund, Jonathan Woetzel, and Mina Mutafchieva. "Debt and (Not Much) Deleveraging". McKinsey Global Institute, 2015.

Dodwell, C. H. "The Copper Eskimos of Coronation Gulf". *Canadian Geographical Journal* 13, no. 2 (1936): 61–82.

Donoghue, John David, Daniel D. Whitney and Iwao Ishino. *People in the Middle*: *The Rhadé of South Vietnam*. Ann Arbor, MI: Michigan State University, 1962.

Donohue, Laura K. "Transplantation". In *Global Anti-Terrorism Law and Policy*, *edited by* Victor V. Ramraj, Michael Hor, Kent Roach, and George Williams, 67–88, 2nd ed. New York: Cambridge University Press, 2012, https://doi.org/10.1017/cbo9781139043793.005

Doosje, Bertjan, Annemarie Loseman, and Kees Bos. "Determinants of Radicalization of Islamic Youth in the Netherlands: Personal Uncertainty, Perceived Injustice, and Perceived Group Threat". *Journal of Social Issues* 69, no. 3 (2013): 586–604, https://doi.org/10.1111/josi.12030

Douglas, Roger Neil. *Law, Liberty, and the Pursuit of Terrorism*. Ann Arbor, MI: University of Michigan Press, 2014.

Dow, James W. "Using R for Cross-Cultural Research". *World Cultures* 14 (2003): 144–54.

Draper, Patricia. "!Kung Women: Contrasts in Sexual Egalitarianism in Foraging and Sedentary Contexts". In *Toward an Anthropology of Women*, edited by Rayna R. Reiter, 77–109. New York: Monthly Review Press, 1975.

Dry, D. P. L. "Some Aspects of Hausa Family Structure". In *Proceedings of the International West African Conference Held at Ibadan*, 158–63, 1949.

Duanmu, Jing-Lin. "A Race to Lower Standards? Labor Standards and Location Choice of Outward FDI from the BRIC Countries". *International Business Review* 23, no. 3 (2014): 620–34, https://doi.org/10.1016/j.ibusrev.2013.10.004

Duckitt, John. "Authoritarianism and Group Identification: A New View of an Old Construct". *Political Psychology*, 1989, 63–84, https://doi.org/10.2307/3791588

— "A Dual-Process Cognitive-Motivational Theory of Ideology and Prejudice". *Advances in Experimental Social Psychology* 33 (2001): 41–113, https://doi.org/10.1016/s0065-2601(01)80004-6

— "Introduction to the Special Section on Authoritarianism in Societal Context: The Role of Threat". *International Journal of Psychology* 48, no. 1 (2013): 1–5, https://doi.org/10.1080/00207594.2012.738298

— "Authoritarian Personality". In *International Encyclopedia of the Social & Behavioral Sciences*, by James D. Wright, 2nd ed. Amsterdam: Elsevier, 2015, 255–61, https://doi.org/10.1016/b978-0-08-097086-8.24042-7

Duckitt, John and Kirstin Fisher. "The Impact of Social Threat on Worldview and Ideological Attitudes". *Political Psychology* 24, no. 1 (2003): 199–222, https://doi.org/10.1111/0162-895x.00322

Duke, Steven B. and Albert C. Gross. *America's Longest War*: *Rethinking Our Tragic Crusade against Drugs*. Los Angeles, CA: Tarcher, 1994.

Dunbar, Robin, Clive Gamble, and John Gowlett, eds. *Social Brain, Distributed Mind*. Oxford: Oxford University Press for the British Academy, 2010.

Dunlop, John. *The 2002 Dubrovka and 2004 Beslan Hostage Crises: A Critique of Russian Counter-Terrorism*. Ibidem-Verlag, 2006.

— *The Moscow Bombings of September 1999: Examinations of Russian Terrorist Attacks at the Onset of Vladimir Putin's Rule*. Stuttgart: Ibidem, 2012.

Durkheim, Emile. *Suicide: A Study in Sociology*. Glencoe: Free Press, 1951 [1897].

Durrant, Russil. "Collective Violence: An Evolutionary Perspective". *Aggression and Violent Behavior* 16, no. 5 (2011): 428–36, https://doi.org/10.1016/j.avb.2011.04.014

Dye, David H. "Trends in Cooperation and Conflict in Native Eastern North America". In *War, Peace, and Human Nature: The Convergence of Evolutionary and Cultural Views, edited by* Douglas P. Fry, 132–50. Oxford University Press, 2013, https://doi.org/10.1093/acprof:oso/9780199858996.003.0008

Eberle, Paul and Shirley Eberle. *The Politics of Child Abuse*. Secaucus, NJ: Lyle Stuart, 1986.

Echebarria-Echabe, Agustin and Emilia Fernández-Guede. "Effects of Terrorism on Attitudes and Ideological Orientation". *European Journal of Social Psychology* 36, no. 2 (2006): 259–65, https://doi.org/10.1002/ejsp.294

Eckhardt, William. "Authoritarianism". *Political Psychology* 12 (1991): 97–124, https://doi.org/10.2307/3791348

— *Civilizations, Empires, and Wars: A Quantitative History of War*. Jefferson, NC: McFarland & Company Inc., 1992.

Edgar, Frank and Neil Skinner. *Hausa Tales and Traditions: An English Translation of Tatsuniyoyi Na Hausa*. London: Frank Cass, 1969.

Edmonds, Bruce. "The Revealed Poverty of the Gene-Meme Analogy—Why Memetics per se has Failed to Produce Substantive Results". *Journal of Memetics* 9, no. 1 (2005), http://jom-emit.cfpm.org/2005/vol9/edmonds_b.html

Edwards, Robert. *White Death: Russia's War on Finland, 1939–40*. Phoenix, AZ: Orion, 2006.

Eibl-Eibesfeldt, Irenäus. "Early Socialization in the !xõ Bushmen". In *The Bushmen: San Hunters and Herders of Southern Africa*, edited by Phillip V. Tobias and Megan Biesele, 130–36. Cape Town: Human & Rousseau, 1978.

— *Human Ethology*. New York: Aldine de Gruyter, 1989.

Eisner, Peter and Knut Royce. *The Italian Letter: How the Bush Administration Used a Fake Letter to Build the Case for War in Iraq*. New York: Rodale Books, 2007.

Ellingson, Terry Jay. *The Myth of the Noble Savage*. Berkeley, CA: University of California Press, 2001.

Ellis, Bruce J., Aurelio José Figueredo, Barbara H. Brumbach, and Gabriel L. Schlomer. "Fundamental Dimensions of Environmental Risk". *Human Nature* 20, no. 2 (2009): 204–68, https://doi.org/10.1007/s12110-009-9063-7

Elliston, Jon. *Psywar on Cuba: The Declassified History of US Anti-Castro Propaganda.* Melbourne: Ocean Press, 1999.

Ember, Carol R. and Melvin Ember. "Resource Unpredictability, Mistrust, and War". *Journal of Conflict Resolution* 36, no. 2 (1992): 242–62, https://doi.org/10.1177/0022002792036002002

—"War, Socialization, and Interpersonal Violence: A Cross-Cultural Study". *Journal of Conflict Resolution* 38, no. 4 (1994): 620–46, https://doi.org/10.1177/0022002794038004002

Endicott, Kirk. "Peaceful Foragers: The Significance of the Batek and Moriori for the Question of Innate Human Violence". In *War, Peace, and Human Nature: The Convergence of Evolutionary and Cultural Views*, edited by Douglas P. Fry, 243–61. New York: Oxford University Press, 2013, https://doi.org/10.1093/ac prof:oso/9780199858996.003.0012

Erickson, Edwin E. "Tradition and Evolution in Song Style: A Reanalysis of Cantometric data". *Behavior Science Research* (*Cross-Cultural Research*) 11, no. 4 (1976): 277–308, https://doi.org/10.1177/106939717601100403

Ericson, Richard, Patricia M. Baranek, and Janet BL Chan. *Negotiating Control: A Study of News Sources.* Milton Keynes: Open University Press, 1989.

Esser, Frank. "Tabloidization of News: A Comparative Analysis of Anglo-American and German Press Journalism". *European Journal of Communication* 14, no. 3 (1999): 291–324, https://doi.org/10.1177/0267323199014003001

Esty, Daniel C., Jack A. Goldstone, Ted R. Gurr, Pamela Surko, and Alan N. Unger. "State Failure Task Force Report (Phase I)". Science Applications International Corporation, 1995.

Fasanella, Giovanni, Giovanni Pellegrino, and Claudio Sestieri. *Segreto Di Stato: La Verità Da Gladio Al Caso Moro.* Torino: Einaudi, 2000.

Fearon, James D. and David D. Laitin. "Ethnicity, Insurgency, and Civil War". *American Political Science Review* 97, no. 1 (2003): 75–90, https://doi.org/10.1017/s0003055403000534

Feldbauer, Gerhard. *Agenten, Terror, Staatskomplott: Der Mord an Aldo Moro, Rote Brigaden, Und CIA.* Cologne: PapyRossa, 2000.

Feldman, Stanley and Karen Stenner. "Perceived Threat and Authoritarianism". *Political Psychology* 18, no. 4 (1997): 741–70, https://doi.org/10.1111/0162-895x.00077

Ferguson, R. Brian. *Yanomami Warfare: A Political History.* Santa Fe: School of American Research Press, 1995.

Ferraresi, Franco. *Threats to Democracy: The Radical Right in Italy after the War*. Princeton, NJ: Princeton University Press, 1996.

Fischer-Shofty, Meytal, Yechiel Levkovitz, and Simone G. Shamay-Tsoory. "Oxytocin Facilitates Accurate Perception of Competition in Men and Kinship in Women". *Social Cognitive and Affective Neuroscience* 8, no. 3 (2013): 313–17, https://doi.org/10.1093/scan/nsr100

Fog, Agner. "Cultural r/k Selection". *Journal of Memetics* 1 (1997): 14–28, http://cfpm.org/jom-emit/1997/vol1/fog_a.html

—*Cultural Selection*. Dordrecht: Kluwer, 1999.

—"An Evolutionary Theory of Cultural Differentiation". In *Proceedings of the XV World Conference of the International Union of Prehistoric and Protohistoric Scientists*, 31–34. Lisbon, 2006.

—"The Supposed and the Real Role of Mass Media in Modern Democracy". Researchgate, July 3, 2013, http://www.researchgate.net/publication/245021478_The_supposed_and_the_real_role_of_mass_media_in_modern_democracy

Fokken, Heinrich. "Gottesanschauungen und religiöse Überlieferungen der Massai". *Archiv für Anthropologie* 15 (1917): 237–52.

Foran, John. *Taking Power: On the Origins of Third World Revolutions*. Cambridge and New York: Cambridge University Press, 2005, https://doi.org/10.1017/cbo9780511488979

Fortna, Virginia Page. "Do Terrorists Win? Rebels' Use of Terrorism and Civil War Outcomes". *International Organization* 69, no. 3 (2015): 519–56, https://doi.org/10.1017/s0020818315000089

Fosbrooke, Henry A. "An Administrative Survey of the Maasai Social System". *Tanganyika Notes and Records* 26 (1948): 1–50.

Foucault, Michel. *The History of Sexuality. Vol. 1: An Introduction*. New York: Vintage Books, 1978.

—*Power/Knowledge: Selected Interviews and Other Writings, 1972–1977*. Brighton: Harvester Press, 1980.

Fowler, James H. "Altruistic Punishment and the Origin of Cooperation". *Proceedings of the National Academy of Sciences of the USA* 102, no. 19 (2005): 7047–49, https://doi.org/10.1073/pnas.0500938102

Francioni, Francesco, and Federico Lenzerini. "The Obligation to Prevent and Avoid Destruction of Cultural Heritage: From Bamiyan to Iraq". In *Art and Cultural Heritage: Law, Policy and Practice*, edited by Barbara T Hoffman, 28–40. Cambridge and New York: Cambridge University Press, 2006.

Francis, David. "Asymmetrical Warfare and the Regionalisation of Domestic Civil Wars in Africa: Implications for EU Interventions". In *Winning the Asymmetric War*, edited by Josef Schröfl, Sean Michael Cox, and Thomas Pankratz, 237–60. Frankfurt: P. Lang, 2009.

Fritsche, Immo, Eva Jonas, and Thomas Kessler. "Collective Reactions to Threat: Implications for Intergroup Conflict and for Solving Societal Crises". *Social Issues and Policy Review* 5, no. 1 (2011): 101–36, https://doi.org/10.1111/j.1751-2409.2011.01027.x

Fry, Douglas P. *Beyond War: The Human Potential for Peace.* New York: Oxford University Press, 2009.

Fry, Douglas P. and Patrik Söderberg. "Lethal Aggression in Mobile Forager Bands and Implications for the Origins of War". *Science* 341, no. 6143 (2013): 270–73, https://doi.org/10.1126/science.1235675

Fuller, Jack. *What Is Happening to News: The Information Explosion and the Crisis in Journalism.* Chicago, IL: University of Chicago Press, 2010, https://doi.org/10.7208/chicago/9780226268996.001.0001

Galbraith, James K. *Inequality and Instability: A Study of the World Economy Just Before the Great Crisis.* New York: Oxford University Press, 2012, https://doi.org/10.1093/acprof:osobl/9780199855650.001.0001

Galor, Oded. *Unified Growth Theory.* Princeton, NJ: Princeton University Press, 2011.

Galpin, Francis W. "The Music of the Sumerians and their immediate Successors the Babylonians and Assyrians". *Sammlung Musikwissenschaftlicher Abhandlungen*, Vol. 33. Strasbourg: Librairie Heitz, 1955.

Ganguly, Pranab. "Religious Beliefs of the Negritos of Little Andaman". *Eastern Anthropologist* 14 (1961): 243–48.

Ganor, Boaz. "Defining Terrorism: Is One Man's Terrorist Another Man's Freedom Fighter?" *Police Practice and Research* 3, no. 4 (2002): 287–304, https://doi.org/10.1080/1561426022000032060

Ganser, Daniele. *NATO's Secret Armies: Operation Gladio and Terrorism in Western Europe.* London: Frank Cass, 2005.

Garcia, Justin R., and Daniel J. Kruger. "Unbuckling in the Bible Belt: Conservative Sexual Norms Lower Age at Marriage", *Journal of Social, Evolutionary, and Cultural Psychology* 4, no. 4 (2010): 206–14, https://doi.org/10.1037/h0099288

Gat, Azar. *War in Human Civilization.* New York: Oxford University Press, 2006.

—"Proving Communal Warfare among Hunter-Gatherers: The Quasi-Rousseauan Error". *Evolutionary Anthropology: Issues, News, and Reviews* 24, no. 3 (2015): 111–26, https://doi.org/10.1002/evan.21446

—*The Causes of War and the Spread of Peace: Will War Rebound?* New York: Oxford University Press, 2017.

Gavrilets, Sergey and Laura Fortunato. "A Solution to the Collective Action Problem in between-Group Conflict with within-Group Inequality". *Nature Communications* 5, no. 3526 (2014), https://doi.org/10.1038/ncomms4526

Geary, David C. *Male, Female*: *The Evolution of Human Sex Differences*. 2nd ed. Washington, D.C.: American Psychological Association, 2010.

Geiss, Robin. "Asymmetric Conflict Structures". *International Review of the Red Cross* 88, no. 864 (2006): 757–77, https://doi.org/10.1017/s1816383107000781

Gelfand, Michele J., J. L. Raver, L. Nishii, L. M. Leslie, J. Lun, B. C. Lim, L. Duan, et al. "Differences between Tight and Loose Cultures: A 33-Nation Study". *Science* 332, no. 6033 (2011): 1100–04, https://doi.org/10.1126/science.1197754

Gibler, Douglas M. and Andrew Owsiak. "Democracy and the Settlement of International Borders, 1919–2001". *Journal of Conflict Resolution*, Forthcoming (2017), https://doi.org/10.1177/0022002717708599

Gifford, Edward Winslow. "Culture Element Distributions: XII Apache-Pueblo". *Anthropological Records* 4 (1940): 1–207.

Gino, Francesca, and Lamar Pierce. "The Abundance Effect: Unethical Behavior in the Presence of Wealth". *Organizational Behavior and Human Decision Processes* 109, no. 2 (2009): 142–55, https://doi.org/10.1016/j.obhdp.2009.03.003

Giustozzi, Antonio. "Auxiliary Irregular Forces in Afghanistan: 1978–2008". In *Making Sense of Proxy Warfare*: *States, Surrogates, and the Use of Force*, edited by Michael A. Innes, 89–107. Washington, D.C.: Potomac, 2012.

Glassner, Barry. *The Culture of Fear*: *Why Americans Are Afraid of the Wrong Things*. New York: Basic Books, 1999.

Glowacki, Luke and Chris von Rueden. "Leadership Solves Collective Action Problems in Small-Scale Societies". *Philosophical Transactions of the Royal Society*, B 370, no. 1683 (2015), https://doi.org/10.1098/rstb.2015.0010

Glowacki, Luke and Richard W. Wrangham. "The Role of Rewards in Motivating Participation in Simple Warfare". *Human Nature* 24, no. 4 (2013): 444–60, https://doi.org/10.1007/s12110-013-9178-8

— "Warfare and Reproductive Success in a Tribal Population". *Proceedings of the National Academy of Sciences* 112, no. 2 (2015): 348–53, https://doi.org/10.1073/pnas.1412287112

Goggin, Mary-Margaret. "'Decent' vs. 'Degenerate' Art: The National Socialist Case". *Art Journal* 50, no. 4 (1991): 84–92, https://doi.org/10.1080/00043249.1991.10791484

Goldberg, Steven. *Why Men Rule*: *A Theory of Male Dominance*. Chicago, IL: Open Court, 1993.

Goldfarb, Alex. *Death of a Dissident*: *The Poisoning of Alexander Litvinenko and the Return of the KGB*. New York: Simon and Schuster, 2007.

Goldstone, Jack A. "Toward a Fourth Generation of Revolutionary Theory". *Annual Review of Political Science* 4 (2001): 139–87, https://doi.org/10.1146/annurev.polisci.4.1.139

— "Comparative Historical Analysis and Knowledge Accumulation in the Study of Revolutions". In *Comparative Historical Analysis in the Social Sciences*, edited by James Mahoney and Dietrich Rueschemeyer, 41–90. New York: Cambridge University Press, 2003, https://doi.org/10.1017/cbo9780511803963.003

— "A Theory of Political Demography". In *Political Demography: How Population Changes Are Reshaping International Security and National Politics*, edited by Jack A. Goldstone, E. P. Kaufmann, and M. D. Toft, 10–28. Oxford University Press, 2012.

Goldstone, Jack A., et al. "State Failure Task Force Report: Phase III Findings". Science Applications International Corporation, 2000.

— "A Global Model for Forecasting Political Instability". *American Journal of Political Science* 54, no. 1 (2010): 190–208, https://doi.org/10.1111/j.1540-5907.2009.00426.x

González, Karina Velasco, Maykel Verkuyten, Jeroen Weesie, and Edwin Poppe. "Prejudice towards Muslims in the Netherlands: Testing Integrated Threat Theory". *British Journal of Social Psychology* 47, no. 4 (2008): 667–85, https://doi.org/10.1348/014466608x284443

Good, Kenneth R. "Limiting Factors in Amazonian Ecology". In *Food and evolution: Toward a theory of human food habits*, edited by Marvin Harris and Eric B. Ross, 407–22. Philadelphia, PA: Temple University Press, 1987.

Goode, Erich and Nachman Ben-Yehuda. *Moral Panics: The Social Construction of Deviance*. 2nd ed. Chichester: John Wiley & Sons, 2009, https://doi.org/10.1002/9781444307924

Gössner, Rolf. *Das Anti-Terror-System. Politische Justiz im präventiven Sicherheitsstaat*. Hamburg: VSA-Verlag, 1991.

Graber, Doris A. *Processing the News: How People Tame the Information Tide*. 2nd ed. University Press of America, 1993.

Green, Elliott. "The Political Demography of Conflict in Modern Africa". *Civil Wars* 14, no. 4 (2012): 477–98, https://doi.org/10.1080/13698249.2012.740198

Greenaway, Katharine H., Winnifred R. Louis, Matthew J. Hornsey, and Janelle M. Jones. "Perceived Control Qualifies the Effects of Threat on Prejudice". *British Journal of Social Psychology* 53, no. 3 (2014): 422–42, https://doi.org/10.1111/bjso.12049

Greenberg, Joseph. *The Influence of Islam on a Sudanese Religion*. New York: J. J. Augustin, 1946.

— "Islam and Clan Organization among the Hausa". *Southwestern Journal of Anthropology* 3, no. 3 (1947): 193–211, https://doi.org/10.1086/soutjanth.3.3.3628775

Greenberg, Jeff and Jamie Arndt. "Terror Management Theory". In *Handbook of Theories of Social Psychology*, edited by Paul A. M. van Lange, A. W. Kruglanski, and E. T. Higgins, 398–415, 2011, https://doi.org/10.4135/9781446249215.n20

Greer, Chris. *Sex Crime and the Media*: *Sex Offending and the Press in a Divided Society*. London: Routledge, 2012.

Griskevicius, Vladas, Andrew W. Delton, Theresa E. Robertson, and Joshua M. Tybur. "Environmental Contingency in Life History Strategies: The Influence of Mortality and Socioeconomic Status on Reproductive Timing". *Journal of Personality and Social Psychology* 100, no. 2 (2011): 241–54, https://doi.org/10.1037/a0021082

Grohs-Paul, Waltraud. *Familiale und schulische Sozialisation bei den Warao-Indianern des Orinoko-Delta, Venezuela*. Stuttgart: Hochschulverlag, 1979.

Guenther, Mathias. *Tricksters and Trancers*: *Bushman Religion and Society*. Bloomington, IN: Indiana University Press, 1999.

Guilaine, Jean and Jean Zammit. *The Origins of War*: *Violence in Prehistory*. Malden, MA: Blackwell, 2008.

Gunther, R. and A. Mugham, eds. *Democracy and the Media*: *A Comparative Perspective*. Cambridge and New York: Cambridge University Press, 2000, https://doi.org/10.1017/cbo9781139175289

Gusfield, Joseph R. "Constructing the Ownership of Social Problems: Fun and Profit in the Welfare State". *Social Problems* 36, no. 5 (1989): 431–41, https://doi.org/10.2307/3096810

Gusinde, Martin. *Die Feuerland Indianer (Vol. II)*: *Die Yamana*. Vienna: Moedling, 1937.

—*Nordwind-Südwind*: *Mythen und Märchen der Feuerlandindianer*. Kassel: Erich Röth Verlag, 1966.

—*Die Feuerland Indianer*: *Vom Leben und Denken der Wassernomaden in West-Patagonien*. Vienna: Moedling, 1937.

Haggenmacher, G. A. "G. A. Haggenmacher's Reise im Somali-Lande 1874". In *Mittheilungen aus Justus Perthes' Geographischer Anstalt über wichtige neue Erforschungen auf dem Gesamtgebiete der Geographie*. 22. Band, Ergänzungs-Hefte Nr. 47, edited by A. Peterman. Gotha: Justus Perthes, 1876.

Hajek, Andrea. "Teaching the History of Terrorism in Italy: The Political Strategies of Memory Obstruction". *Behavioral Sciences of Terrorism and Political Aggression* 2, no. 3 (2010): 198–216, https://doi.org/10.1080/19434471003597456

Hanyok, Robert J. "Skunks, Bogies, Silent Hounds, and the Flying Fish: The Gulf of Tonkin Mystery, 2–4 August 1964". *Cryptologic Quarterly*. 2001. Released 2005 by National Security Archive, https://doi.org/10.1163/9789004346185.usao-11_098

Harako, Reizo. "The Mbuti as Hunters". *Kyoto University African Studies* 10 (1976): 37–99.

Hardin, Garrett. "The Competitive Exclusion Principle". *Science* 131, no. 3409 (1960): 1292–97, https://doi.org/10.1126/science.131.3409.1292

Harding, David J. and Kristin S. Seefeldt. "Mixed Methods and Causal Analysis". In *Handbook of Causal Analysis for Social Research*, by Stephen L. Morgan, 91–108. Dordrecht: Springer, 2013, https://doi.org/10.1007/978-94-007-6094-3_6

Harrell, Stevan. *Perspectives on the Yi of Southwest China*. Berkeley, CA: University of California Press, 2001, https://doi.org/10.1525/california/9780520219885.001.0001

Harrington, Jesse R. and Michele J. Gelfand. "Tightness-looseness across the 50 United States". *Proceedings of the National Academy of Sciences* 111, no. 22 (2014): 7990–95, https://doi.org/10.1073/pnas.1317937111

Harrit, Niels H., et. al. "Active Thermitic Material Discovered in Dust from the 9/11 World Trade Center Catastrophe". *The Open Chemical Physics Journal* 2, no. 1 (2009): 7–31, https://doi.org/10.2174/1874412500902010007

Harzing, Anne-Wil. "Response Styles in Cross-National Survey Research. A 26-Country Study". *International Journal of Cross Cultural Management* 6, no. 2 (2006): 243–66, https://doi.org/10.1177/1470595806066332

Haslam, Nick, and Steve Loughnan. "Dehumanization and Infrahumanization". *Annual Review of Psychology* 65 (2014): 399–423, https://doi.org/10.1146/annurev-psych-010213-115045

Hastings, Brad M. and Barbara Shaffer. "Authoritarianism. The Role of Threat, Evolutionary Psychology, and the Will to Power". *Theory & Psychology* 18, no. 3 (2008): 423–40, https://doi.org/10.1177/0959354308089793

Hayek, Friedrich A. "The Theory of Complex Phenomena". In *Studies in Philosophy, Politics and Economics*, 22–42. Chicago, IL: University of Chicago Press, 1967.

Healy, Amy Erbe and Michael Breen. "Religiosity in Times of Insecurity: An Analysis of Irish, Spanish and Portuguese European Social Survey Data, 2002–12". *Irish Journal of Sociology* 22, no. 2 (2014): 4–29, https://doi.org/10.7227/ijs.22.2.2

Hegre, Håvard and Nicholas Sambanis. "Sensitivity Analysis of Empirical Results on Civil War Onset". *Journal of Conflict Resolution* 50, no. 4 (2006): 508–35, https://doi.org/10.1177/0022002706289303

Hegre, Håvard, Joakim Karlsen, Håvard M. Nygård, Håvard Strand, and Henrik Urdal. "Predicting Armed Conflict, 2010–2050". *International Studies Quarterly* 57, no. 2 (2013): 250–70, https://doi.org/10.1111/isqu.12007

Heine, Lazar, Jihyun Lee, and Fons van de Vijver. "Two Dimensions of Psychological Country-Level Differences: Conservatism/Liberalism and Harshness/Softness". *Learning and Individual Differences* 30 (2014): 22–33, https://doi.org/10.1016/j.lindif.2013.12.001

Heine, Steven J., Darrin R. Lehman, Kaiping Peng, and Joe Greenholtz. "What's Wrong with Cross-Cultural Comparisons of Subjective Likert Scales?: The Reference-Group Effect". *Journal of Personality and Social Psychology* 82, no. 6 (2002): 903–18, https://doi.org/10.1037//0022-3514.82.6.903

Heinen, H. Dieter. "Oko Warao, We are Canoe People". *Acta Ethnologica et Linguistica* 59 (1985): 1–140.

Heinen, H. Dieter. "Warao". In *Encyclopedia of World Cultures: South America*, Vol. 7, edited by Johannes Wilbert, 356–59. Boston, MA: Massachusetts: G. K. Hall & Co., 1994.

Heinsohn, Gunnar. *Söhne und Weltmacht*. Zürich: Orell Füssli, 2003.

Henrich, Joseph. "Cultural Group Selection, Coevolutionary Processes and large-scale Cooperation". *Journal of Economic Behavior & Organization* 53, no. 1 (2004): 3–35, https://doi.org/10.1016/s0167-2681(03)00094-5

Hetherington, Marc and Elizabeth Suhay. "Authoritarianism, Threat, and Americans' Support for the War on Terror". *American Journal of Political Science* 55, no. 3 (2011): 546–60, https://doi.org/10.1111/j.1540-5907.2011.00514.x

Hett, Benjamin Carter. *Burning the Reichstag: An Investigation Into the Third Reich's Enduring Mystery*. Oxford and New York: Oxford University Press, 2014.

Hiatt, Steven W. *A Game as Old as Empire: The Secret World of Economic Hit Men and the Web of Global Corruption*. San Francisco, CA: Berrett-Koehler Publishers, 2007.

Hill, Ann Maxwell and Eric Diehl. "A Comparative Approach to Lineages among the Xiao Liangshan Nuosu and Han". In *Perspectives on the Yi of Southwest China*, edited by Stevan Harrell, 51–67. Berkeley, CA: University of California Press, 2001, https://doi.org/10.1525/california/9780520219885.003.0004

Hirsch-Hoefler, Sivan, Daphna Canetti, Carmit Rapaport, and Stevan E. Hobfoll. "Conflict will Harden Your Heart: Exposure to Violence, Psychological Distress, and Peace Barriers in Israel and Palestine". *British Journal of Political Science* 46, no. 4 (2014), 845–59, https://doi.org/10.1017/s0007123414000374

Hiskett, Mervyn. *A History of Hausa Islamic Verse*. London: School of Oriental and African Studies, University of London, 1975.

Hofstede, Geert, Gert Jan Hofstede, and Michael Minkov. *Cultures and Organizations: Software of the Mind*. 3rd ed. New York: McGraw-Hill, 2010.

Hogben, Sidney John and Anthony H. M. Kirk-Greene. *The Emirates of Northern Nigeria: A Preliminary Survey of Their Historical Traditions*. London: Oxford University Press, 1966.

Hogg, Michael A. *The Social Psychology Of Group Cohesiveness*. New York: Harvester Wheatsheaf, 1992.

—"From Uncertainty to Extremism: Social Categorization and Identity Processes". *Current Directions in Psychological Science* 23, no. 5 (2014): 338–42, https://doi.org/10.1177/0963721414540168

Hogg, Michael A., Christie Meehan, and Jayne Farquharson. "The Solace of Radicalism: Self-Uncertainty and Group Identification in the Face of Threat". *Journal of Experimental Social Psychology* 46, no. 6 (2010): 1061–66, https://doi.org/10.1016/j.jesp.2010.05.005

Hogg, Michael A., Daan van Knippenberg, and David E. Rast III. "The Social Identity Theory of Leadership: Theoretical Origins, Research Findings, and Conceptual Developments". *European Review of Social Psychology* 23, no. 1 (2012): 258–304, https://doi.org/10.1080/10463283.2012.741134

Hogg, Michael A., and Janice Adelman. "Uncertainty-Identity Theory: Extreme Groups, Radical Behavior, and Authoritarian Leadership". *Journal of Social Issues* 69, no. 3 (2013): 436–54, https://doi.org/10.1111/josi.12023

Hollis, Alfred Claud. *The Masai: Their Language and Folklore*. Oxford: Clarendon Press, 1905.

Hooper, Paul L., Hillard S. Kaplan, and James L. Boone. "A Theory of Leadership in Human Cooperative Groups". *Journal of Theoretical Biology* 265, no. 4 (2010): 633–46, https://doi.org/10.1016/j.jtbi.2010.05.034

Horowitz, Mark, William Yaworsky, and Kenneth Kickham. "Whither the Blank Slate? A Report on the Reception of Evolutionary Biological Ideas among Sociological Theorists". *Sociological Spectrum* 34, no. 6 (2014): 489–509, https://doi.org/10.1080/02732173.2014.947451

Huddy, Leonie and Stanley Feldman. "Americans Respond Politically to 9/11: Understanding the Impact of the Terrorist Attacks and their Aftermath". *American Psychologist* 66, no. 6 (2011): 455–67, https://doi.org/10.1037/a0024894

Huddy, Leonie, Stanley Feldman, and Christopher Weber. "The Political Consequences of Perceived Threat and Felt Insecurity". *The Annals of the American Academy of Political and Social Science* 614, no. 1 (2007): 131–53, https://doi.org/10.1177/0002716207305951

Hudes, Karen and Sabine Schlemmer-Schulte. "Accountability in Bretton Woods". *ILSA Journal of International and Comparative Law* 15, no. 2 (2009): 501.

Hughes, Geraint. *My Enemy's Enemy: Proxy Warfare in International Politics*. Eastbourne: Sussex Academic Press, 2012.

Hui, Lu. "Preferential Bilateral-Cross-Cousin Marriage among the Nuosu in Liangshan". In *Perspectives on the Yi of Southwest China*, edited by Stevan Harrell, 68–80. Berkeley, CA: University of California Press, 2001, https://doi.org/10.1525/california/9780520219885.003.0005

Human Rights Watch. "Illusion of Justice: Human Rights Abuses in US Terrorism Prosecutions". Columbia Law School, 2014, https://www.hrw.org/report/2014/07/21/illusion-justice/human-rights-abuses-us-terrorism-prosecutions

Huntingford, G. W. B. *The Southern Nilo-Hamites*. London: International African Institute, 1953.

Hvistendahl, Mara. "Young and Restless Can Be a Volatile Mix". *Science* 333, no. 6042 (2011): 552–54, https://doi.org/10.1126/science.333.6042.552

Hyades, P. and J. Deniker. *Mission Scientifique du Cap Horn, 1882–1883. (Tome VII): Anthropologie, Ethnographie*. Paris: Gauthier-Villars et Fils, 1891.

Hyslop, John. *The Inka Road System*. Orlando, FL: Academic Press, 1984.

Ichikawa, Mitsuo. "The Japanese Tradition in Central African Hunter-gatherer Studies, with Comparative Observations on the French and American Traditions". In *Hunter-gatherers in History, Archaeology and Anthropology*, edited by Alan Barnard, 103–14. Oxford: Berg, 2004.

Igel, Regine. *Terrorismus-Lügen: Wie die Stasi im Untergrund agierte*. Munich: Herbig, 2012.

Inglehart, Ronald. "Mapping Global Values". In *Measuring and Mapping Cultures: 25 Years of Comparative Value Surveys*, edited by Yilmaz R. Esmer and Thorleif Pettersson, 11–32. Leiden: Brill, 2007, https://doi.org/10.1163/ej.9789004158207.i-193.7

Inglehart, Ronald and Pippa Norris. *Sacred and Secular: Religion and Politics Worldwide*. Cambridge and New York: Cambridge University Press, 2004, https://doi.org/10.1017/cbo9780511791017

Inglehart, Ronald and Christian Welzel. *Modernization, Cultural Change, and Democracy: The Human Development Sequence*. Cambridge and New York: Cambridge University Press, 2005, https://doi.org/10.1017/cbo9780511790881

Inglehart, Ronald, and Christian Welzel. Supplementary data, http://www.worldvaluessurvey.org/wvs/articles/folder_published/article_base_54

International Monetary Fund. "World Economic Outlook Database", 2012, https://www.imf.org/external/pubs/ft/weo/2015/01/weodata/index.aspx

Iversen, Reidun and Wenche Stray. *Kvinner i Afghanistan*. Oslo: Afghanistankomiteen i Norge, 1985.

Iyengar, Shanto. "Framing Responsibility for Political Issues". *Annals of the American Academy of Political and Social Science*, no. 546 (1996): 59–70, https://doi.org/10.1177/0002716296546001006

Iyengar, Shanto and Donald R. Kinder. *News That Matters: Television and American Opinion*. 2nd ed. Chicago, IL: University of Chicago Press, 2010, https://doi.org/10.7208/chicago/9780226388601.001.0001

Jackson, Jay W. "Realistic Group Conflict Theory: A Review and Evaluation of the Theoretical and Empirical Literature". *The Psychological Record* 43, no. 3 (1993): 395–413.

Jackson, Richard, Marie Breen-Smyth, Jeroen Gunning, and Lee Jarvis. *Terrorism: A Critical Introduction*. Palgrave Macmillan, 2011.

Jacur, Francesca Romanin, Angelica Bonfanti, and Francesco Seatzu. *Natural Resources Grabbing: An International Law Perspective*. Brill Nijhoff, 2015.

Jaffe, Klaus and Luis Zaballa. "Co-Operative Punishment Cements Social Cohesion". *Journal of Artificial Societies and Social Simulation* 13, no. 3 (2010): 4, https://doi.org/10.18564/jasss.1568

Jenkins, Philip. "Strategy of Tension: The Belgian Terrorist Crisis 1982–1986". *Studies in Conflict & Terrorism* 13, nos. 4–5 (1990): 299–309, https://doi.org/10.1080/10576109008435838

Jenness, Diamond. "The Copper Eskimos". *Geographical Review* 4, no. 2 (1917): 81–91, https://doi.org/10.2307/207288

—*Report of the Canadian Arctic Expedition 1913–18, Vol. XII. The life of the Copper Eskimos*. Ottawa: F. A. Acland, 1922.

—*Report of the Canadian Arctic Expedition, 1913–1918. Vol. XIV: Eskimo Songs. Songs of the Copper Eskimos*. Ottawa: F. A. Acland, 1925.

—*The People of the Twilight*. New York: Macmillan, 1928.

—*Report of the Canadian Arctic Expedition, 1913–1918. Vol. XVI: Material Culture of the Copper Eskimo*. Ottawa: Edmond Cloutier, 1946.

Jingzhong, Wu. "Nzymo as seen in Some Yi Classical Books". In *Perspectives on the Yi of Southwest China*, edited by Stevan Harrell, 35–48. Berkeley, CA: University of California Press, 2001, https://doi.org/10.1525/california/9780520219885.003.0003

Johnson, Dominic D. P. "Leadership in War: Evolution, Cognition, and the Military Intelligence Hypothesis". In *The Handbook of Evolutionary Psychology: Foundations*, edited by David M. Buss, 2nd ed., vol. 2, part V: 722–44. New York: John Wiley & Sons Inc., 2015.

Jones, Steven E., et. al. "Extremely High Temperatures during the World Trade Center Destruction". *Journal of 9/11 Studies* 2008, no. 1 (2008), http://journalof911studies.com/articles/WTCHighTemp2.pdf

Jordan, Javier and Nicola Horsburgh. "Politics vs Terrorism: The Madrid Case". In *Playing Politics with Terrorism*, edited by George Kassimeris, 203–19. New York: Columbia University Press, 2008.

Jost, John T., Jaime L. Napier, Hulda Thorisdottir, Samuel D. Gosling, Tibor P. Palfai, and Brian Ostafin. "Are Needs to Manage Uncertainty and Threat Associated with Political Conservatism or Ideological Extremity?" *Personality and Social Psychology Bulletin* 33, no. 7 (2007): 989–1007, https://doi.org/10.1177/0146167207301028

Jugert, Philipp and John Duckitt. "A motivational model of authoritarianism: Integrating personal and situational determinants". *Political Psychology* 30, no. 5 (2009): 693–719, https://doi.org/10.1111/j.1467-9221.2009.00722.x

Kagwa, Apolo. *The Customs of the Baganda*. New York: Columbia University Press, 1934.

Kakkar, Hemant and Niro Sivanathan, "When the Appeal of a Dominant Leader Is Greater than a Prestige Leader", *Proceedings of the National Academy of Sciences*, June 12, 2017, https://doi.org/10.1073/pnas.1617711114

Karmin, Monika, Lauri Saag, Mário Vicente, Melissa A. Wilson Sayres, Mari Järve, Ulvi Gerst Talas, Siiri Rootsi, et al. "A Recent Bottleneck of Y Chromosome Diversity Coincides with a Global Change in Culture". *Genome Research* 25, no. 4 (2015): 459–66, https://doi.org/10.1101/gr.186684.114

Kassimeris, George. "Case Study–The 17th November Group: Europe's Last Revolutionary Terrorists". In *Playing Politics with Terrorism, edited by* George Kassimeris, 97–121. New York: Columbia University Press, 2008, https://doi.org/10.1002/9780470744499.ch6

Katz, Richard and Megan Biesele. "!Kung healing: The Symbolism of Sex Roles and Culture Change". In *The Past and Future of !Kung Ethnography: Critical Reflections and Symbolic Perspectives. Essays in Honour of Lorna Marshall*, edited by Megan Biesele, R. Gordon R and R. Lee, 195–230. Hamburg: Helmut Buske Verlag, 1986.

Keeley, Lawrence H. *War Before Civilization: The Myth of the Peaceful Savage.* Oxford and New York: Oxford University Press, 1996.

Keen, David. *Endless War? Hidden Functions of the "War on Terror"*. London: Pluto Press, 2006.

— "Greed and Grievance in Civil War". *International Affairs* 88, no. 4 (2012): 757–77, https://doi.org/10.1111/j.1468-2346.2012.01100.x

— *Useful Enemies: When Waging Wars Is More Important Than Winning Them.* New Haven, CT: Yale University Press, 2012.

Kellner, Douglas. "War Correspondents, the Military, and Propaganda: Some Critical Reflections". In *The Propaganda Society: Promotional Culture and Politics in Global Context*, edited by Gerald Sussman, 179–92. New York: Peter Lang Publishing, 2011.

Kelly, Raymond C. *Warless Societies and the Origin of War.* Ann Arbor, MI: University of Michigan Press, 2000, https://doi.org/10.3998/mpub.11589

Kesaeva, Ella. "Terroristy-Agenty. Neizvestnye podrobnosti beslanskoi Tragedii". *Novaia Gazeta*, No. 86, November 20, 2008, http://www.novayagazeta.ru/politics/37948.html.   English   translation: https://larussophobe.wordpress.com/2008/11/29/another-original-lr-translation-beslan-and-the-kgb

Kessler, Thomas and J. Christopher Cohrs. "The evolution of authoritarian processes: Fostering cooperation in large-scale groups". *Group Dynamics: Theory, Research, and Practice* 12, no. 1 (2008): 73–84, https://doi.org/10.1037/1089-2699.12.1.73

Kim, Heejung S. and Joni Y. Sasaki. "Cultural Neuroscience: Biology of the Mind in Cultural Contexts". *Annual Review of Psychology* 65 (2014): 487–514, https://doi.org/10.1146/annurev-psych-010213-115040

Kincaid, Harold. *Philosophical Foundations of the Social Sciences: Analyzing Controversies in Social Research*. Cambridge University Press, 1996.

Kingston, Shane. "Terrorism, the Media, and the Northern Ireland Conflict". *Studies in Conflict & Terrorism* 18, no. 3 (1995): 203–31, https://doi.org/10.1080/10576109508435980

Kirch, Patrick Vinton. *The Evolution of the Polynesian Chiefdoms*. Cambridge University Press, 1989.

Kirchoff, Paul. "The Warrau". In *Handbook of South American Indians: The Tropical Forest Tribes*, Vol. 3, edited by Julian Haynes Steward, 869–81. Bureau of American Ethnology Bulletin, 143–3. Washington, D.C.: US Government Printing Office, 1948.

Kiyonari, Toko, Shigehito Tanida, and Toshio Yamagishi. "Social Exchange and Reciprocity: Confusion or a Heuristic?" *Evolution and Human Behavior* 21, no. 6 (2000): 411–27, https://doi.org/10.1016/s1090-5138(00)00055-6

Klare, Michael T. "The Changing Geopolitics of Oil". In *Handbook of Oil Politics*, edited by Robert E. Looney, 30–44. London: Routledge, 2012.

Klein, Eberhard C. *Sozialer Wandel in Kiteezi/Buganda: Ein Dorf im Einflußbereich der Stadt Kampala*. Munich: Weltforum Verlag, 1969.

Ko, Siang-Feng. "Marriage among the Independent Lolos of Western China". *The American Journal of Sociology* 54 (1949): 487–96, https://doi.org/10.1086/220412

Kottak, Conrad P. "Ecological Variables in the Origin and Evolution of African States: The Buganda Example". *Comparative Studies in Society and History* 14, no. 3 (1972): 351–80, https://doi.org/10.1017/s0010417500006721

Krapf, J. Lewis. *Travels, Researches, and Missionary Labors during an Eighteen Years' Residence in Eastern Africa*. 2nd ed. London: Frank Cass & Co., 1968 [1860], https://doi.org/10.4324/9780203042403

Kraus, Michael W., Paul K. Piff, Rodolfo Mendoza-Denton, Michelle L. Rheinschmidt, and Dacher Keltner. "Social Class, Solipsism, and Contextualism: How the Rich Are Different from the Poor". *Psychological Review* 119, no. 3 (2012): 546–72, https://doi.org/10.1037/a0028756

Krause, Peter. "The Political Effectiveness of Non-State Violence: A Two-Level Framework to Transform a Deceptive Debate". *Security Studies* 22, no. 2 (2013): 259–94, https://doi.org/10.1080/09636412.2013.786914

Krueger, Frank, Raja Parasuraman, Lara Moody, Peter Twieg, Ewart de Visser, Kevin McCabe, Martin O'Hara, and Mary R. Lee. "Oxytocin Selectively Increases Perceptions of Harm for Victims but Not the Desire to Punish Offenders of Criminal Offenses". *Social Cognitive and Affective Neuroscience* 8, no. 5 (2013): 494–98, https://doi.org/10.1093/scan/nss026

Kuhn, Thomas S. *The Structure of Scientific Revolutions*. Chicago, IL: University of Chicago Press, 1962.

Ladd, Jonathan McDonald. "Predispositions and Public Support for the President during the War on Terrorism". *Public Opinion Quarterly* 71, no. 4 (2007): 511–38, https://doi.org/10.1093/poq/nfm033

LaGraffe, Daniel. "The Youth Bulge in Egypt: An Intersection of Demographics, Security, and the Arab Spring". *Journal of Strategic Security* 5, no. 2 (2012): 65–80, https://doi.org/10.5038/1944-0472.5.2.4

Lahr, M. Mirazón, et. al. "Inter-Group Violence among Early Holocene Hunter-gatherers of West Turkana, Kenya". *Nature* 529, no. 7586 (2016): 394–98, https://doi.org/10.1038/nature16477

Lakatos, Imre. "Falsification and the Methodology of Scientific Research Programmes". In *Criticism and the Growth of Knowledge*, edited by Imre Lakatos and Alan Musgrave, 91–196. Cambridge: Cambridge University Press, 1974, https://doi.org/10.1017/cbo9781139171434.009

Lammers, Joris, Janka I. Stoker, Jennifer Jordan, Monique Pollmann, and Diederik A. Stapel. "Power Increases Infidelity Among Men and Women". *Psychological Science* 22, no. 9 (2011): 1191–97, https://doi.org/10.1177/0956797611416252

Lammers, Joris, and Huadong Yang. "Feelings of Power Lead People to Take Sides with Other Powerful Parties". *Revista de Psicología Social* 27, no. 3 (2012): 337–46, https://doi.org/10.1174/021347412802845612

Laszlo, Ervin, and Alexander Laszlo. "The Contribution of the Systems Sciences to the Humanities". *Systems Research and Behavioral Science* 14, no. 1 (1997): 5–19, https://doi.org/10.1002/(sici)1099-1743(199701/02)14:1%3C5::aid-sres150%3E3.0.co;2-m

"Lavaan: Latent Variable Analysis". The Comprehensive R Archive Network. http://cran.r-project.org/web/packages/lavaan/index.html

Le Billon, Philippe. "The Geopolitical Economy of 'Resource Wars'." *Geopolitics* 9, no. 1 (2004): 1–28, https://doi.org/10.1080/14650040412331307812

Lê; Sébastien, Julie Josse, François Husson, et. al. "FactoMineR: An R Package for Multivariate Analysis". *Journal of Statistical Software* 25, no. 1 (2008): 1–18, https://doi.org/10.18637/jss.v025.i01

LeBar, Frank M., Gerald C. Hickey, and John K. Musgrave. *Ethnic Groups of Mainland Southeast Asia*. New Haven, CT: HRAF Press, 1964.

LeBlanc, Steven A. "Forager Warfare and Our Evolutionary Past". In *Violence and Warfare among Hunter-Gatherers, edited by* Mark W. Allen and Terry L. Jones, 26–46. Walnut Creek, CA: Left Coast Press, 2014.

Lechler, Marie. *An Econometrical Analysis of the Interdependencies between the Demographic Transition and Democracy*. Hamburg: Anchor Academic Publishing, 2015.

Lee, Richard B. *The !Kung San: Men, Women, and Work in a Foraging Society*. Cambridge: Cambridge University Press, 1979.

Lehmann, Laurent and Marcus W. Feldman. "War and the Evolution of Belligerence and Bravery". *Proceedings of the Royal Society B: Biological Sciences* 275, no. 1653 (2008): 2877–85, https://doi.org/10.1098/rspb.2008.0842

Leino, Petteri. "Zhdanovin Provokaatio—Mainilan Laukaukset". *Ammattisotilas* 2009, no. 4: 7–10.

Levy, Jack S. and William R. Thompson. *Causes of War*. Malden, MA: John Wiley & Sons, 2010.

Lewis, Ioan Myrddin. *Peoples of the horn of Africa: Somali, Afar and Saro*. London: International African Institute, 1955.

—*A Pastoral Democracy*. London: Oxford University Press, 1961.

—*Marriage and the Family in Northern Somaliland*. Kampala: East African Institute of Social Research, 1962.

Lieberman, Evan S. "Nested Analysis: Toward the Integration of Comparative-Historical Analysis with Other Social Science Methods". In *Advances in Comparative-Historical Analysis*, edited by James Mahoney and Kathleen Thelen, 240–63. Cambridge: Cambridge University Press, 2015, https://doi.org/10.1017/cbo9781316273104.010

Liétard, Alfred. *Au Yun-Nan: Les Lo-lo P'o, Une Tribu des Aborigènes de la Chine Méridionale*. Münster: Aschendorffsche Verlagsbuchhandlung, 1913. Anthropos Bibliothek, vol. 1, no. 5.

Litvinenko, Alexander and Yuri Felshtinsky. *Blowing Up Russia: The Secret Plot to Bring Back KGB Terror*. London: Gibson Square, 2007.

Lizot, Jacques. *Le Cercle des Feux: Faits et Dits des Indiens Yanomami*. Paris: Éditions du Seuil, 1976.

Lomax, Alan. *Folk Song Style and Culture*. Washington, D.C.: American Association for the Advancement of Science, 1968.

Looney, Robert E. *Handbook of Oil Politics*. Routledge, 2012.

Lothrop, Samuel K. *The Indians of Tierra del Fuego*. New York: Museum of the American Indian Heye Foundation, 1928, https://doi.org/10.5479/sil.472342.39088016090599

Low, Bobbi S. "Human Responses to Environmental Extremeness and Uncertainty: A Cross-Cultural Perspective". in *Risk and Uncertainty in Tribal and Peasant Economies*, edited by Elizabeth A. Cashdan, 229–55. Boulder: Westview Press, 1990.

Luard, Evan. *War in International Society: A Study in International Sociology*. London: I. B. Tauris, 1986.

Lucas, Lisa. "'To Them That Have Shall Be Given, But…': The Future of Funding and Evaluating Research in UK Universities". In *Beyond Mass Higher Education: Building On Experience*, edited by Ian McNay, 120–33. Maidenhead and New York: The Society for Research into Open Education and Open University Press, 2006.

Lugira, Aloysius Muzzanganda. *Ganda Art: A Study of the Ganda Mentality with Respect to Possibilities of Acculturation in Christian Art.* Kampala: Osasa Publications, 1970.

Luhmann, Niklas. *Social Systems.* Stanford, CA: Stanford University Press, 1995.

Luling, Virginia. *Somali Sultanate: The Geledi City-State Over 150 Years.* London: Haan, 2002.

Lumsden, Charles J. and Edward O. Wilson. *Genes, Mind, and Culture: The Coevolutionary Process.* Cambridge, MA: Harvard University Press, 1981.

Lutz, Brenda J., and James M. Lutz. *Global Terrorism.* 2nd ed. Routledge, 2008.

Lynch, Andrew. "Legislating Anti-Terrorism: Observations on Form and Process". In *Global Anti-Terrorism Law and Policy, edited by* Victor V. Ramraj, Michael Hor, Kent Roach, and George Williams, 151–82, 2nd ed. Cambridge University Press, 2012, https://doi.org/10.1017/cbo9781139043793.009

Maffi, Luisa, "Linguistic, Cultural, and Biological Diversity". *Annual Reviews of Anthropology.* 34 (2005): 599–617, https://doi.org/10.1146/annurev.anthro.34.081804.120437

Mahoney, James and Kathleen Thelen. *Advances in Comparative-Historical Analysis.* Cambridge University Press, 2015.

Mair, Lucy P. *An African People in the Twentieth Century.* London: Routledge, 1934.

Malpass, Michael A. *Daily Life in the Inca Empire.* Westport: Greenwood Press, 1996.

Man, Edward Horace. *Aboriginal Inhabitants of the Andaman Islands.* Delhi: Sanskaran Prakashak, 1883. Reprint 1975.

Marshall, Lorna. "!Kung Bushman Religious Beliefs". *Africa* 32 (1962): 221–52, https://doi.org/10.2307/1157541

—*Nyae Nyae !Kung: Beliefs and Rites.* Cambridge, MA: Peabody Museum of Archaeology and Ethnology, Harvard University, 1999.

—*The !Kung of Nyae Nyae.* Cambridge, MA: Harvard University Press, 1976.

Martens, Andy, Brian L. Burke, Jeff Schimel, and Erik H. Faucher. "Same but Different: Meta-Analytically Examining the Uniqueness of Mortality Salience Effects". *European Journal of Social Psychology* 41, no. 1 (2011): 6–10, https://doi.org/10.1002/ejsp.767

Martin, Debra L. and David W. Frayer, eds. *Troubled Times: Violence and Warfare in the Past.* Amsterdam: Gordon and Breach, 1997.

Mason, J. Alden. *The Ancient Civilizations of Peru.* Middlesex: Penguin, 1957.

Mayer, Arno J. *Dynamics of Counterrevolution in Europe, 1870–1956: An Analytic Framework.* New York: Harper & Row, 1971.

Mayer, Jane. *The Dark Side*: *The inside Story of how the War on Terror turned into a War on American Ideals*. New York: Doubleday, 2008.

McAllister, Lisa S., Gillian V. Pepper, Sandra Virgo, and David A. Coall. "The Evolved Psychological Mechanisms of Fertility Motivation: Hunting for Causation in a Sea of Correlation". *Philosophical Transactions of the Royal Society*, B 371, no. 1692 (2016): 20150151, https://doi.org/10.1098/rstb.2015.0151

McCauley, John F. and Daniel N. Posner, "The Political Sources of Religious Identification: Evidence from the Burkina Faso-Côte d'Ivoire Border". *British Journal of Political Science*, 2017, 1–21, https://doi.org/10.1017/s0007123416000594

McDonald, Melissa M., Carlos David Navarrete, and Mark van Vugt. "Evolution and the Psychology of Intergroup Conflict: The Male Warrior Hypothesis". *Philosophical Transactions of the Royal Society B: Biological Sciences* 367, no. 1589 (2012): 670–79, https://doi.org/10.1098/rstb.2011.0301

McFarland, Sam, Vladimir Ageyev and Marina Abalakina. "The Authoritarian Personality in the United States and the Former Soviet Union: Comparative Studies". In *Strength and Weakness*: *The Authoritarian Personality Today*, edited by William F. Stone, Gerda Lederer, and Richard Christie, 199–225. New York: Springer, 1993, https://doi.org/10.1007/978-1-4613-9180-7_10

McFarland, Sam. "Authoritarianism, Social Dominance, and Other Roots of Generalized Prejudice". *Political Psychology* 31, no. 3 (2010): 453–77, https://doi.org/10.1111/j.1467-9221.2010.00765.x

McGhee, Robert. "Copper Eskimo Prehistory". *Publications in Archaeology* 2 (1972): 1–141.

McLeay, Michael, Amar Radia, and Ryland Thomas. "Money Creation in the Modern Economy". *Bank of England Quarterly Bulletin* 54, no. 1 (2014): 14–27, http://www.bankofengland.co.uk/publications/Documents/quarterlybulletin/2014/qb14q1prereleasemoneycreation.pdf

McManus, John. "A Market-Based Model of News Production". *Communication Theory* 5, no. 4 (1995): 301–38, https://doi.org/10.1111/j.1468-2885.1995.tb00113.x

—"The Commercialization of News". In *The Handbook of Journalism Studies*, edited by Karin Wahl-Jorgensen and Thomas Hanitzsch, 218–35. New York: Routledge, 2009.

McNamara, Rita Anne, Ara Norenzayan, and Joseph Henrich. "Supernatural Punishment, in-Group Biases, and Material Insecurity: Experiments and Ethnography from Yasawa, Fiji". *Religion, Brain & Behavior*, 6, no. 1 (2014): 34–55, https://doi.org/10.1080/2153599x.2014.921235

Means, Philip A. *Ancient Civilizations of the Andes*. New York: Charles Scribner's Sons, 1931.

Mehlum, Halvor, Karl Moene, and Ragnar Torvik. "Institutions and the Resource Curse". *The Economic Journal* 116, no. 508 (2006): 1–20, https://doi.org/10.1111/j.1468-0297.2006.01045.x

Mellor, Mary. *The Future of Money: From Financial Crisis to Public Resource*. Pluto Press, 2010.

Mercader, Julio, Freya Runge, Luc Vrydaghs, Hughes Doutrelepont, Corneille EN Ewango, et al. "Phytoliths from Archaeological Sites in the Tropical Forest of Ituri, Democratic Republic of Congo". *Quaternary Research* 54, no. 1 (2000): 102–12, https://doi.org/10.1006/qres.2000.2150

Merker, Moritz. *Die Masai: Ethnographische Monographie eines ostafrikanischen Semitenvolkes*. Berlin: Dietrich Reimer, 1904.

Meyer, Philip. "The Influence Model and Newspaper Business". *Newspaper Research Journal* 25, no. 1 (2004): 66–83, https://doi.org/10.1177/073953290402500106

Minkov, Michael, Vesselin Blagoev, and Geert Hofstede. "The Boundaries of Culture: Do Questions about Societal Norms Reveal Cultural Differences?" *Journal of Cross-Cultural Psychology* 44, no. 7 (2012): 1094–1106, https://doi.org/10.1177/0022022112466942

Moghadam, Assaf. "Failure and Disengagement in the Red Army Faction". *Studies in Conflict & Terrorism* 35, no. 2 (2012): 156–81, https://doi.org/10.1080/1057610x.2012.639062

Moïse, Edwin E. *Tonkin Gulf and the Escalation of the Vietnam War*. University of North Carolina Press, 1996.

Moller, Herbert. "Youth as a Force in the Modern World". *Comparative Studies in Society and History* 10, no. 3 (1968): 237–60, https://doi.org/10.1017/s0010417500004898

Moor, Nienke, Wout Ultee, and Ariana Need. "Analogien, Subsistenztechnologien und (nicht-)moralische allmächtige Schöpfergötter in vorindustriellen Gesellschaften". *Kölner Zeitschrift für Soziologie und Sozialpsychologie* 59, no. 3 (2007): 383–409, https://doi.org/10.1007/s11577-007-0054-6

Mueller, John E. *Overblown: How Politicians and the Terrorism Industry Inflate National Security Threats, and Why We Believe Them*. Simon and Schuster, 2006.

Mueller, Laurence D. "Theoretical and Empirical Examination of Density-dependent Selection". *Annual Review of Ecology and Systematics* no. 28 (1997): 269–88, https://doi.org/10.1146/annurev.ecolsys.28.1.269

Müller, Michael, and Andreas Kanonenberg. *Die RAF-Stasi Connection*. Berlin: Rowohlt, 1992.

Mumford, Andrew and Bruno C. Reis. "Constructing and Deconstructing Warrior-Scholars". In *The Theory and Practice of Irregular Warfare: Warrior-Scholarship in Counter-Insurgency*, edited by Andrew Mumford and Bruno C. Reis, 4–17. London: Routledge, 2014.

Mumford, Andrew. *Proxy Warfare*. Cambridge: Polity Press, 2013.

Münkler, Herfried. *The New Wars.* Cambridge: Polity, 2005.

Murdock, George P. and Douglas R. White. "Standard Cross-Cultural Sample". *Ethnology* 8, no. 4 (1969): 329–69, https://doi.org/10.2307/3772907

— "Standard Cross-Cultural Sample: On-line Edition". *Social Dynamics and Complexity* (2006), http://repositories.cdlib.org/imbs/socdyn/wp/Standard_Cross-Cultural_Sample

Murshed, S. Iftikhar. *Afghanistan: The Taliban Years.* London: Bennett & Bloom, 2006.

Nacos, Brigitte L., Yaeli Bloch-Elkon, and Robert Y. Shapiro. *Selling Fear: Counterterrorism, the Media, and Public Opinion.* Chicago, IL: University of Chicago Press, 2011, https://doi.org/10.7208/chicago/9780226567204.001.0001

National Institute of Standards and Technology. "Final Report on the Collapse of the World Trade Center Towers". 2005, http://nvlpubs.nist.gov/nistpubs/Legacy/NCSTAR/ncstar1.pdf

— "Structural Fire Response and Probable Collapse Sequence of World Trade Center Building 7". 2008, https://www.nist.gov/publications/structural-fire-response-and-probable-collapse-sequence-world-trade-center-building-7

National Security Archive. "Pentagon Proposed Pretexts for Cuba Invasion in 1962", 2001, http://nsarchive.gwu.edu/news/20010430

Navarrete, C. David, Robert Kurzban, Daniel M.T. Fessler, and Lee A. Kirkpatrick. "Anxiety and Intergroup Bias: Terror Management or Coalitional Psychology?" *Group Processes & Intergroup Relations* 7, no. 4 (2004): 370–97, https://doi.org/10.1177/1368430204046144

Nelson, Barbara J. *Making an Issue of Child Abuse: Political Agenda Setting for Social Problems.* Chicago, IL: University of Chicago Press, 1984.

Nemet-Nejat, Karen R. *Daily Life in Ancient Mesopotamia.* Westport, Connecticut: Greenwood Press, 1998.

Nersessian, Nancy J. "Kuhn, Conceptual Change, and Cognitive Science". In *Thomas Kuhn,* edited by Thomas Nickles, 178–211. Cambridge and New York: Cambridge University Press, 2003.

New World Encyclopedia. "List of Countries by Human Development Index", https://www.newworldencyclopedia.org/entry/List_of_countries_by_Human_Development_Index

Nicolas, Guy. *Dynamique Sociale et Appréhension du Monde au sein d'une Société Hausa.* Paris: Institut d'ethnologie, Musée de l'Homme, 1975.

Nippold, Walter. *Rassen—und Kulturgeschichte der Negrito-Völker Südost-Asiens.* Leipzig: Verlag Jordan & Gramberg, 1936.

Nojumi, Neamatollah. *The Rise of the Taliban in Afghanistan: Mass Mobilization, Civil War, and the Future of the Region.* New York: Palgrave MacMillan, 2002.

Nordenskjöld, Otto. *Från Eldslandet: Skildringer från den Svenska Expeditionen till Magellansländerna 1895–97.* Stockholm: P. A. Norstedt & Söners Förlag, 1898.

Norris, Jesse. "Entrapment and Terrorism on the Left: An Analysis of Post-9/11 Cases". *New Criminal Law Review*, 19, no. 2 (2016): 236–78, https://doi.org/10.1525/nclr.2016.19.2.236

Norris, Pippa, and Ronald Inglehart. *Sacred and Secular: Religion and Politics Worldwide*. 2nd ed. Cambridge and New York: Cambridge University Press, 2011.

North, Adrian C. and David J. Hargreaves. "Lifestyle Correlates of Musical Preference: 1, 2, 3". *Psychology of Music* 35 (2007): 58–87, 179–20, 473–497, https://doi.org/10.1177/0305735607068888 https://doi.org/10.1177/0305735607070302 https://doi.org/10.1177/0305735607072656

Nowak, Martin A. "Five Rules for the Evolution of Cooperation". *Science* 314, no. 5805 (2006): 1560–63, https://doi.org/10.1126/science.1133755

Nuti, Leopoldo. "The Italian 'Stay-Behind' network–The Origins of Operation 'Gladio'." *Journal of Strategic Studies* 30, no. 6 (2007): 955–80, https://doi.org/10.1080/01402390701676501

O'Brien, Kevin A. "Surrogate Agents: Private Military and Security Operators in an Unstable World". In *Making Sense of Proxy Wars: States, Surrogates & the Use of Force*, edited by Michael A. Innes, 109–36. Washington, D.C.: Potomac, 2012.

O'Day, Alan. *Dimensions of Irish Terrorism*. Aldershot: Dartmouth, 1993.

O'Malley, Maureen A. "Evolutionary Approaches in the Social Sciences". In *The SAGE Handbook of Social Science Methodology*, edited by William Outhwaite and Stephen P. Turner, 333–57. Adelrshot: Sage, 2007, https://doi.org/10.4135/9781848607958.n19

Oates, Sarah, Lynda Lee Kaid, and Mike Berry. *Terrorism, Elections, and Democracy: Political Campaigns in the United States, Great Britain, and Russia*. New York: Palgrave Macmillan, 2010.

OECD. "Harmful Tax Competition: An Emerging Global Issue". Paris: OECD Publishing, 1998, http://www.oecd-ilibrary.org/taxation/harmful-tax-competition_9789264162945-en

Olivas-Luján, Miguel R., Anne-Wil Harzing, and Scott McCoy. "September 11, 2001. Two Quasi-Experiments on the Influence of Threats on Cultural Values and Cosmopolitanism". *International Journal of Cross Cultural Management* 4, no. 2 (2004): 211–28, https://doi.org/10.1177/1470595804044750

Olsen, Dale A. *Music of the Warao of Venezuela: Song People of the Rain Forest*. Gainesville, FL: University Press of Florida, 1996.

Opler, Morris E. "An Outline of Chiricahua Apache Social Organization". In *Social Anthropology of North American Tribes*, edited by Fred Eggan, 173–239. Chicago, IL: University of Chicago Press, 1937.

Opler, Morris E. *An Apache Life-Way: The Economic, Social, and Religious Institutions of the Chiricahua Indians*. Chicago, IL: University of Chicago Press, 1941.

Owsiak, Andrew P. "Democratization and International Border Agreements". *The Journal of Politics* 75 (July 2013): 717–29, https://doi.org/10.1017/s0022381613000364

Oxfam. "An Economy for the 99%". Oxford: Oxfam International, 2017, https://www.oxfam.org/sites/www.oxfam.org/files/file_attachments/bp-economy-for-99-percent-160117-en.pdf

Padgett, Vernon R. and Dale O. Jorgenson. "Superstition and Economic Threat: Germany, 1918–1940". *Personality and Social Psychology Bulletin* 8, no. 4 (1982): 736–41, https://doi.org/10.1177/0146167282084021

Padilla, Art, Robert Hogan, and Robert B. Kaiser. "The Toxic Triangle: Destructive Leaders, Susceptible Followers, and Conducive Environments". *The Leadership Quarterly* 18, no. 3 (2007): 176–94, https://doi.org/10.1016/j.leaqua.2007.03.001

Pagès, Jérôme. "Analyse Factorielle de Données Mixtes". *Revue de Statistique Appliquée* 52, no. 4 (2004): 93–111.

Paletz, David L. and Alex P. Schmid. *Terrorism and the Media*. London: Sage, 1992.

Pallis, Svend Aage. *The Antiquity of Iraq: A Handbook of Assyriology*. Copenhagen: Ejnar Munksgaard, 1956.

Palmer, Herbert R. *Sudanese Memoirs: Being Mainly Translations of a Number of Arabic Manuscripts Relating to the Central and Western Sudan*. Vol. 3. Frank Cass & Co., 1967 [1928].

Pandya, Vishvajit. "Making of the Other. Vignettes of Violence in Andamanese Culture". *Critique of Anthropology* 20, no. 4 (2000): 359–391, https://doi.org/10.1177/0308275x0002000403

—*Above the Forest: A Study of Andamanese Ethnoanemology, Cosmology, and the Power of Ritual*. Delhi: Oxford University Press, 1993.

Pape, R. A. "The Strategic Logic of Suicide Terrorism". *American Political Science Review* 97, no. 3 (2003): 343–61, https://doi.org/10.1017/s000305540300073x

Peck, Sarah and Sarah Chayes. "The Oil Curse. A Remedial Role for the Oil Industry". Carnegie Endowment for International Peace, 2015.

Pellegrino, Giovanni. "Commissione parlamentare d'inchiesta sul terrorismo in Italia e sulle cause della mancata individuazione dei responsabili delle stragi. 9ª seduta", February 12, 1997.

Pelto, Pertii J. "The Differences between 'Tight' and 'Loose' Societies". *Trans-Action* 5, no. 5 (1968): 37–40, https://doi.org/10.1007/bf03180447

Peretz, Don. *The Arab-Israel Dispute*. New York: Facts On File, 1996.

Perez, Carlota. *Technological Revolutions and Financial Capital: The Dynamics of Bubbles and Golden Ages*. Cheltenham: Edward Elgar Publishing, 2002, https://doi.org/10.4337/9781781005323

— "Technological Revolutions and Techno-Economic Paradigms". *Cambridge Journal of Economics* 34, no. 1 (2010): 185–202, https://doi.org/10.1093/cje/bep051

Perrin, Andrew J. "National Threat and Political Culture: Authoritarianism, Antiauthoritarianism, and the September 11 Attacks". *Political Psychology* 26, no. 2 (2005): 167–94, https://doi.org/10.1111/j.1467-9221.2005.00414.x

Perry, Ryan, Chris G. Sibley, and John Duckitt. "Dangerous and Competitive Worldviews: A Meta-Analysis of Their Associations with Social Dominance Orientation and Right-Wing Authoritarianism". *Journal of Research in Personality* 47, no. 1 (2013): 116–27, https://doi.org/10.1016/j.jrp.2012.10.004

Peters, Butz. *Tödlicher Irrtum: Die Geschichte der RAF*. Frankfurt: Fischer, 2004.

Peters, Gretchen. "The Afghan Insurgency and Organized Crime". In *The Rule of Law in Afghanistan: Missing in Inaction*, edited by Whit Mason, 99–122. Cambridge and New York: Cambridge University Press, 2011, https://doi.org/10.1017/cbo9780511760082.008

Peters, John F. *Life Among the Yanomami*. Peterborough, Ontario: Broadview Press, 1998.

Pettersson, Therése and Peter Wallensteen. "Armed Conflicts, 1946–2014". *Journal of Peace Research* 52, no. 4 (2015): 536–50, https://doi.org/10.1177/0022343315595927

Phayal, Anup. "Cost of Violence and the Peaceful Way out: A Comparison of the Provisional Irish Republican Army and the Communist Party of Nepal (Maoists)". *Dynamics of Asymmetric Conflict* 4, no. 1 (2011): 1–20, https://doi.org/10.1080/17467586.2011.575170

Picard, Robert. "Commercialism and Newspaper Quality". *Newspaper Research Journal* 25, no. 1 (2004): 54–65, https://doi.org/10.1177/073953290402500105

Pierson, Paul. "Power and Path Dependence". In *Advances in Comparative-Historical Analysis*, edited by James Mahoney and Kathleen Thelen, 123–46. Cambridge University Press, 2015, https://doi.org/10.1017/cbo9781316273104.006

Piff, Paul K. "Wealth and the Inflated Self: Class, Entitlement, and Narcissism". *Personality and Social Psychology Bulletin* 40, no. 1 (2014): 34–43, https://doi.org/10.1177/0146167213501699

Piff, Paul K., Daniel M. Stancato, Stéphane Côté, Rodolfo Mendoza-Denton, and Dacher Keltner. "Higher Social Class Predicts Increased Unethical Behavior". *Proceedings of the National Academy of Sciences* 109, no. 11 (2012): 4086–91, https://doi.org/10.1073/pnas.1118373109

Pillar, Paul R. *Intelligence and US Foreign Policy: Iraq, 9/11, and Misguided Reform*. New York: Columbia University Press, 2011.

Plummer, Ken. *Telling Sexual Stories: Power, Change and Social Worlds*. London: Routledge, 1995, https://doi.org/10.4324/9780203425268

Pollard, Samuel. *In Unknown China*. London: Seeley, Service & Co. Ltd., 1921.

Pop-Eleches, Grigore and Graeme Robertson. "After the Revolution". *Problems of Post-Communism* 61, no. 4 (2014): 3–22, https://doi.org/10.2753/ppc1075-8216610401

Popper, Karl R. *The Open Society and Its Enemies*. London: Routledge, 1966 [1945].

—*Conjectures and Refutations: The Growth of Scientific Knowledge*. London: Routledge, 1963.

Portman, Maurice Vidal (1888a). "Andamanese Music, with Notes on Oriental Music and Musical Instruments". *Journal of the Royal Asiatic Society of Great Britain and Ireland*, New Series 20 (1888): 181–218, https://doi.org/10.1017/s0035869x00020025

—(1888b). "The Exploration and Survey of the Little Andamans". *Proceedings of the Royal Geographical Society* 10 (1888): 567–76, https://doi.org/10.2307/1800974

Price, Michael E., and Mark Van Vugt. "The Evolution of Leader-follower Reciprocity: The Theory of Service-for-Prestige". *Frontiers in Human Neuroscience* 8 (2014): 363, https://doi.org/10.3389/fnhum.2014.00363

—"The Service-for-Prestige Theory of Leader-Follower Relations: A Review of the Evolutionary Psychology and Anthropology Literatures". In *The Biological Foundations of Organizational Behavior*, edited by Stephen M. Colarelli and Richard D. Arvey, 169–201. Chicago, IL: University of Chicago Press, 2015.

"Profile: Dr David Kelly". *BBC News*, January 27, 2004, http://news.bbc.co.uk/2/hi/uk_news/politics/3076869.stm

Puccioni, Nello. *Antropologia e Etnografia delle Genti della Somalia*. Vol. 3. Bologna: Nicola Zanichelli, 1936.

Putnam, Patrick. "The Pygmies of the Ituri Forest". In *A Reader in General Anthropology*, edited by Carleton Stevens Coon, 322–42. London: Jonathan Cape, 1950.

Quinlan, Robert J. "Human Parental Effort and Environmental Risk". *Proceedings of the Royal Society B: Biological Sciences* 274, no. 1606 (2007): 121–25, https://doi.org/10.1098/rspb.2006.3690

Radcliffe-Brown, Alfred. *The Andaman Islanders: A Study in Social Anthropology*. Cambridge: Cambridge University Press, 1922.

Rahimi, Fahima. *Women in Afghanistan/Frauen in Afghanistan*. Kabul: Grauwiler, 1977, Liestal 1986: Bibliotheca Afghanica.

Rashid, Ahmed. *Taliban: Islam, Oil and the New Great Game in Central Asia*. London: I. B. Tauris, 2002.

Rasmussen, Knud. *Intellectual Culture of the Copper Eskimos. Report of the Fifth Thule Expedition 1921–24*. Copenhagen: Gyldendalske boghandel, 1932.

Rast, David E., III; Michael A. Hogg, and Steffen R. Giessner. "Self-Uncertainty and Support for Autocratic Leadership". *Self and Identity* 12, no. 6 (2013): 635–49, https://doi.org/10.1080/15298868.2012.718864

Ravndal, Jacob Aasland. "Thugs or Terrorists? A Typology of Right-Wing Terrorism and Violence in Western Europe". *Journal for Deradicalization*, no. 3 (2015): 1–38.

Ray, Benjamin C. *Myth, Ritual and Kingship in Buganda*. New York: Oxford University Press, 1991.

Ray, J. J. "The Scientific Study of Ideology is Too Often More Ideological than Scientific". *Personality and Individual Differences* 10, no. 3 (1989): 331–36, https://doi.org/10.1016/0191-8869(89)90106-2

Redlich, Fritz. *Hitler: Diagnosis of a Destructive Prophet*. New York: Oxford University Press, 1998.

Regan, Patrick M. and Daniel Norton. "Greed, Grievance, and Mobilization in Civil Wars". *Journal of Conflict Resolution* 49, no. 3 (2005): 319–36, https://doi.org/10.1177/0022002704273441

Reid, Edna O. F. "Evolution of a Body of Knowledge: An Analysis of Terrorism Research". *Information Processing & Management* 33, no 1 (1997): 91–106, https://doi.org/10.1016/s0306-4573(96)00052-0

Reilly, Marie. "Data Analysis Using Hot Deck Multiple Imputation". *The Statistician* 42 (1993): 307–13, https://doi.org/10.2307/2348810

Reinhart, Carmen M. and Kenneth Rogoff. *This Time Is Different: Eight Centuries of Financial Folly*. Princeton, NJ: Princeton University Press, 2009.

Reith, Margaret. "Viewing of Crime Drama and Authoritarian Aggression: An Investigation of the Relationship between Crime Viewing, Fear, and Aggression". *Journal of Broadcasting & Electronic Media* 43, no. 2 (1999): 211, https://doi.org/10.1080/08838159909364485

Richards, Audrey I. and Priscilla Reining. "Report on Fertility Surveys in Buganda and Buhaya, 1952". In *Culture and Human Fertility: A Study of the Relation of Cultural Conditions to Fertility in Non-industrial and Transitional Societies*, edited by Frank Lorimer, 351–403. Zürich: UNESCO, 1954.

Richardson, George P. *Feedback Thought in Social Science and Systems Theory*. Philadelphia, PA: University of Pennsylvania Press, 1991.

Richardson, Louise. *What Terrorists Want: Understanding the Terrorist Threat*. New York: Random House, 2006.

Richerson, Peter J. and Robert Boyd. *Not by Genes Alone: How Culture Transformed Human Evolution*. Chicago, IL: University of Chicago Press, 2008.

Rickert, Edward J. "Authoritarianism and Economic Threat: Implications for Political Behavior". *Political Psychology* 19, no. 4 (1998): 707–20, https://doi.org/10.1111/0162-895x.00128

Rid, Thomas and Marc Hecker. *War 2.0*: *Irregular Warfare in the Information Age*. Westport, CT: Praeger, 2009.

Riek, Blake M., Eric W. Mania and Samuel L. Gaertner. "Intergroup Threat and Outgroup Attitudes: A Meta-Analytic Review". *Personality and Social Psychology Review* 10, no. 4 (2006): 336–53, https://doi.org/10.1207/s15327957pspr1004_4

Rigney, Daniel. *The Matthew Effect*: *How Advantage Begets Further Advantage*. New York: Columbia University Press, 2010.

Risen, James. *State of War*: *The Secret History of the CIA and the Bush Administration*. New York: Free Press, 2006.

Ritter, Scott. *Iraq Confidential*: *The Untold Story of America's Intelligence Conspiracy*. London: IB Tauris, 2005.

Ritter, Scott and William Rivers Pitt. *War on Iraq*: *What Team Bush Doesn't Want You to Know*. London: Profile Books, 2002.

Roes, Frans L. and Michel Raymond. "Belief in Moralizing Gods". *Evolution and Human Behavior* 24, no. 2 (2003): 126–35, https://doi.org/10.1016/s1090-5138(02)00134-4

Roos, Patrick, Michele Gelfand, Dana Nau, and Janetta Lun. "Societal Threat and Cultural Variation in the Strength of Social Norms: An Evolutionary Basis". *Organizational Behavior and Human Decision Processes* 129 (2015): 14–23, https://doi.org/10.1016/j.obhdp.2015.01.003

Roscoe, John. *The Baganda*: *An Account of their Native Customs and Beliefs*. London: Frank Cass & Co., 1911.

Roscoe, Paul. "Foragers and War in Contact-Era New Guinea". In *Violence and Warfare among Hunter-Gatherers*, edited by Mark W. Allen and Terry L. Jones, 223–40. Walnut Creek: Left Coast Press, 2014.

Rosen, Stephen Peter. *War and Human Nature*. Princeton, NJ: Princeton University Press, 2005.

Rosenfeld, Alan. "Militant Democracy: The Legacy of West Germany's War on Terror in the 1970s". *The European Legacy* 19, no. 5 (2014): 568–89, https://doi.org/10.1080/10848770.2014.943531

Ross, Marc Howard. *The Culture of Conflict*: *Interpretations and Interests in Comparative Perspective*. New Haven, CT: Yale University Press, 1993.

Ross, Michael. *The Oil Curse*: *How Petroleum Wealth Shapes the Development of Nations*. Princeton, NJ: Princeton University Press, 2012.

Roth, Martha T. "Age at Marriage and the Household: A Study of Neo-Babylonian and Neo-Assyrian Forms". *Comparative Studies in Society and History* 29, no. 4 (1987): 715–47, https://doi.org/10.1017/s0010417500014857

Rothe, Dawn and Stephen L. Muzzatti. "Enemies Everywhere: Terrorism, Moral Panic, and US Civil Society". *Critical Criminology* 12, no. 3 (2004): 327–50, https://doi.org/10.1007/s10612-004-3879-6

Rouhana, Nadim. "Key Issues in Reconciliation: Challenging Traditional Assumptions on Conflict Resolution and Power Dynamics". In *Intergroup Conflicts and Their Resolution: A Social Psychological Perspective*, edited by Daniel Bar-Tal, 291–314, 2011.

Rowse, Arthur E. "Gladio: The Secret US War to Subvert Italian Democracy". *Covert Action Quarterly* 49 (1994): 20–27.

Rummel, Rudolph J. *The Dimensions of Nations*. Beverly Hills, CA: Sage, 1972.

Rusch, Walter. *Klassen und Staat in Buganda vor der Kolonialzeit*. Berlin: Akademie Verlag, 1975.

Russell, Elbert W. "Factors of Human Aggression: A Cross-Cultural Factor Analysis of Characteristics Related to Warfare and Crime". *Behavior Science Notes (Cross-Cultural Research)* 7, no. 4 (1972): 275–312, https://doi.org/10.1177/106939717200700401

Sabatier, Leopold. *Recueil des coutumes Rhadées du Darlac*. Hanoi: Imprimerie d'Extrême-Orient, 1940.

Saggs, Henry William Frederick. *The Greatness that was Babylon*. London: Sidgwick and Jackson, 1962.

— *Everyday Life in Babylonia & Assyria*. London: B. T. Batsford, 1965.

Saitoti, Tepilit Ole. *The Worlds of a Maasai Warrior: An Autobiography*. New York: Random House, 1986.

Sambanis, Nicholas. "Using Case Studies to Refine and Expand the Theory of Civil War". In *Understanding Civil War: Evidence and Analysis*, edited by Paul Collier and Nicholas Sambanis, 303–34. World Bank, 2005.

Sankan, S. S. Ole. *The Maasai*. Nairobi: East African Literature Bureau, 1971.

Satter, David. *Darkness at Dawn: The Rise of the Russian Criminal State*. New Haven, CT: Yale University Press, 2003.

Saucier, Gerard, et al. "Cross-Cultural Differences in a Global 'Survey of World Views'." *Journal of Cross-Cultural Psychology* 46, no. 1 (2015): 53–70.

Sawyer, Jack and Robert A. LeVine. "Cultural Dimensions: A Factor Analysis of the World Ethnographic Sample". *American Anthropologist* 68, no. 3 (1966): 708–31, https://doi.org/10.1525/aa.1966.68.3.02a00060

Schack, Todd. "Perpetual Media Wars: The Cultural Front in the Wars on Terror and Drugs". In *The War on Terror and American Popular Culture. September 11 and Beyond*, edited by Andrew Schopp and Matthew B. Hill. Cranbury, NJ: Associated University Press, 2009.

Schaeffer, Jörg. *Traditionelle Gesellschaft und Geschichte der Rhadé im südvietnamesischen Hochland*. Freiburg: Johannes Krause, 1979.

Schebesta, Paul. *Die Bambuti-Pygmäen vom Ituri: Ergebnisse zweier Forschungsreisen zu den zentralafrikanischen Pygmäen*, Vol. I. Bruxelles: Georges van Campenhout, 1938.

—*Die Bambuti-Pygmäen vom Ituri*: *Ergebnisse zweier Forschungsreisen zu den zentralafrikanischen Pygmäen*, Vol. II–I. Bruxelles: Georges van Campenhout, 1941.

—*Die Bambuti-Pygmäen vom Ituri*: *Ergebnisse zweier Forschungsreisen zu den zentralafrikanischen Pygmäen*, Vol. II–II. Bruxelles: Georges van Campenhout, 1948.

—*Die Bambuti-Pygmäen vom Ituri*: *Ergebnisse zweier Forschungsreisen zu den zentralafrikanischen Pygmäen*, Vol. II–III. Bruxelles: Georges van Campenhout, 1950.

Schetsche, Michael. *Wissenssoziologie sozialer Probleme*: *Grundlegung einer relativistischen Problemtheorie*. Wiesbaden: Westdeutscher Verlag, 2000.

Schmeidel, John. "My Enemy's Enemy: Twenty Years of Co-Operation between West Germany's Red Army Faction and the GDR Ministry for State Security". *Intelligence and National Security* 8, no. 4 (1993): 59–72, https://doi.org/10.1080/02684529308432225

Schmid, Katharina and Orla T. Muldoon. "Perceived Threat, Social Identification, and Psychological Well-Being: The Effects of Political Conflict Exposure". *Political Psychology* 36, no. 1 (2015): 75–92, https://doi.org/10.1111/pops.12073

Schoel, Christiane, Matthias Bluemke, Patrick Mueller, and Dagmar Stahlberg. "When Autocratic Leaders Become an option–Uncertainty and Self-Esteem Predict Implicit Leadership Preferences". *Journal of Personality and Social Psychology* 101, no. 3 (2011): 521, https://doi.org/10.1037/a0023393

Schrenck, Leopold von. *Reisen und Forschungen im Amur-Lande (Vol. 3)*: *Die Völker des Amur-Landes*. Commissionäre der Kaiserlichen Akademie der Wissenschaften: St. Petersburg, 1881.

Schroeder, Albert H. *A Study of the Apache Indians* (Parts 4 and 5). New York: Garland Publishing, 1974.

Schurr, Theodore G. and Stephen T. Sherry. "Mitochondrial DNA and Y Chromosome Diversity and the Peopling of the Americas: Evolutionary and Demographic Evidence". *American Journal of Human Biology* 16, no. 4 (2004): 420–39, https://doi.org/10.1002/ajhb.20041

Scott, Peter Dale. *Drugs, Oil, and War*: *The United States in Afghanistan, Colombia, and Indochina*. Lenham, MD: Rowman & Littlefield, 2003.

Seeland, Nicolas. "Die Ghiliaken". *Russische Revue* 21 (1882): 97–130, 222–254.

Sheng, Feng, Yi Liu, Bin Zhou, Wen Zhou, and Shihui Han. "Oxytocin Modulates the Racial Bias in Neural Responses to Others' Suffering". *Biological Psychology* 92, no. 2 (2013): 380–86, https://doi.org/10.1016/j.biopsycho.2012.11.018

Shenk, Mary K., "Testing Three Evolutionary Models of the Demographic Transition: Patterns of Fertility and Age at Marriage in Urban South India", *American Journal of Human Biology* 21, no. 4 (2009): 501–11, https://doi.org/10.1002/ajhb.20943

Shimei, Qubi and Ma Erzi. "Homicide and Homicide Cases in Old Liangshan". In *Perspectives on the Yi of Southwest China*, edited by Stevan Harrell, 94–103. Berkeley, CA: University of California Press, 2001.

Shinar, Dov and Vladimir Bratic. "Asymmetric War and Asymmetric Peace: Real Realities and Media Realities in the Middle East and the Western Balkans". *Dynamics of Asymmetric Conflict* 3, no. 2 (2010): 125–42, https://doi.org/10.108 0/17467586.2010.531034

Shirer, William L. *The Rise and Fall of the Third Reich. A History of Nazi Germany*. New York: Simon and Schuster, 1960.

Shiva, Vandana. *Making Peace with the Earth*. London: Pluto Press, 2013.

Shoemaker, Pamela J. and Stephen D. Reese. *Mediating the Message in the 21st Century: A Media Sociology Perspective*. 3rd ed. New York: Routledge, 2014.

—"Hardwired for News: Using Biological and Cultural Evolution to Explain the Surveillance Function". *Journal of Communication* 46, no. 3 (1996): 32–47, https://doi.org/10.1111/j.1460-2466.1996.tb01487.x

Shternberg, L. Ya. *Sem'ia i rod u narodov Severo-vostochnoi Azii*. Leningrad: Izdatel'stvo Instituta Narodov Severa TsIK SSSR, 1933.

Sibley, Chris G., Marc S. Wilson and John Duckitt. "Effects of Dangerous and Competitive Worldviews on Right-Wing Authoritarianism and Social Dominance Orientation over a Five-Month Period". *Political Psychology* 28, no. 3 (2007): 357–71, https://doi.org/10.1111/j.1467-9221.2007.00572.x

Sibley Chris G. and John Duckitt. "Personality and Prejudice: A Meta-Analysis and Theoretical Review". *Personality and Social Psychology Review* 12, no. 3 (2008): 248–79, https://doi.org/10.1177/1088868308319226

Sibley, Chris G. and Joseph Bulbulia. "Faith after an Earthquake: A Longitudinal Study of Religion and Perceived Health before and after the 2011 Christchurch New Zealand Earthquake". *PloS One* 7, no. 12 (2012): e49648, https://doi.org/10.1371/journal.pone.0049648

Sigmund, Karl, Hannelore De Silva, Arne Traulsen, and Christoph Hauert. "Social Learning Promotes Institutions for Governing the Commons". *Nature* 466, no. 7308 (2010): 861–63, https://doi.org/10.1038/nature09203

Signorielli, Nancy. "Television's Mean and Dangerous World: A Continuation of the Cultural Indicators Perspective". In *Cultivation Analysis: New Directions in Media Effects Research*, edited by Nancy Signorielli and Michael Morgan, 85–106. Newbury Park: Sage, 1990.

Silberbauer, George B. *Hunter and Habitat in the Central Kalahari Desert*. Cambridge: Cambridge University Press, 1981.

Simon, Joel. "Look Who's Inspiring Global Censorship". *Columbia Journalism Review* 40, no. 5 (2002): 64–65.

Simonton, Dean Keith. "After Einstein: Scientific Genius Is Extinct". *Nature* 493, no. 7434 (2013): 602, https://doi.org/10.1038/493602a

Sipes, Richard G. "War, Combative Sports, and Aggression: A Preliminary Causal Model of Cultural Patterning". In *War, Its Causes and Correlates*, edited by Martin A. Nettleship and Dale Givens, 749–64. The Hague: Mouton, 1975.

Skocpol, Theda. *Social Revolutions in the Modern World.* Cambridge: Cambridge University Press, 1994.

Smith, Eric Alden. "Why Do Good Hunters Have Higher Reproductive Success?" *Human Nature* 15, no. 4 (2004): 343–64, https://doi.org/10.1007/s12110-004-1013-9

Smith, Mary Felice. *Baba of Karo, a Woman of the Muslim Hausa.* New Haven, CT: Yale University Press, 1954.

Smith, Michael G. *The Economy of Hausa Communities of Zaria.* London: Her Majesty's Stationery Office for the Colonial Office, 1955.

—*Government in Zazzau, 1800–1950.* London: Oxford University Press, 1960.

—"The Hausa of Northern Nigeria". In *Peoples of Africa*, edited by James L. Gibbs Jr., 119–55. New York: Holt, Rinehart & Winston, 1965.

—*The Affairs of Daura.* Berkeley, CA: University of California Press, 1978.

Smole, William J. *The Yanoama Indians: A Cultural Geography.* Austin, TX: University of Texas Press, 1976.

Snow, Douglas. *Distant Thunder: Patterns of Conflict in the Developing World.* Abington and New York: Routledge, 2015.

Sokolov, Boris V. *Tainy Finskoi Voiny (Voennye tainy XX veka).* Moscow: Veche, 2000.

Soule, John W. "Problems in Applying Counterterrorism to Prevent Terrorism: Two Decades of Violence in Northern Ireland Reconsidered". *Studies in Conflict & Terrorism* 12, no. 1 (1989): 31–46, https://doi.org/10.1080/10576108908435759

Southwold, Martin. "The Ganda of Uganda". In *Peoples of Africa*, edited by James Lowell Gibbs, 81–118. New York: Rinehart and Winston, 1969.

Special Operations Research Office. *Selected Groups in the Republic of Vietnam: The Rhade.* Washington, D.C.: The American University, 1965.

Speckhard, Anne. *Talking to Terrorists. Understanding the Psycho-Social Motivations of Militant Jihadi Terrorists, Mass Hostage Takers, Suicide Bombers & Martyrs.* Mclean, VA: Advances Press, 2012.

Spencer, B. and F. J. Gillen. *The Arunta: A Study of a Stone Age People.* Vols 1 and 2. London: MacMillan, 1927.

Spencer, Baldwin and Francis James Gillen. *Across Australia.* Vol. 2. London: Macmillan, 1912.

Spencer, Paul. *The Maasai of Matapato: A Study of Rituals of Rebellion.* London: Manchester University Press, 1988.

Sperber, Dan. *Explaining Culture*. Oxford: Blackwell Publishers, 1996.

Spisak, Brian R., Peter H. Dekker, Max Krüger, and Mark van Vugt. "Warriors and Peacekeepers: Testing a Biosocial Implicit Leadership Hypothesis of Intergroup Relations Using Masculine and Feminine Faces". *PloS One* 7, no. 1 (2012): e30399, https://doi.org/10.1371/journal.pone.0030399

St. Marie, Joseph J. and Shahdad Naghshpour. *Revolutionary Iran and the United States: Low-Intensity Conflict in the Persian Gulf*. Farnham: Ashgate, 2011.

Stallen, Mirre, Carsten K. W. de Dreu, Shaul Shalvi, Ale Smidts, and Alan G. Sanfey. "The Herding Hormone: Oxytocin Stimulates In-Group Conformity". *Psychological Science* 23, no. 11 (2012): 1288–92, https://doi.org/10.1177/0956797612446026

Stankov, Lazar, Jihyun Lee, and Fons van de Vijver. "Two Dimensions of Psychological Country-Level Differences: Conservatism/ Liberalism and Harshness/Softness". *Learning and Individual Differences* 30 (2014): 22–33, https://doi.org/10.1016/j.lindif.2013.12.001

Stankov, Lazar, and Jihyun Lee. "Toward a Psychological Atlas of the World with Mixture Modeling". *Journal of Cross-Cultural Psychology*, 2015, 1–14, https://doi.org/10.1177/0022022115611749

Stearns, Stephen C. *The Evolution of Life Histories*. Vol. 249. Oxford: Oxford University Press, 1992.

Stefánsson, Vilhjálmur. *The Stefánsson-Anderson Arctic Expedition of the American Museum: Preliminary Ethnological Report* (Anthropological Papers of the American Museum of Natural History, Vol. 14, Part 1). New York, 1914.

Steffen, Will, Katherine Richardson, Johan Rockström, et al. "Planetary Boundaries: Guiding Human Development on a Changing Planet". *Science* 347, no. 6223 (2015): 1259855, https://doi.org/10.1126/science.1259855

Stellmacher, Jost and Thomas Petzel. "Authoritarianism as a Group Phenomenon". *Political Psychology* 26, no. 2 (2005): 245–74, https://doi.org/10.1111/j.1467-9221.2005.00417.x

Stempel, Carl, Thomas Hargrove, and Guido H. Stempel. "Media Use, Social Structure, and Belief in 9/11 Conspiracy Theories". *Journalism & Mass Communication Quarterly* 84, no. 2 (2007): 353–72, https://doi.org/10.1177/107769900708400210

Stewart, Robert A. C. *Cultural Dimensions: a Factor Analysis of Textor's A Cross-Cultural Summary*. New Haven, CT: Human Relations Area Files, 1971.

Stol, Marten. "Private Life in Ancient Mesopotamia". In *Civilizations of the Ancient Near East*. Vol. 1, edited by Jack M. Sasson, 485–501. New York: Simon & Schuster, 1995.

Strehlow, Carl. *Die Aranda—und Loritja-Stämme in Zentral-Australien*, Vols 1–5. Frankfurt am Main: Städtlischen Völker-Museum, 1907–1920.

Strehlow, Theodor George Henry. *Aranda Traditions*. Melbourne: Melbourne University Press, 1947.

Suarez, Maria Matilde. *Los Warao: Indigenas del Delta del Orinoco*. Caracas: Venegafrica, 1968.

Summers, Kyle. "The Evolutionary Ecology of Despotism". *Evolution and Human Behavior* 26, no. 1 (2005): 106–35, https://doi.org/10.1016/j.evolhumbehav.2004.09.001

Sütterlin, Christa. "Art and Indoctrination: From the Biblia Pauperum to the Third Reich". In *Indoctrinability, Ideology and Warfare: Evolutionary Perspectives*, edited by Irenäus Eibl-Eibesfeldt and Frank Kemp Salter, 279–300. New York: Berghahn, 1998.

Tewksbury, David, Jennifer Jones, Matthew W. Peske, Ashlea Raymond, and William Vig. "The Interaction of News and Advocate Frames: Manipulating Audience Perceptions of a Local Public Policy Issue". *Journalism & Mass Communication Quarterly* 77, no. 4 (2000): 804–29, https://doi.org/10.1177/107769900007700406

Thangaraj, Kumarasamy, Lalji Singh, Alla G. Reddy, V. Raghavendra Rao, Subhash C. Sehgal, et al. "Genetic Affinities of the Andaman Islanders, A Vanishing Human Population". *Current Biology* 13, no. 2 (2003): 86–93, https://doi.org/10.1016/s0960-9822(02)01336-2

Thayer, Bradley A. *Darwin and International Relations: On the Evolutionary Origins of War and Ethnic Conflict*. Lexington, KT: University Press of Kentucky, 2004.

Thies, Cameron G. "A Social Psychological Approach to Enduring Rivalries". *Political Psychology* 22, no. 4 (2001): 693–725, https://doi.org/10.1111/0162-895x.00259

Thomas, Elizabeth Marshall. *The Harmless People*. New York: Alfred A. Knopf, 1959).

Thorpe, Ian J. N. "Anthropology, Archaeology, and the Origin of Warfare". *World Archaeology* 35, no. 1 (2003): 145–65, https://doi.org/10.1080/0043824032000079198

Tilly, Charles. *Coercion, Capital, and European States, AD 990–1992*. Oxford: Blackwell, 1992.

Tindale, Norman Barnett. *Aboriginal Tribes of Australia: Their Terrain, Environmental Controls, Distribution, Limits, and Proper Names*. Berkeley, CA: University of California Press, 1974.

Toft, Monica Duffy. "Getting Religion? The Puzzling Case of Islam and Civil War", *International Security* 31, no. 4 (2007): 97–131, https://doi.org/10.1162/isec.2007.31.4.97

Tooby, John and Leda Cosmides. "The Evolution of War and its Cognitive Foundations". *Institute for Evolutionary Studies Technical Report* 88, no. 1 (1988): 1–15.

—"Groups in Mind: The Coalitional Roots of War and Morality". In *Human Morality and Sociality: Evolutionary and Comparative Perspectives*, edited by Henrik Høgh-Olesen, 91–234. New York: Palgrave MacMillan, 2010.

Trautmann, Stefan T., Gijs van de Kuilen, and Richard J. Zeckhauser. "Social Class and (Un)Ethical Behavior: A Framework, With Evidence From a Large Population Sample". *Perspectives on Psychological Science* 8, no. 5 (2013): 487–97, https://doi.org/10.1177/1745691613491272

Tremearne, A. J. N. *Hausa Superstitions and Customs: An Introduction to the Folk-Lore and the Folk*. London: Frank Cass & Co., 1970 [1913].

— *Ban of the Bori: Demons and Demon-Dancing in West and North Africa*. New York: Routledge, 2013 [1914].

Turchin, Peter. *War and Peace and War: The Rise and Fall of Empires*. New York: Penguin Books, 2006.

—"Dynamics of Political Instability in the United States, 1780–2010". *Journal of Peace Research* 49, no. 4 (2012): 577–91, https://doi.org/10.1177/0022343312442078

Turchin, Peter and Sergey Gavrilets. "Evolution of Complex Hierarchical Societies". *Social Evolution and History* 8, no. 2 (2009): 167–98.

Turchin, Peter and Sergey A. Nefedov. *Secular Cycles*. Princeton, NJ: Princeton University Press, 2009.

Turnbull, Colin M. *The Forest People*. London: Jonathan Cape, 1961.

— *Wayward Servants: The Two Worlds of the African Pygmies*. New York: Natural History Press, 1965. Reprint: Westport, CT: Greenwood Press, 1976.

— *The Mbuti Pygmies: Change and Adaptation*. New York: Holt, Rinehart and Winston, 1983.

Turse, Nick. *Tomorrow's Battlefield: US Proxy Wars and Secret Ops in Africa*. Chicago, IL: Haymarket Books, 2015.

Turton, David. *War and Ethnicity: Global Connections and Local Violence*. Woodbridge: Boydell & Brewer Ltd, 2003.

Tversky, Amos and Daniel Kahneman. "Advances in Prospect Theory: Cumulative Representation of Uncertainty". *Journal of Risk and Uncertainty* 5, no. 4 (1992): 297–323, https://doi.org/10.1007/bf00122574

Tyler, Tim, *Memetics: Memes and the Science of Cultural Evolution*. Mersenne Publishing, 2011, https://books.google.co.uk/books?id=OcJY5Dgtz2EC

Ullrich, Johannes and J. Christopher Cohrs. "Terrorism Salience Increases System Justification: Experimental Evidence". *Social Justice Research* 20, no. 2 (2007): 117–39, https://doi.org/10.1007/s11211-007-0035-y

United Nations Data Retrieval System. "World Population Prospects: The 2012 Revision", http://data.un.org/

United Nations, Department of Economic and Social Affairs, Population Division. "World Population Prospects: The 2015 Revision". United Nations, 2015.

United Nations Development Programme. "Human Development Report", 2013.

Uppsala Conflict Data Program. "UCDP/PRIO Armed Conflict Dataset version 17.1". Uppsala University, 2017, http://ucdp.uu.se/downloads/

Urdal, Henrik. "People vs. Malthus: Population Pressure, Environmental Degradation, and Armed Conflict Revisited". *Journal of Peace Research* 42, no. 4 (2005): 417–34, https://doi.org/10.1177/0022343305054089

— "A Clash of Generations? Youth Bulges and Political Violence". *International Studies Quarterly* 50, no. 3 (2006): 607–29, https://doi.org/10.1111/j.1468-2478.2006.00416.x

Uz, Irem. "The Index of Cultural Tightness and Looseness Among 68 Countries". *Journal of Cross-Cultural Psychology* 46, no. 6 (2015): 319–35, https://doi.org/10.1177/0022022114563611

Valero, Helena and Ettore Biocca. *Yanoáma: The Narrative of a White Girl Kidnapped by Amazonian Indians*. New York: E. P. Dutton & Co., 1970.

Van Arensbergen, Pleun, Inge van der Weijden, and Peter van den Besselaar. "The Selection of Talent as a Group Process. A Literature Review on the Social Dynamics of Decision Making in Grant Panels". *Research Evaluation* 23, no. 4 (2014): 298–311, https://doi.org/10.1093/reseval/rvu017

Van der Dennen, Johan M. G. *The Origin of War: The Evolution of a Male-Coalitional Reproductive Strategy*. Groningen: Origin Press, 1995.

Van Hiel, Alain, Mario Pandelaere, and Bart Duriez. "The Impact of Need for Closure on Conservative Beliefs and Racism: Differential Mediation by Authoritarian Submission and Authoritarian Dominance". *Personality and Social Psychology Bulletin* 30, no. 7 (2004): 824–37, https://doi.org/10.1177/0146167204264333

Van Ussel, Jos. *Sexualunterdrückung. Geschichte der Sexualfeindschaft*. Hamburg: Rowohlt, 1970.

Van Vugt, Mark. "Evolutionary origins of leadership and followership". *Personality and Social Psychology Review* 10, no. 4 (2006): 354–71, https://doi.org/10.1207/s15327957pspr1004_5

Van Vugt, Mark, David de Cremer, and Dirk P. Janssen. "Gender Differences in Cooperation and Competition: The Male-Warrior Hypothesis". *Psychological Science* 18, no. 1 (2007): 19–23, https://doi.org/10.1111/j.1467-9280.2007.01842.x

Van Vugt, Mark and Brian R. Spisak. "Sex Differences in the Emergence of Leadership during Competitions within and between Groups". *Psychological Science* 19, no. 9 (2008): 854–58, https://doi.org/10.1111/j.1467-9280.2008.02168.x

Verschueren, Herwig. "The European Internal Market and the Competition between Workers". *European Labour Law Journal* 6, no. 2 (2015): 128–51, https://doi.org/10.1177/201395251500600203

Vitali, Stefania, James B. Glattfelder, and Stefano Battiston. "The Network of Global Corporate Control". *PloS One* 6, no. 10 (2011): e25995, https://doi.org/10.1371/journal.pone.0025995

Vohs, Kathleen D., Nicole L. Mead, and Miranda R. Goode. "The Psychological Consequences of Money". *Science* 314, no. 5802 (2006): 1154–56, https://doi.org/10.1126/science.1132491

Von Rueden, Christopher R. and Adrian V. Jaeggi. "Men's Status and Reproductive Success in 33 Nonindustrial Societies: Effects of Subsistence, Marriage System, and Reproductive Strategy". *Proceedings of the National Academy of Sciences*, 113, no. 39 (2016): 10824–29. https://doi.org/10.1073/pnas.1606800113

Von Rueden, Christopher, Michael Gurven, and Hillard Kaplan. "Why Do Men Seek Status? Fitness Payoffs to Dominance and Prestige". *Proceedings of the Royal Society B: Biological Sciences* 278 (2010), 2223–32, https://doi.org/10.1098/rspb.2010.2145

Von Saldern, Adelheid. "*Volk* and *Heimat* Culture in Radio Broadcasting during the Period of Transition from Weimar to Nazi Germany". *The Journal of Modern History* 76, no. 2 (2004): 312–46, https://doi.org/10.1086/422932

Wahl-Jorgensen, Karin and Thomas Hanitzsch, eds. *The Handbook of Journalism Studies*. New York: Routledge, 2009.

Watts, Joseph, Simon J. Greenhill, Quentin D. Atkinson, Thomas E. Currie, Joseph Bulbulia, and Russell D. Gray. "Broad Supernatural Punishment but Not Moralizing High Gods Precede the Evolution of Political Complexity in Austronesia". *Proceedings of the Royal Society of London B: Biological Sciences* 282, no. 1804 (2015): 20142556, https://doi.org/10.1098/rspb.2014.2556

Weber, Hannes. "Demography and Democracy: The Impact of Youth Cohort Size on Democratic Stability in the World". *Democratization* 20, no. 2 (2013): 335–57, https://doi.org/10.1080/13510347.2011.650916

Weimann, Gabriel and Conrad Winn. *The Theater of Terror: Mass Media and International Terrorism*. New York: Longman, 1994.

Weinberg, Leonard. "Italian Neo-Fascist Terrorism: A Comparative Perspective". *Terrorism and Political Violence* 7, no. 1 (1995): 221–38, https://doi.org/10.1080/09546559508427290

Weinstein, Jeremy M. "Resources and the Information Problem in Rebel Recruitment". *Journal of Conflict Resolution* 49, no. 4 (2005): 598–624, https://doi.org/10.1177/0022002705277802

Welch, Michael. *Scapegoats of September 11th: Hate Crimes & State Crimes in the War on Terror*. New Brunswick, NJ: Rutgers University Press, 2006.

Welsh, Brandon C. and Rebecca D. Pfeffer. "Reclaiming Crime Prevention in an Age of Punishment: An American History". *Punishment & Society* 15, no. 5 (2013): 534–53, https://doi.org/10.1177/1462474513504798

Welzel, Christian. *Freedom Rising: Human Empowerment and the Quest for Emancipation*. New York: Cambridge University Press, 2013.

—"Freedom Rising. Resources. Tables and Figures", http://www.cambridge.org/dk/academic/subjects/politics-international-relations/comparative-politics/freedom-rising-human-empowerment-and-quest-emancipation

Werner, Richard A. "How Do Banks Create Money, and Why can Other Firms Not do the Same? An Explanation for the Coexistence of Lending and Deposit-Taking". *International Review of Financial Analysis* 36 (2014): 71–77, https://doi.org/10.1016/j.irfa.2014.10.013

West, Stuart A., Claire El Mouden, and Andy Gardner. "Sixteen Common Misconceptions about the Evolution of Cooperation in Humans". *Evolution and Human Behavior* 32, no. 4 (2011): 231–62, https://doi.org/10.1016/j.evolhumbehav.2010.08.001

White, Douglas R. "Focused Ethnographic Bibliography: Standard Cross-Cultural Sample". *Cross-Cultural Research* 23, no. 1–4 (1989): 1–145, https://doi.org/10.1177/106939718902300102

White, Frances J., Michel T. Waller, and Klaree J. Boose. "Evolution of Primate Peace". In *War, Peace, and Human Nature: The Convergence of Evolutionary and Cultural Views*, edited by Douglas P. Fry, chapter 19. New York: Oxford University Press, 2013.

Whitehead, Dennis. "The Gleiwitz Incident". *After the Battle*, no. 142 (2008): 3–23.

Whitehouse, L. E. "Maasai Social Customs". *Journal of the East Africa and Uganda Natural History Society* 47–48 (1933): 146–53.

Wilbert, Johannes. "Die soziale und politische Organisation der Warrau". *Kölner Zeitshcrift für Soziologie und Sozialpsychologie* 10 (1958): 272–91.

—*Folk Literature of the Warao Indians: Narrative Material and Motif Content*. Los Angeles, CA: University of California, Latin American Center, 1970.

—*Mystic Endowment: Religious Ethnography of the Warao Indians*. Cambridge, MA: Harvard University Press, 1993.

—*Mindful of Famine: Religious Climatology of the Warao Indians*. Cambridge, MA: Harvard University Press, 1996.

Wilbert, J. and M. Layrisse, editors. *Demographic and Biological Studies of the Warao Indians*. Los Angeles, CA: University of California, Latin American Center Publications, 1980.

Wilbert, Johannes and Karin Simoneau. *Folk Literature of the Yanomami Indians*. Los Angeles, CA: University of California, Latin American Center Publications, 1990.

Wilkie, Andrew. *Axis of Deceit*: *The Extraordinary Story of an Australian Whistleblower*. Collingwood Vic: Black Inc. Agenda, 2010.

Wilkie, David S. and Bryan Curran. "Historical Trends in Forager and Farmer Exchange in the Ituri Rain Forest of Northeastern Zaire". *Human Ecology* 21, no. 4 (1993): 389–417, https://doi.org/10.1007/bf00891141

Willan, Philip P. *Puppetmasters*: *The Political Use of Terrorism in Italy*. San Jose, CA: Authors Choice Press, 1991.

Willer, Robb. "The Effects of Government-Issued Terror Warnings on Presidential Approval Ratings". *Current Research in Social Psychology* 10, no. 1 (2004): 1–12.

Williams, Brian Glyn. "Fighting with a Double-Edged Sword: Proxy Militias in Iraq, Afghanistan, Bosnia, and Chechnya". In *Making Sense of Proxy Wars*: *States, Surrogates & the Use of Force*, edited by Michael A. Innes, 61–88. Washington, D.C.: Potomac, 2012.

Wilson, Sigismond Ayodele. "Diamond Exploitation in Sierra Leone 1930 to 2010: A Resource Curse?" *GeoJournal* 78, no. 6 (2013): 997–1012, https://doi.org/10.1007/s10708-013-9474-1

Wilson, Margo and Martin Daly. "Competitiveness, Risk Taking, and Violence: The Young Male Syndrome". *Ethology and Sociobiology* 6, 1 (1985): 59–73, https://doi.org/10.1016/0162-3095(85)90041-x

Wilson, Andrew S., Timothy Taylor, Maria Constanza Ceruti, Jose Antonio Chavez, Johan Reinhard, et al. "Stable Isotope and DNA Evidence for Ritual Sequences in Inca Child Sacrifice". *Proceedings of the National Academy of Sciences* 104, no. 42 (2007): 16456–61, https://doi.org/10.1073/pnas.0704276104

Wilson, Michael L., Christophe Boesch, Barbara Fruth, Takeshi Furuichi, Ian C. Gilby, et al. "Lethal Aggression in *Pan* Is Better Explained by Adaptive Strategies than Human Impacts". *Nature* 513, no. 7518 (2014): 414–17, https://doi.org/10.1038/nature13727

Winkler, Willi. "Ein ZEIT-Gespräch mit Ex-Terroristen Horst Mahler über die Apo, den Weg in den Terror und die Versöhnung mit dem Grundgesetz". *Die Zeit*, No. 19, May 2, 1997, http://www.glasnost.de/hist/apo/97mahler.html

Wisnewski, Gerhard, Wolfgang Landgraeber, and Ekkehard Sieker. *Das RAF Phantom*: *Neue Ermittlungen in Sachen Terror*. Munich: Droemer Knaur, 2008.

Wohl, Michael J.A., Nyla R. Branscombe and Stephen Reysen, "Perceiving Your Group's Future to be in Jeopardy: Extinction Threat Induces Collective Angst and the Desire to Strengthen the Ingroup". *Personality and Social Psychology Bulletin* 36, no. 7 (2010): 898–910, https://doi.org/10.1177/0146167210372505

Woods, Joshua. "The 9/11 Effect: Toward a Social Science of the Terrorist Threat". *The Social Science Journal* 48, no. 1 (2011): 213–33, https://doi.org/10.1016/j.soscij.2010.06.001

Woodward, Bob. *Plan of Attack: The Road to War*. New York: Simon and Schuster, 2004.

Wrangham, Richard W. "Evolution of Coalitionary Killing". *Yearbook of Physical Anthropology* 42 (1999): 1–30.

Wrangham, Richard W., and Luke Glowacki. "Intergroup Aggression in Chimpanzees and War in Nomadic Hunter-gatherers". *Human Nature* 23, no. 1 (2012): 5–29, https://doi.org/10.1007/s12110-012-9132-1

Yair, Omer and Dan Miodownik. "Youth Bulge and Civil War: Why a Country's Share of Young Adults Explains Only Non-Ethnic Wars". *Conflict Management and Peace Science* 33, no. 1 (2016): 25–44, https://doi.org/10.1177/0738894214544613

Yu, Liu. "Searching for the Heroic Age of the Yi People of Liangshan". In *Perspectives on the Yi of Southwest China*, edited by Stevan Harrell, 104–17. Berkeley and Los Angeles, CA: University of California Press, 2001, https://doi.org/10.1525/california/9780520219885.003.0008

Yueh-Hua, Lin. *The Lolo of Liang Shan*. New Haven: HRAF Press, 1961.

Zarlenga, Stephen. *The Lost Science of Money: The Mythology of Money–The Story of Power*. Valatie, NY: American Monetary Institute, 2002.

Zerries, Otto. *Waika: Die kulturgeschichtliche Stellung der Waika-Indianer des oberen Orinoco im Rahmen der Völkerkunde Südamerikas*. Munich: Klaus Renner Verlag, 1964.

Zimmermann, Clemens. *Medien im Nationalsozialismus. Deutschland 1933–1945, Italien 1922–1943, Spanien 1936–1951*. Wien: Böhlau Verlag, 2007.

Zodrow, George R. "Capital Mobility and Capital Tax Competition". *National Tax Journal* 63, no. 4 (2010): 865–902, https://doi.org/10.17310/ntj.2010.4s.03

Zweigenhaft, Richard L. "A Do Re Mi Encore: A Closer Look at the Personality Correlates of Music Preferences". *Journal of Individual Differences* 29, no. 1 (2008): 45–55, https://doi.org/10.1027/1614-0001.29.1.45

11. Illustrations

This book is published under a Creative Commons Attribution license 4.0, except for some of the photos, as listed below.

Index

This book need not end here...

At Open Book Publishers, we are changing the nature of the traditional academic book. The title you have just read will not be left on a library shelf, but will be accessed online by hundreds of readers each month across the globe. OBP publishes only the best academic work: each title passes through a rigorous peer-review process. We make all our books free to read online so that students, researchers and members of the public who can't afford a printed edition will have access to the same ideas.

This book and additional content is available at:
https://www.openbookpublishers.com/product/657

Customize

Personalize your copy of this book or design new books using OBP and third-party material. Take chapters or whole books from our published list and make a special edition, a new anthology or an illuminating coursepack. Each customized edition will be produced as a paperback and a downloadable PDF. Find out more at:
https://www.openbookpublishers.com/section/59/1

Donate

If you enjoyed this book, and feel that research like this should be available to all readers, regardless of their income, please think about donating to us. We do not operate for profit and all donations, as with all other revenue we generate, will be used to finance new Open Access publications.
https://www.openbookpublishers.com/section/13/1/support-us

⬜ Like Open Book Publishers

🐦 Follow @OpenBookPublish

BLOG Read more at the OBP Blog

You may also be interested in:

Peace and Democratic Society

Edited by Amartya Sen

https://www.openbookpublishers.com/product/78

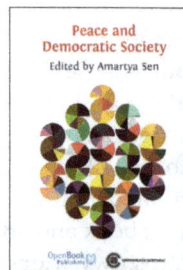

Democracy and Power
The Delhi Lectures

By Noam Chomsky. Introduction by Jean Drèze

https://www.openbookpublishers.com/product/300

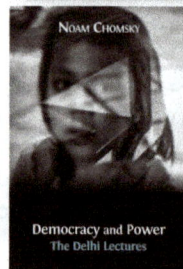

The Universal Declaration of Human Rights in the 21st Century

Edited by Gordon Brown

https://www.openbookpublishers.com/product/467

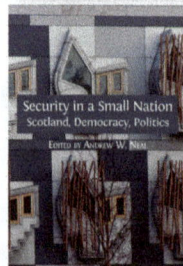

Security in a Small Nation
Scotland, Democracy, Politics

Edited by Andrew W. Neal

https://www.openbookpublishers.com/product/467

www.ingramcontent.com/pod-product-compliance
Lightning Source LLC
Chambersburg PA
CBHW070239290326
41929CB00046B/1966